MicroRNA: Regulation and Clinical Applications

MicroRNA: Regulation and Clinical Applications

Editor: Shane Eakin

MURPHY & MOORE
www.murphy-moorepublishing.com

www.murphy-moorepublishing.com

MURPHY & MOORE

Cataloging-in-Publication Data

MicroRNA : regulation and clinical applications / edited by Shane Eakin.
p. cm.
Includes bibliographical references and index.
ISBN 978-1-63987-790-4
1. MicroRNA. 2. Small interfering RNA. 3. RNA--Metabolism. I. Eakin, Shane.
QP623.5.S63 M53 2023
572.88--dc23

Murphy & Moore Publishing
1 Rockefeller Plaza,
New York City,
NY 10020, USA

ISBN 978-1-63987-790-4

Contents

Preface

MicroRNAs (miRNAs) are a class of non-coding RNAs found in plants, animals and some viruses. They play a significant role in the regulation of gene expressions and are involved in all physiological processes. These physiological processes include development and differentiation, heart function, metabolism, hemostasis, and apoptosis. The alteration of the miRNA function is associated with some diseases and disorders such as cancer, inherited diseases, cardiovascular diseases and neurodegenerative diseases. Most of the miRNAs are transcribed from DNA sequences to produce primary miRNAs, which are then processed into precursor miRNAs. These finally turn into mature miRNAs. This book provides comprehensive insights into the regulation and clinical applications of microRNA. A number of latest researches have been included to keep the readers up-to-date with the global concepts in this area of study. With state-of-the-art inputs by acclaimed experts of this field, this book targets students and professionals.

After months of intensive research and writing, this book is the end result of all who devoted their time and efforts in the initiation and progress of this book. It will surely be a source of reference in enhancing the required knowledge of the new developments in the area. During the course of developing this book, certain measures such as accuracy, authenticity and research focused analytical studies were given preference in order to produce a comprehensive book in the area of study.

This book would not have been possible without the efforts of the authors and the publisher. I extend my sincere thanks to them. Secondly, I express my gratitude to my family and well-wishers. And most importantly, I thank my students for constantly expressing their willingness and curiosity in enhancing their knowledge in the field, which encourages me to take up further research projects for the advancement of the area.

Editor

1

Targeting MicroRNA in Cancer using Plant-Based Proanthocyanidins

Rishipal R. Bansode [1,*], Janak R. Khatiwada [1], Jack N. Losso [2] and Leonard L. Williams [1]

1 Center for Excellence in Post-Harvest Technologies, North Carolina Research Campus, North Carolina Agricultural and Technical State University, Kannapolis, NC 28081, USA; jrkhatiw@ncat.edu (J.R.K.); llw@ncat.edu (L.L.W.)

2 School of Nutrition & Food Sciences, Louisiana State University, Baton Rouge, LA 70803, USA; jlosso@lsu.edu

* Correspondence: rbansode@ncat.edu.

Academic Editor: Maurizio Battino

Abstract: Proanthocyanidins are oligomeric flavonoids found in plant sources, most notably in apples, cinnamon, grape skin and cocoa beans. They have been also found in substantial amounts in cranberry, black currant, green tea, black tea and peanut skins. These compounds have been recently investigated for their health benefits. Proanthocyanidins have been demonstrated to have positive effects on various metabolic disorders such as inflammation, obesity, diabetes and insulin resistance. Another upcoming area of research that has gained widespread interest is microRNA (miRNA)-based anticancer therapies. MicroRNAs are short non-coding RNA segments, which plays a crucial role in RNA silencing and post-transcriptional regulation of gene expression. Currently, miRNA based anticancer therapies are being investigated either alone or in combination with current treatment methods. In this review, we summarize the current knowledge and investigate the potential of naturally occurring proanthocyanidins in modulating miRNA expression. We will also assess the strategies and challenges of using this approach as potential cancer therapeutics.

Keywords: proanthocyanidins; microRNA; cancer; polyphenols

1. Introduction

Research related to microRNAs (miRNAs) has gained prominence as a viable cancer therapy [1]. These are short non-encoding RNAs of 20–24 nucleotides that are involved in gene regulation and cell signaling [2].

They have been shown to contain gene-expression regulatory activities, and they can function both as promoters and inhibitors of oncogenesis [3]. The inhibitory activities of miRNA have led researchers to investigate its potential in cancer therapeutics. MicroRNAs are known to achieve this inhibitory role by binding to *cis*-elements in the 3′-untranslated region (UTR) of mRNAs [4]. Following the binding of miRNAs, inhibition occurs by one of three mechanisms: (1) target mRNA site-specific cleavage; (2) repression of target mRNA translation; and (3) mRNA degradation by cytoplasmic processing bodies [5].

MicroRNAs are integral parts of feedback circuits in normal physiology. They play a role in cell differentiation, regulating cellular pathway and exhibiting buffering effects to key biological processes [2,6–12]. On the other hand, deregulation of miRNA results in enhanced cellular plasticity, dedifferentiation and increased translation of oncogenes [13–15].

Over 1400 human miRNAs have been identified, and many of these miRNAs are found to be strongly conserved across species [16]. It is estimated that more than 60% of all mRNAs are regulated by miRNAs via posttrancriptional mode [17]. It is for this reason that recent research has been directed towards understanding these miRNAs in diverse biological processes, including in a variety of diseases.

2. Biogenesis of miRNA

The canonical pathway of miRNA processing includes transcription by RNA polymerase II to form a primary miRNA (pri-miRNA) transcript with 5′-end caps and 3′-end poly-A tails (Figure 1). This transcript consists of more than one hairpin structure. It is subsequently cleaved by the nuclease, Drosha and its cofactor DiGeorge syndrome critical region 8 (DGCR8), thus generating a precursor miRNA (pre-miRNA). This pre-miRNA is transported from the nucleus to the cytoplasm by exportin-5 (Exp5) using RanGTP as a cofactor. It undergoes further processing in the cytoplasm by RNase III Dicer in a complex with trans-activation response RNA-binding protein (TRBP) to generate approximately 22-nucleotide duplex mature miRNA. Finally, the miRNA duplex unwinds and the functional strand of mature miRNA complexes with Argonaute (Ago2), forming an RNA-induced silencing complex (RISC). This RISC complex directs silencing via sequence-specific mRNAs target cleavage, translational repression and mRNA deadenylation [18].

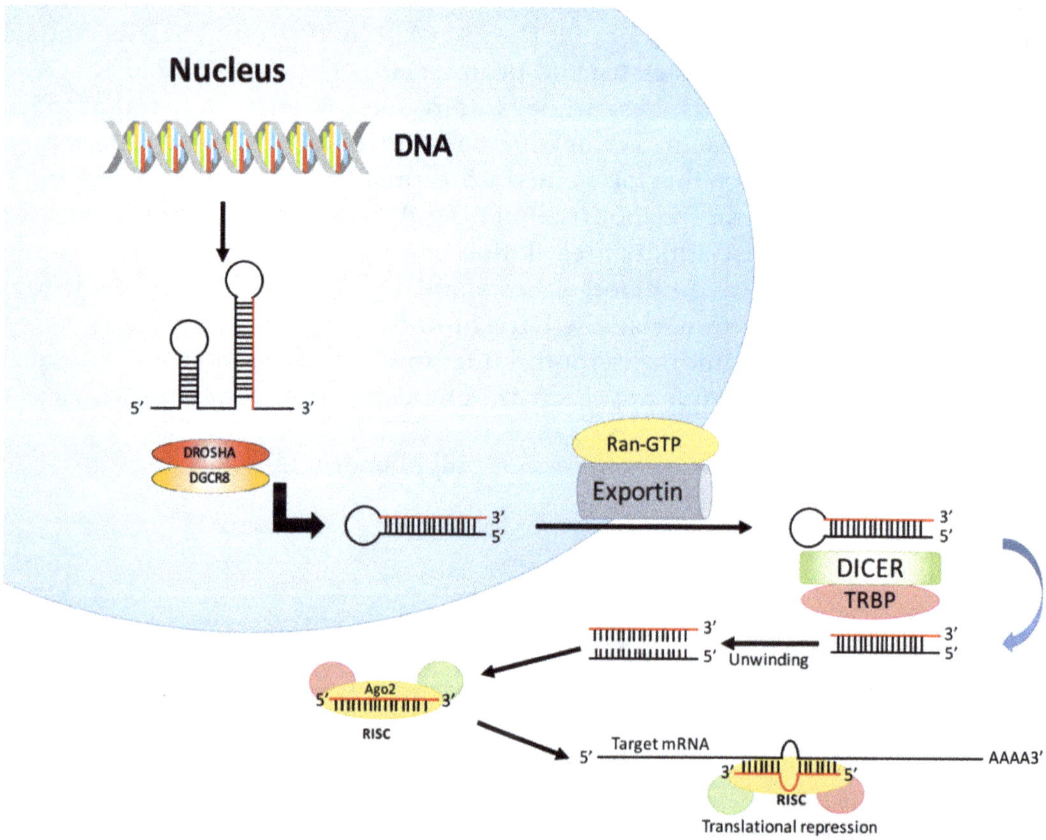

Figure 1. The biogenesis of microRNA.

3. Mechanism of Action of miRNAs

The main mechanism with which miRNA effects protein-coding genes is by interaction with the 3′-UTR of target mRNA, subsequently leading to mRNA degradation or translation repression [19]. The interaction of miRNA can also occur at the 5′-untranslated region (5′-UTR) of the protein coding sequence, thereby affecting translation repression or activation of target proteins [20,21]. Translation repression can also occur when miRNA targets the coding sequence of the target protein [22]. Alternately, miRNAs can up-regulate the translation of target genes by interacting with regulatory complexes [23]. The function of miRNAs is illustrated in Figure 2.

Figure 2. Function of miRNA in translational repression or activation of mRNAs.

4. Implication of miRNAs in Cancer

Recent studies have identified deregulation miRNAs in various tumors (Table 1). These miRNAs have been determined to be located in the proximity of the breakpoints, regions of amplification or loss of heterozygocity [24]. Alternately, they can lead to carcinogenesis by altered miRNA expression due to defective miRNA processing, post transcriptional regulation, mutation and epigenetic changes [11,25–27]. Their main function in tumorigenesis is to regulate oncogenes or tumor-suppressor genes. These miRNAs that are involved in tumorigenesis are known as oncomirs. They can act as oncogenes by down-regulating genes involved in cell differentiation and apoptosis, or can act as tumor suppressors by down-regulating oncogenes [28]. For example, miR-21 is widely expressed in various types of cancers, and it hinders the activity of tumor-suppressor genes [29,30].

Table 1. Overview of key microRNAs associated with cancer.

MicroRNA	Up- (↑)/Down- (↓) regulation	Cancer type	References
miR-21	↑	Lung cancer	[31]
let-7 family	↓	Lung cancer cells	[32]
miR-125b, mir-145	↓	Breast cancer	[33]
miR-21, mir-155	↑		
miR-27a	↑	Breast cancer	[34,35]
miR-143, mir-145	↓	Colorectal cancer	[36]
miR-106a, mir-21	↑	Colorectal cancer	[37]
miR-15, mir-16	↓	Chronic lymphocytic leukemia (CLL)	[38]
let-7b, let-7c miR-24	↓ ↑	Acute myeloid leukemia (AML)	[39,40]
miR-29a miR-155	↑ ↑	B-cell chronic lymphocytic leukemia (CLL)	[41] [42]
miR-125b	↑	B-cell acute lymphocytic leukemia (ALL)	[43]
miR-19	↑	T-cell acute lymphocytic leukemia (ALL)	[44]
miR-21	↑	pre-B malignant lymphoid-like phenotype in mice	[45]
miR-122	↓	Liver cancer	[46]
miR-221, miR-222, miR-21	↑	Hepatocellular carcinoma (HCC)	[47]
miR-103, miR-107	↑	Pancreatic cancer	[48]
miR-141	↑	Prostate cancer	[49]
miR-221	↑	Ovarian cancer	[50]
miR-21, let-7 family	↓		
miR-17-92 cluster	↑	Prostate cancer cells	[51]

The miR-17-92 cluster (miR-17, miR-18a, miR-19a, miR-19b-1, miR-20a and miR-92-1) also known as oncomiR-1, is aberrantly expressed in various solid tumors [52]. In humans, miR 17-92 is over-expressed in several hematopoietic malignancies and solid tumors [53]. The sequences are highly conserved across species and also within the miR 17-92 family [54]. The expression of miR 17-92 has been studied in animals and human cancer and has been found to show pleiotropic functions [55,56].

Under both normal and malignant transformations, miR 17-92 stimulates proliferation, inhibit differentiation, initiate angiogenesis and promote cell survival.

Members of the let-7 family, mir-15a and mir-16-1, are well-known tumor suppressors that can affect proto-oncogenes like RAS, HMGA2, MYC and IMP-1 [57–60]. Generally, these miRNAs are down-regulated in most cancers. The oncogenic property of miR 17-92 mainly results in elevated c-myc expression. Other modes of action in carcinogenesis are through the modulation of cell signaling, cell adhesion, hypoxia-related genes and angiogenesis [61–64].

5. Dietary Phytochemical and Chemoprevention

Phytochemicals are important constituents of fruits, vegetables and legumes. These naturally occurring bioactive compounds represent a rich source of micronutrients in the human diet [65]. Epidemiological studies have shown that regular consumption of fruits and vegetables reduce the risk of cancer, metabolic syndrome and age-related diseases [66]. Phytochemicals such as isothiocyanates are known to prevent cancer by affecting cellular mechanisms [67]. A majority of these compounds exhibit antioxidant properties and can regulate carcinogenesis by scavenging reactive oxygen species (ROS) and inhibiting other oxidative stress [68]. Other molecular pathways with which phytochemicals inhibit carcinogenesis are by their anti-inflammatory and pro-apoptotic activity [69]. Recent studies have highlighted that the mechanism of chemoprevention also involves modulating miRNAs that regulate cellular pathways associated with inflammation or apoptosis [70].

6. Classification of Polyphenols

Polyphenols represent the most abundant phytochemicals in the human diet. Depending on their chemical structure, they are classified into several groups. Most polyphenols occur in nature as conjugates with sugar or organic acids [71]. The chemical structure consists of a benzene ring attached to one or more hydroxyl groups. They can also form polymers as in the case of flavonoids [72]. Polyphenols and their metabolites can regulate cell functionality by binding to target proteins or by affecting histone modification [73–75]. The role of polyphenols in regulating miRNA is not clearly understood. However, recent studies show that polyphenols may be involved in miRNA modulations by multiple mechanisms. Polyphenols can directly bind to miRNAs posttranscriptionally, and they can affect their functionality [76]. They can also bind to mRNAs and proteins, thereby affecting the miRNA biogenesis [72]. The latter involves inhibition of cell signaling by either phosphorylating or dephosphorylating proteins involved in miRNA maturation such as the Dicer complex [77]. Polyphenols can also inhibit transcription factors that bind to the promoter regions of miRNAs [78].

7. Structure of Proanthocyanidins

Proanthocyanidins are condensed tannins that are the most abundant phytochemicals in nature. Among these, the group that mainly contains (epi)-catechins is known as procyanidins. These procyanidins exist in numerous stereo-chemical isomeric structures (Figure 3). Proanthocyanidins containing (epi)-afzelechin or (epi)-gallocatechin are known as prodelphinidins [79]. They can be broadly differentiated into two subgroups with linkages of catechin or epicatechin at C4-C8 and/or C2-O7 (Type A), or C4-C8 and/or C4-C6 (Type B). They can form complexes with multiple catechins and epicatechins, and with their gallic acid esters to form polymers. The type of polymers are classified as monomers, dimers, trimers, tetramers or oligomers (>4 polymers) [80]. Proanthocyanidins are found in various foods which include their most common B-type polymers. However, some A-type procyanidins predominantly occur in cranberries, plums, avocados, peanuts and cinnamon [79].

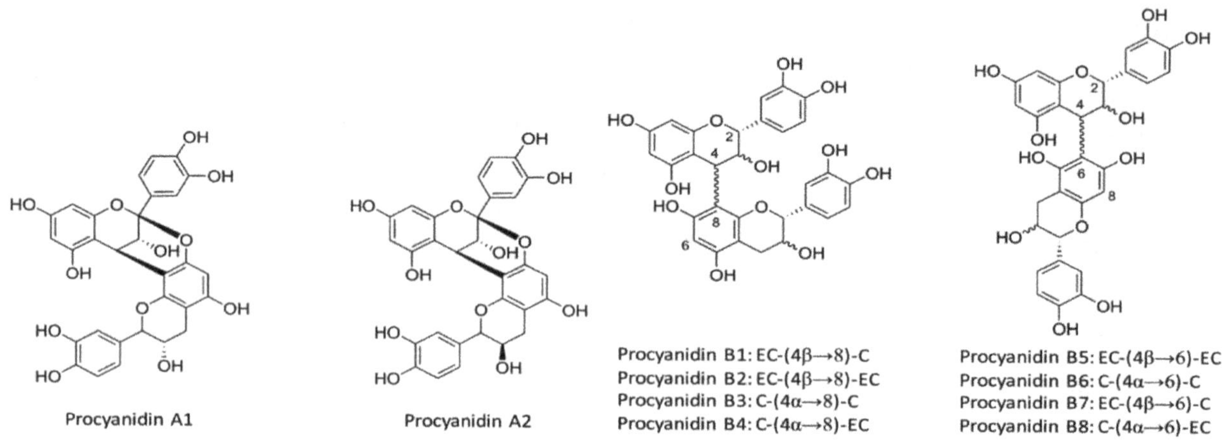

Figure 3. Structure of Type-A and Type-B proanthocyanidins.

Proanthocyanidins like most other flavonoids, exists in glycosylated forms [81]. The type of glycosylation and its associated sugar moiety greatly influence its absorption [82]. The glycosylated proanthocyanidins are stable during digestion and resist acid hydrolysis [83]. They are catabolized by colonic microflora into metabolites that are absorbed into the circulatory system [84]. The metabolites then reach target tissues and undergo glucuronidation, methylation and sulfation [85]. Some studies also reported the bioavailability of unconjugated B-type procyanidins in plasma reaching a peak concentration within two hours of ingestion [86–90].

8. Proanthocyanidins Influence Regulation of microRNAs *in Vitro*

Research related to the benefits of proanthocyanidins in the regulation of miRNA is relatively new and is being actively investigated. Recent studies have shown that they can modulate miRNAs that play an important role in cancer, glucose and lipid homeostasis [91–93]. Current studies establishing the role of procyanidins in miRNAs regulation are presented in the Table 2. One study involved cranberry extract treatment to esophaegeal adenocarcinoma (EAC) cell lines and its precursor Barrett's esophagus (BE). This study found that EAC treated with cranberry procyanidin rich extract (CPE) induced miR-410 and miR-520d-5p and reduced miR-202, miR-516a-3p and miR-586 levels [91]. They showed that these five miRNAs are linked to the regulation of target genes that are involved in tumor suppression (p53 and p16), oncogenesis (Rb and Erb8) and inflammation (NFkB). The overall treatment affected 26 genes that play key role in angiogenesis, T-cell activation and apoptosis [91].

Proanthocyanidins extracts from grape seed (GSPE) and cocoa (CPE) differentially expressed miR-30b *, miR-197, miR-532-3p and miR-1224-3p in HepG2 hepatoma cells [93]. This study suggested that the degree of polymerization, botanical origin and growth conditions could affect miRNA regulation of proanthocyanidins. In another study using GSPE in HepG2 cells, miR-122 and miR-33a levels were reduced [76]. These two miRNAs are known to mediate dyslipedimia and insulin resistance [94]. miR-33a/b regulate cholesterol homeostasis and fatty acid β-oxidation, while miR-122 play a vital role in liver homeostasis [95–97]. Because miR-33a/b occurs in the intronic region of the sterol response element proteins (SREBF2 and SREBF1), the authors further investigated if the host gene expression would affect the intronic miRNA levels [98]. Their results showed that host gene SREBP2 expression was unaltered while miRNA expression of miR-33 was modulated.

The authors using ^{1}H-NMR spectroscopy reported that EGCG and resveratrol could bind to miRNA and thus influence its target genes. This study directly implicates the role of polyphenols at a posttranscriptional level. This outcome inferred that polyphenols can directly bind to miRNAs and can inhibit their posttranscriptional regulation [99].

9. Animal Model Studies Establish the Role of Proanthocyanidins in miRNA Regulation

The health-promoting effects of grape proanthocyanidins have been previously studied using animal models [100–104]. Procyanidins are known to exert beneficial effects by regulating lipids homeostasis and reducing atherogenic risk [96]. Studies involving ApoE knockout mice fed with proanthocyanidins for two weeks at a dosage equivalent to human intake of 300 mg/day showed modulation of a total of 55 miRNAs in the livers of Apoe −/− compared to C57BL/6 mice [105]. The highest upregulation was for mmu-miR-133b, a miRNA involved in cell differentiation [106]. Procyanidins exposure at a concentration of 25 mg/kg of bodyweight showed significant downregulation of miR-1249, miR-483, and miR-30c-1 * and upregulation of miR-3544 in pancreatic islets [107]. Proanthocyanidins are also shown to inhibit AsPC-1 pancreatic cancer cell growth and migration by upregulating let-7a [108].

Grape seed proanthocyanidins (GSP) were also shown to suppress miR-106b in melanoma cells (A375, Hs294t) by arresting G1-phase and increasing p21/WAF1/Cip1 protein levels and affecting its downstream signaling by inhibiting pRb, E2F1 and E2F2 [109]. They also corroborated this finding *in vivo* following A375 cells tumor xenograft in nude mice that showed administration of GSP in the diet reduced tumor growth and suppressed miR-106b expression in the tumors [109].

Table 2. List of miRNAs regulated by proanthocyanidins.

MicroRNA	Up- (↑)/Down- (↓) regulation	Source	Experimental Model	References
miR-30b * miR-1224-3p, miR-197, miR-532-3p	↓ ↑	GSPE [1], CPE [2]	HepG2 cells	[93]
miR-33a, miR-122	↓	GSPE	Rats, HepG2 cells	[76,99,110]
miR-410, miR-520d-5p	↑	Cranberry-PE [3]	JHAD1 and OE33 cells	[91]
miR-202, miR-516a-3p, miR-586	↓			
let-7a	↑	GSPE	AsPC-1 cells	[108]
let-7c-2 *, miR-125a-3p, miR-146b, miR-190, miR-190b, miR-196a, miR-196b, miR-197, miR-291b-3p, miR-292, miR-294, miR-297a, miR-29a *, miR-302a, miR-302b, miR-302c, miR-302d, miR-30c-1 *, miR-335-3p, miR-374 *, miR-450a-3p, miR-450b-5p, miR-455 *, miR-464, miR-7b, miR-469, miR-483 *, miR-487b, miR-505, miR-539, miR-542-3p, miR-551b, miR-669a, miR-676, miR-698, miR-7b *, miR-881	↑	GSPE	$Apoe^{-/-}$ mice	[105]
let-7b *, miR-1, miR-106a, miR-133a, miR-133b, miR-17, miR-181a, miR-191 *, miR-200c, miR-291b-5p, miR-296-5p, miR-324-5p, miR-374, miR-486, miR-671-5p, miR-801, miR-878-3p, miR-99b	↓			
miR-3544 miR-1249, miR-483, miR-30c1 *	↑ ↓	GSPE	Pancreatic islets of Wistar rats	[107]
miR-106b	↓	GSPE	A375 cells tumor xenograft in nude mice	[109]

[1] GSPE: Grape seed proanthocyanidins extract; [2] CPE: Cocoa proanthocyanidins extract; [3] Cranberry-PE: Cranberry proanthocyanidins extract; * miRNA originated from the opposite arm of the precursor miRNA

10. Conclusions

There is a growing amount of evidence that dietary polyphenols are important micronutrients that can play a role in preventing degenerative diseases. Fruits-and-vegetables-based proanthocyanidins are one of the most abundant phytochemicals. They could serve as a tool to benefit human health.

Although studies have been undertaken to establish the role of proanthocyanidins in various cancers using *in vitro* and *in vivo* methods, large scale epidemiological data is lacking. Unlike pharmacological drugs, polyphenols are complex compounds and their chemical and biochemical characteristics are dependent on source and cultivar. Clear evidence of their mechanistic function in regulating cellular responses is still unknown. It is hypothecated that polyphenols could modulate miRNA by directly binding it or interacting with key effector proteins involved in miRNA biogenesis. They can also affect miRNA by indirectly regulating its target genes or by modifying host gene expression as some miRNAs are present in intronic regions of the genes. Given the complexity of their structures and their role in affecting multiple cellular pathways simultaneously, attributing their beneficial effects to a distinctive molecular pathway is challenging. Despite these limitations, recent studies have clearly established their role in cancer-prevention. On a similar level, miRNA's role in human epigenetics is just being investigated. There are formidable clinical implications of using miRNAs as an effective chemotherapy. These, combined with advances in nutritional diets designed to address specific disease conditions, will serve to improve the fight against cancer.

Author Contributions: R.R.B. wrote the manuscript. J.R.K. performed the literature review. J.N.L. and L.L.W. provided supervision and assistance with manuscript revision.

References

1. Cheng, C.J.; Bahal, R.; Babar, I.A.; Pincus, Z.; Barrera, F.; Liu, C.; Svoronos, A.; Braddock, D.T.; Glazer, P.M.; Engelman, D.M.; *et al.* MicroRNA silencing for cancer therapy targeted to the tumour microenvironment. *Nature* **2015**, *518*, 107–110. [CrossRef] [PubMed]
2. Jansson, M.D.; Lund, A.H. MicroRNA and cancer. *Mol. Oncol.* **2012**, *6*, 590–610. [CrossRef] [PubMed]
3. Shukla, S.; Meeran, S.M.; Katiyar, S.K. Epigenetic regulation by selected dietary phytochemicals in cancer chemoprevention. *Cancer Lett.* **2014**, *355*, 9–17. [CrossRef] [PubMed]
4. Lynn, F.C. Meta-regulation: MicroRNA regulation of glucose and lipid metabolism. *Trends Endocrinol. MeTable* **2009**, *20*, 452–459. [CrossRef] [PubMed]
5. Carthew, R.W.; Sontheimer, E.J. Origins and mechanisms of miRNAs and siRNAs. *Cell* **2009**, *136*, 642–655. [CrossRef] [PubMed]
6. Herranz, H.; Cohen, S.M. MicroRNAs and gene regulatory networks: Managing the impact of noise in biological systems. *Genes Dev.* **2010**, *24*, 1339–1344. [CrossRef] [PubMed]
7. Hornstein, E.; Shomron, N. Canalization of development by microRNAs. *Nat. Genet.* **2006**, *38*, S20–S24. [CrossRef] [PubMed]
8. Levine, E.; McHale, P.; Levine, H. Small regulatory RNAs may sharpen spatial expression patterns. *PLoS Comput. Biol.* **2007**, *3*, e233. [CrossRef] [PubMed]
9. Reinhart, B.J.; Slack, F.J.; Basson, M.; Pasquinelli, A.E.; Bettinger, J.C.; Rougvie, A.E.; Horvitz, H.R.; Ruvkun, G. The 21-nucleotide *let*-7 RNA regulates developmental timing in *Caenorhabditis elegans*. *Nature* **2000**, *403*, 901–906. [PubMed]
10. Yi, R.; Poy, M.N.; Stoffel, M.; Fuchs, E. A skin microRNA promotes differentiation by repressing 'stemness'. *Nature* **2008**, *452*, 225–229. [CrossRef] [PubMed]
11. Cordes, K.R.; Sheehy, N.T.; White, M.P.; Berry, E.C.; Morton, S.U.; Muth, A.N.; Lee, T.H.; Miano, J.M.; Ivey, K.N.; Srivastava, D. miR-145 and miR-143 regulate smooth muscle cell fate and plasticity. *Nature* **2009**, *460*, 705–710. [CrossRef] [PubMed]
12. Lim, L.P.; Lau, N.C.; Garrett-Engele, P.; Grimson, A.; Schelter, J.M.; Castle, J.; Bartel, D.P.; Linsley, P.S.; Johnson, J.M. Microarray analysis shows that some microRNAs downregulate large numbers of target mRNAs. *Nature* **2005**, *433*, 769–773. [CrossRef] [PubMed]
13. Kumar, M.S.; Lu, J.; Mercer, K.L.; Golub, T.R.; Jacks, T. Impaired microRNA processing enhances cellular transformation and tumorigenesis. *Nat. Genet.* **2007**, *39*, 673–677. [CrossRef] [PubMed]
14. Peter, M.E. Let-7 and miR-200 microRNAs: Guardians against pluripotency and cancer progression. *Cell Cycle* **2009**, *8*, 843–852. [CrossRef] [PubMed]

15. Palmero, E.I.; de Campos, S.G.; Campos, M.; de Souza, N.C.; Guerreiro, I.D.; Carvalho, A.L.; Marques, M.M. Mechanisms and role of microRNA deregulation in cancer onset and progression. *Genet. Mol. Biol.* **2011**, *34*, 363–370. [CrossRef] [PubMed]
16. Griffiths-Jones, S. Mirbase: MicroRNA sequences and annotation. In *Curr Protoc Bioinformatics*; John Wiley & Sons, Inc.: Manchester, UK, 2010.
17. Lewis, B.P.; Burge, C.B.; Bartel, D.P. Conserved seed pairing, often flanked by adenosines, indicates that thousands of human genes are microRNA targets. *Cell* **2005**, *120*, 15–20. [CrossRef] [PubMed]
18. Winter, J.; Jung, S.; Keller, S.; Gregory, R.I.; Diederichs, S. Many roads to maturity: MicroRNA biogenesis pathways and their regulation. *Nat. Cell Biol.* **2009**, *11*, 228–234. [CrossRef] [PubMed]
19. Ling, H.; Fabbri, M.; Calin, G.A. MicroRNAs and other non-coding RNAs as targets for anticancer drug development. *Nat. Rev. Drug Discov.* **2013**, *12*, 847–865. [CrossRef] [PubMed]
20. Lytle, J.R.; Yario, T.A.; Steitz, J.A. Target mRNAs are repressed as efficiently by microRNA-binding sites in the 5′ UTR as in the 3′ UTR. *Proc. Natl. Acad. Sci. USA* **2007**, *104*, 9667–9672. [CrossRef] [PubMed]
21. Eiring, A.M.; Harb, J.G.; Neviani, P.; Garton, C.; Oaks, J.J.; Spizzo, R.; Liu, S.; Schwind, S.; Santhanam, R.; Hickey, C.J.; *et al.* miR-328 functions as an RNA decoy to modulate hnRNP E2 regulation of mRNA translation in leukemic blasts. *Cell* **2010**, *140*, 652–665. [CrossRef] [PubMed]
22. Tay, Y.; Zhang, J.; Thomson, A.M.; Lim, B.; Rigoutsos, I. MicroRNAs to Nanog, Oct4 and Sox2 coding regions modulate embryonic stem cell differentiation. *Nature* **2008**, *455*, 1124–1128. [CrossRef] [PubMed]
23. Vasudevan, S.; Tong, Y.; Steitz, J.A. Switching from repression to activation: MicroRNAs can up-regulate translation. *Science* **2007**, *318*, 1931–1934. [CrossRef] [PubMed]
24. Budhu, A.; Ji, J.; Wang, X.W. The clinical potential of microRNAs. *J. Hematol. Oncol.* **2010**, *3*, 37. [CrossRef] [PubMed]
25. Deng, S.; Calin, G.A.; Croce, C.M.; Coukos, G.; Zhang, L. Mechanisms of microRNA deregulation in human cancer. *Cell Cycle* **2008**, *7*, 2643–2646. [CrossRef] [PubMed]
26. Schmittgen, T.D. Regulation of microRNA processing in development, differentiation and cancer. *J. Cell Mol. Med.* **2008**, *12*, 1811–1819. [CrossRef] [PubMed]
27. Thomson, J.M.; Newman, M.; Parker, J.S.; Morin-Kensicki, E.M.; Wright, T.; Hammond, S.M. Extensive post-transcriptional regulation of microRNAs and its implications for cancer. *Genes Dev.* **2006**, *20*, 2202–2207. [CrossRef] [PubMed]
28. Zhang, B.; Pan, X.; Cobb, G.P.; Anderson, T.A. MicroRNAs as oncogenes and tumor suppressors. *Dev. Biol.* **2007**, *302*, 1–12. [CrossRef] [PubMed]
29. Yang, C.H.; Pfeffer, S.R.; Sims, M.; Yue, J.; Wang, Y.; Linga, V.G.; Paulus, E.; Davidoff, A.M.; Pfeffer, L.M. The oncogenic microRNA-21 inhibits the tumor suppressive activity of FBXO11 to promote tumorigenesis. *J. Biol. Chem.* **2015**, *290*, 6037–6046. [CrossRef] [PubMed]
30. Zhou, X.; Wang, X.; Huang, Z.; Wang, J.; Zhu, W.; Shu, Y.; Liu, P. Prognostic value of miR-21 in various cancers: An updating meta-analysis. *PLoS ONE* **2014**, *9*, e102413. [CrossRef] [PubMed]
31. Seike, M.; Goto, A.; Okano, T.; Bowman, E.D.; Schetter, A.J.; Horikawa, I.; Mathe, E.A.; Jen, J.; Yang, P.; Sugimura, H.; *et al.* miR-21 is an EGFR-regulated anti-apoptotic factor in lung cancer in never-smokers. *Proc. Natl. Acad. Sci. USA* **2009**, *106*, 12085–12090. [CrossRef] [PubMed]
32. Takamizawa, J.; Konishi, H.; Yanagisawa, K.; Tomida, S.; Osada, H.; Endoh, H.; Harano, T.; Yatabe, Y.; Nagino, M.; Nimura, Y.; *et al.* Reduced expression of the let-7 microRNAs in human lung cancers in association with shortened postoperative survival. *Cancer Res.* **2004**, *64*, 3753–3756. [CrossRef] [PubMed]
33. Iorio, M.V.; Ferracin, M.; Liu, C.G.; Veronese, A.; Spizzo, R.; Sabbioni, S.; Magri, E.; Pedriali, M.; Fabbri, M.; Campiglio, M.; *et al.* MicroRNA gene expression deregulation in human breast cancer. *Cancer Res.* **2005**, *65*, 7065–7070. [CrossRef] [PubMed]
34. Mertens-Talcott, S.U.; Chintharlapalli, S.; Li, X.; Safe, S. The oncogenic microRNA-27a targets genes that regulate specificity protein transcription factors and the G2-M checkpoint in MDA-MB-231 breast cancer cells. *Cancer Res.* **2007**, *67*, 11001–11011. [CrossRef] [PubMed]
35. Li, X.; Mertens-Talcott, S.U.; Zhang, S.; Kim, K.; Ball, J.; Safe, S. MicroRNA-27a indirectly regulates estrogen receptor α expression and hormone responsiveness in MCF-7 breast cancer cells. *Endocrinology* **2010**, *151*, 2462–2473. [CrossRef] [PubMed]
36. Michael, M.Z.; SM, O.C.; van Holst Pellekaan, N.G.; Young, G.P.; James, R.J. Reduced accumulation of specific microRNAs in colorectal neoplasia. *Mol. Cancer Res.* **2003**, *1*, 882–891. [PubMed]

37. Link, A.; Balaguer, F.; Shen, Y.; Nagasaka, T.; Lozano, J.J.; Boland, C.R.; Goel, A. Fecal microRNAs as novel biomarkers for colon cancer screening. *Cancer Epidemiol. Biomarkers. Prev.* **2010**, *19*, 1766–1774. [CrossRef] [PubMed]
38. Calin, G.A.; Dumitru, C.D.; Shimizu, M.; Bichi, R.; Zupo, S.; Noch, E.; Aldler, H.; Rattan, S.; Keating, M.; Rai, K.; *et al.* Frequent deletions and down-regulation of micro- RNA genes miR15 and miR16 at 13q14 in chronic lymphocytic leukemia. *Proc. Natl. Acad. Sci. USA* **2002**, *99*, 15524–15529. [CrossRef] [PubMed]
39. Jongen-Lavrencic, M.; Sun, S.M.; Dijkstra, M.K.; Valk, P.J.; Lowenberg, B. MicroRNA expression profiling in relation to the genetic heterogeneity of acute myeloid leukemia. *Blood* **2008**, *111*, 5078–5085. [CrossRef] [PubMed]
40. Zaidi, S.K.; Dowdy, C.R.; van Wijnen, A.J.; Lian, J.B.; Raza, A.; Stein, J.L.; Croce, C.M.; Stein, G.S. Altered Runx1 subnuclear targeting enhances myeloid cell proliferation and blocks differentiation by activating a miR-24/MKP-7/MAPK network. *Cancer Res.* **2009**, *69*, 8249–8255. [CrossRef] [PubMed]
41. Pekarsky, Y.; Croce, C.M. Is miR-29 an oncogene or tumor suppressor in CLL? *Oncotarget* **2010**, *1*, 224–227. [CrossRef] [PubMed]
42. Wang, M.; Tan, L.P.; Dijkstra, M.K.; van Lom, K.; Robertus, J.L.; Harms, G.; Blokzijl, T.; Kooistra, K.; van T'veer M, B.; Rosati, S.; *et al.* miRNA analysis in B-cell chronic lymphocytic leukaemia: Proliferation centres characterized by low miR-150 and high BIC/miR-155 expression. *J. Pathol.* **2008**, *215*, 13–20. [CrossRef] [PubMed]
43. Bousquet, M.; Harris, M.H.; Zhou, B.; Lodish, H.F. MicroRNA miR-125B causes leukemia. *Proc. Natl. Acad. Sci. USA* **2010**, *107*, 21558–21563. [CrossRef] [PubMed]
44. Mavrakis, K.J.; Wolfe, A.L.; Oricchio, E.; Palomero, T.; de Keersmaecker, K.; McJunkin, K.; Zuber, J.; James, T.; Khan, A.A.; Leslie, C.S.; *et al.* Genome-wide RNA-mediated interference screen identifies miR-19 targets in Notch-induced T-cell acute lymphoblastic leukaemia. *Nat. Cell Biol.* **2010**, *12*, 372–379. [CrossRef] [PubMed]
45. Medina, P.P.; Nolde, M.; Slack, F.J. Oncomir addiction in an *in vivo* model of microRNA-21-induced pre-B-cell lymphoma. *Nature* **2010**, *467*, 86–90. [CrossRef] [PubMed]
46. Tsai, W.C.; Hsu, P.W.; Lai, T.C.; Chau, G.Y.; Lin, C.W.; Chen, C.M.; Lin, C.D.; Liao, Y.L.; Wang, J.L.; Chau, Y.P.; *et al.* MicroRNA-122, a tumor suppressor microRNA that regulates intrahepatic metastasis of hepatocellular carcinoma. *Hepatology* **2009**, *49*, 1571–1582. [CrossRef] [PubMed]
47. Yoon, S.O.; Chun, S.M.; Han, E.H.; Choi, J.; Jang, S.J.; Koh, S.A.; Hwang, S.; Yu, E. Deregulated expression of microRNA-221 with the potential for prognostic biomarkers in surgically resected hepatocellular carcinoma. *Hum. Pathol.* **2011**, *42*, 1391–1400. [CrossRef] [PubMed]
48. Roldo, C.; Missiaglia, E.; Hagan, J.P.; Falconi, M.; Capelli, P.; Bersani, S.; Calin, G.A.; Volinia, S.; Liu, C.G.; Scarpa, A.; *et al.* MicroRNA expression abnormalities in pancreatic endocrine and acinar tumors are associated with distinctive pathologic features and clinical behavior. *J. Clin. Oncol.* **2006**, *24*, 4677–4684. [CrossRef] [PubMed]
49. Mitchell, P.S.; Parkin, R.K.; Kroh, E.M.; Fritz, B.R.; Wyman, S.K.; Pogosova-Agadjanyan, E.L.; Peterson, A.; Noteboom, J.; O'Briant, K.C.; Allen, A.; *et al.* Circulating microRNAs as stable blood-based markers for cancer detection. *Proc. Natl. Acad. Sci. USA* **2008**, *105*, 10513–10518. [CrossRef] [PubMed]
50. Dahiya, N.; Sherman-Baust, C.A.; Wang, T.L.; Davidson, B.; Shih, I.M.; Zhang, Y.; Wood, W., 3rd; Becker, K.G.; Morin, P.J. MicroRNA expression and identification of putative miRNA targets in ovarian cancer. *PLoS ONE* **2008**, *3*, e2436. [CrossRef] [PubMed]
51. Zhou, P.; Ma, L.; Zhou, J.; Jiang, M.; Rao, E.; Zhao, Y.; Guo, F. miR-17–92 plays an oncogenic role and conveys chemo-resistance to cisplatin in human prostate cancer cells. *Int. J. Oncol.* **2016**, *48*, 1737–1748. [CrossRef] [PubMed]
52. Mogilyansky, E.; Rigoutsos, I. The miR-17/92 cluster: A comprehensive update on its genomics, genetics, functions and increasingly important and numerous roles in health and disease. *Cell Death Differ.* **2013**, *20*, 1603–1614. [CrossRef] [PubMed]
53. Ota, A.; Tagawa, H.; KaRNAn, S.; Tsuzuki, S.; Karpas, A.; Kira, S.; Yoshida, Y.; Seto, M. Identification and characterization of a novel gene, *C13orf25*, as a target for 13q31-q32 amplification in malignant lymphoma. *Cancer Res.* **2004**, *64*, 3087–3095. [CrossRef] [PubMed]
54. Olive, V.; Jiang, I.; He, L. miR-17–92, a cluster of miRNAs in the midst of the cancer network. *Int. J. Biochem. Cell Biol.* **2010**, *42*, 1348–1354. [CrossRef] [PubMed]

55. He, L.; Thomson, J.M.; Hemann, M.T.; Hernando-Monge, E.; Mu, D.; Goodson, S.; Powers, S.; Cordon-Cardo, C.; Lowe, S.W.; Hannon, G.J.; *et al.* A microRNA polycistron as a potential human oncogene. *Nature* **2005**, *435*, 828–833. [CrossRef] [PubMed]
56. Olive, V.; Li, Q.; He, L. miR-17–92: A polycistronic oncomir with pleiotropic functions. *Immunol. Rev.* **2013**, *253*, 158–166. [CrossRef] [PubMed]
57. Johnson, S.M.; Grosshans, H.; Shingara, J.; Byrom, M.; Jarvis, R.; Cheng, A.; Labourier, E.; Reinert, K.L.; Brown, D.; Slack, F.J. Ras is regulated by the let-7 microRNA family. *Cell* **2005**, *120*, 635–647. [CrossRef] [PubMed]
58. Lee, Y.S.; Dutta, A. The tumor suppressor microRNA let-7 represses the HMGA2 oncogene. *Genes Dev.* **2007**, *21*, 1025–1030. [CrossRef] [PubMed]
59. Sampson, V.B.; Rong, N.H.; Han, J.; Yang, Q.; Aris, V.; Soteropoulos, P.; Petrelli, N.J.; Dunn, S.P.; Krueger, L.J. MicroRNA let-7A down-regulates MYC and reverts MYC-induced growth in Burkitt lymphoma cells. *Cancer Res.* **2007**, *67*, 9762–9770. [CrossRef] [PubMed]
60. Boyerinas, B.; Park, S.M.; Shomron, N.; Hedegaard, M.M.; Vinther, J.; Andersen, J.S.; Feig, C.; Xu, J.; Burge, C.B.; Peter, M.E. Identification of let-7-regulated oncofetal genes. *Cancer Res.* **2008**, *68*, 2587–2591. [CrossRef] [PubMed]
61. Mei, Q.; Li, X.; Guo, M.; Fu, X.; Han, W. The miRNA network: Micro-regulator of cell signaling in cancer. *Expert Rev. Anticancer. Ther.* **2014**, *14*, 1515–1527. [CrossRef] [PubMed]
62. Valastyan, S.; Weinberg, R.A. Roles for microRNAs in the regulation of cell adhesion molecules. *J. Cell Sci.* **2011**, *124*, 999–1006. [CrossRef] [PubMed]
63. Favaro, E.; Ramachandran, A.; McCormick, R.; Gee, H.; Blancher, C.; Crosby, M.; Devlin, C.; Blick, C.; Buffa, F.; Li, J.L.; *et al.* MicroRNA-210 regulates mitochondrial free radical response to hypoxia and krebs cycle in cancer cells by targeting iron sulfur cluster protein ISCU. *PLoS ONE* **2010**, *5*, e10345. [CrossRef] [PubMed]
64. Suarez, Y.; Sessa, W.C. MicroRNAs as novel regulators of angiogenesis. *Circ. Res.* **2009**, *104*, 442–454. [CrossRef] [PubMed]
65. Blumberg, J.B.; Camesano, T.A.; Cassidy, A.; Kris-Etherton, P.; Howell, A.; Manach, C.; Ostertag, L.M.; Sies, H.; Skulas-Ray, A.; Vita, J.A. Cranberries and their bioactive constituents in human health. *Adv. Nutr.* **2013**, *4*, 618–632. [CrossRef] [PubMed]
66. Greiner, A.K.; Papineni, R.V.; Umar, S. Chemoprevention in gastrointestinal physiology and disease. Natural products and microbiome. *Am. J. Physiol. Gastrointest. Liver Physiol.* **2014**, *307*, G1–G15. [CrossRef] [PubMed]
67. Russo, M.; Spagnuolo, C.; Tedesco, I.; Russo, G.L. Phytochemicals in cancer prevention and therapy: Truth or dare? *Toxins (Basel)* **2010**, *2*, 517–551. [CrossRef] [PubMed]
68. Su, Z.Y.; Shu, L.; Khor, T.O.; Lee, J.H.; Fuentes, F.; Kong, A.N. A perspective on dietary phytochemicals and cancer chemoprevention: Oxidative stress, Nrf2, and epigenomics. *Top. Curr. Chem.* **2013**, *329*, 133–162. [PubMed]
69. Tan, A.C.; Konczak, I.; Sze, D.M.; Ramzan, I. Molecular pathways for cancer chemoprevention by dietary phytochemicals. *Nutr. Cancer* **2011**, *63*, 495–505. [CrossRef] [PubMed]
70. Milenkovic, D.; Jude, B.; Morand, C. MiRNA as molecular target of polyphenols underlying their biological effects. *Free Radic. Biol. Med.* **2013**, *64*, 40–51. [CrossRef] [PubMed]
71. Clifford, M.N.B.; Brown, J.E.; Andersen, Ø.M.; Markham, K.R. *Dietary Flavonoids and Health-Broadening the Perspective*; CRC Press: Boca Raton, FL, USA, 2006; pp. 319–370.
72. Fraga, C.G.; Galleano, M.; Verstraeten, S.V.; Oteiza, P.I. Basic biochemical mechanisms behind the health benefits of polyphenols. *Mol. Aspects Med.* **2010**, *31*, 435–445. [CrossRef] [PubMed]
73. Blade, C.; Arola, L.; Salvado, M.J. Hypolipidemic effects of proanthocyanidins and their underlying biochemical and molecular mechanisms. *Mol. Nutr. Food Res.* **2010**, *54*, 37–59. [CrossRef] [PubMed]
74. Malireddy, S.; Kotha, S.R.; Secor, J.D.; Gurney, T.O.; Abbott, J.L.; Maulik, G.; Maddipati, K.R.; Parinandi, N.L. Phytochemical antioxidants modulate mammalian cellular epigenome: Implications in health and disease. *Antioxid. Redox Signal.* **2012**, *17*, 327–339. [CrossRef] [PubMed]
75. Hardy, T.M.; Tollefsbol, T.O. Epigenetic diet: Impact on the epigenome and cancer. *Epigenomics* **2011**, *3*, 503–518. [CrossRef] [PubMed]

76. Baselga-Escudero, L.; Arola-ARNAl, A.; Pascual-Serrano, A.; Ribas-Latre, A.; Casanova, E.; Salvado, M.J.; Arola, L.; Blade, C. Chronic administration of proanthocyanidins or docosahexaenoic acid reverses the increase of miR-33a and miR-122 in dyslipidemic obese rats. *PLoS ONE* **2013**, *8*, e69817.
77. Chendrimada, T.P.; Gregory, R.I.; Kumaraswamy, E.; Norman, J.; Cooch, N.; Nishikura, K.; Shiekhattar, R. TRBP recruits the Dicer complex to Ago2 for microRNA processing and gene silencing. *Nature* **2005**, *436*, 740–744. [CrossRef] [PubMed]
78. George, J.; Singh, M.; Srivastava, A.K.; Bhui, K.; Roy, P.; Chaturvedi, P.K.; Shukla, Y. Resveratrol and black tea polyphenol combination synergistically suppress mouse skin tumors growth by inhibition of activated mapks and p53. *PLoS ONE* **2011**, *6*, e23395. [CrossRef] [PubMed]
79. Gu, L.; Kelm, M.A.; Hammerstone, J.F.; Beecher, G.; Holden, J.; Haytowitz, D.; Prior, R.L. Screening of foods containing proanthocyanidins and their structural characterization using LC-MS/MS and thiolytic degradation. *J. Agric. Food Chem.* **2003**, *51*, 7513–7521. [CrossRef] [PubMed]
80. Zhang, L.; Wang, Y.; Li, D.; Ho, C.T.; Li, J.; Wan, X. The absorption, distribution, metabolism and excretion of procyanidins. *Food Funct.* **2016**, *7*, 1273–1281. [CrossRef] [PubMed]
81. Manach, C.; Scalbert, A.; Morand, C.; Remesy, C.; Jimenez, L. Polyphenols: Food sources and bioavailability. *Am. J. Clin. Nutr.* **2004**, *79*, 727–747. [PubMed]
82. Thilakarathna, S.H.; Rupasinghe, H.P. Flavonoid bioavailability and attempts for bioavailability enhancement. *Nutrients* **2013**, *5*, 3367–3387. [CrossRef] [PubMed]
83. Gee, J.M.; DuPont, M.S.; Rhodes, M.J.; Johnson, I.T. Quercetin glucosides interact with the intestinal glucose transport pathway. *Free Radic. Biol. Med.* **1998**, *25*, 19–25. [CrossRef]
84. Tabasco, R.; Sanchez-Patan, F.; Monagas, M.; Bartolome, B.; Victoria Moreno-Arribas, M.; Pelaez, C.; Requena, T. Effect of grape polyphenols on lactic acid bacteria and bifidobacteria growth: Resistance and metabolism. *Food Microbiol.* **2011**, *28*, 1345–1352. [CrossRef] [PubMed]
85. Selma, M.V.; Espin, J.C.; Tomas-Barberan, F.A. Interaction between phenolics and gut microbiota: Role in human health. *J. Agric. Food Chem.* **2009**, *57*, 6485–6501. [CrossRef] [PubMed]
86. Baba, S.; Osakabe, N.; Natsume, M.; Terao, J. Absorption and urinary excretion of procyanidin B2 [epicatechin-(4β-8)-epicatechin] in rats. *Free Radic. Biol. Med.* **2002**, *33*, 142–148. [CrossRef]
87. Holt, R.R.; Lazarus, S.A.; Sullards, M.C.; Zhu, Q.Y.; Schramm, D.D.; Hammerstone, J.F.; Fraga, C.G.; Schmitz, H.H.; Keen, C.L. Procyanidin dimer B2 [epicatechin-(4β-8)-epicatechin] in human plasma after the consumption of a flavanol-rich cocoa. *Am. J. Clin. Nutr.* **2002**, *76*, 798–804. [PubMed]
88. Sano, A.; Yamakoshi, J.; Tokutake, S.; Tobe, K.; Kubota, Y.; Kikuchi, M. Procyanidin B1 is detected in human serum after intake of proanthocyanidin-rich grape seed extract. *Biosci. Biotechnol. Biochem.* **2003**, *67*, 1140–1143. [CrossRef] [PubMed]
89. Shoji, T.; Masumoto, S.; Moriichi, N.; Akiyama, H.; Kanda, T.; Ohtake, Y.; Goda, Y. Apple procyanidin oligomers absorption in rats after oral administration: Analysis of procyanidins in plasma using the porter method and high-performance liquid chromatography/tandem mass spectrometry. *J. Agric. Food Chem.* **2006**, *54*, 884–892. [CrossRef] [PubMed]
90. Tsang, C.; Auger, C.; Mullen, W.; Bornet, A.; Rouanet, J.M.; Crozier, A.; Teissedre, P.L. The absorption, metabolism and excretion of flavan-3-ols and procyanidins following the ingestion of a grape seed extract by rats. *Br. J. Nutr.* **2005**, *94*, 170–181. [CrossRef] [PubMed]
91. Kresty, L.A.; Clarke, J.; Ezell, K.; Exum, A.; Howell, A.B.; Guettouche, T. MicroRNA alterations in barrett's esophagus, esophageal adenocarcinoma, and esophageal adenocarcinoma cell lines following cranberry extract treatment: Insights for chemoprevention. *J. Carcinog.* **2011**, *10*, 34. [CrossRef] [PubMed]
92. Kong, A.N.; Zhang, C.; Su, Z.Y. Targeting epigenetics for cancer prevention by dietary cancer preventive compounds—The case of miRNA. *Cancer Prev. Res. (Phila.)* **2013**, *6*, 622–624. [CrossRef] [PubMed]
93. Arola-Arnal, A.; Blade, C. Proanthocyanidins modulate microRNA expression in human HEPG2 cells. *PLoS ONE* **2011**, *6*, e25982. [CrossRef] [PubMed]
94. Deiuliis, J.A. MicroRNAs as regulators of metabolic disease: Pathophysiologic significance and emerging role as biomarkers and therapeutics. *Int. J. Obes. (Lond.)* **2016**, *40*, 88–101. [CrossRef] [PubMed]
95. Gerin, I.; Clerbaux, L.A.; Haumont, O.; Lanthier, N.; Das, A.K.; Burant, C.F.; Leclercq, I.A.; MacDougald, O.A.; Bommer, G.T. Expression of miR-33 from an SREBP2 intron inhibits cholesterol export and fatty acid oxidation. *J. Biol. Chem.* **2010**, *285*, 33652–33661. [CrossRef] [PubMed]

96. Del Bas, J.M.; Crescenti, A.; Arola-Arnal, A.; Oms-Oliu, G.; Arola, L.; Caimari, A. Intake of grape procyanidins during gestation and lactation impairs reverse cholesterol transport and increases atherogenic risk indexes in adult offspring. *J. Nutr. Biochem.* **2015**, *26*, 1670–1677. [CrossRef] [PubMed]
97. Tsai, W.C.; Hsu, S.D.; Hsu, C.S.; Lai, T.C.; Chen, S.J.; Shen, R.; Huang, Y.; Chen, H.C.; Lee, C.H.; Tsai, T.F.; *et al.* MicroRNA-122 plays a critical role in liver homeostasis and hepatocarcinogenesis. *J. Clin. Invest.* **2012**, *122*, 2884–2897. [CrossRef] [PubMed]
98. Horie, T.; Ono, K.; Horiguchi, M.; Nishi, H.; Nakamura, T.; Nagao, K.; Kinoshita, M.; Kuwabara, Y.; Marusawa, H.; Iwanaga, Y.; *et al.* MicroRNA-33 encoded by an intron of sterol regulatory element-binding protein 2 (SREBP2) regulates hdl *in vivo*. *Proc. Natl. Acad. Sci. USA* **2010**, *107*, 17321–17326. [CrossRef] [PubMed]
99. Baselga-Escudero, L.; Blade, C.; Ribas-Latre, A.; Casanova, E.; Suarez, M.; Torres, J.L.; Salvado, M.J.; Arola, L.; Arola-Arnal, A. Resveratrol and EGCG bind directly and distinctively to miR-33a and miR-122 and modulate divergently their levels in hepatic cells. *Nuclei. Acids Res.* **2014**, *42*, 882–892. [CrossRef] [PubMed]
100. Vinson, J.A.; Mandarano, M.A.; Shuta, D.L.; Bagchi, M.; Bagchi, D. Beneficial effects of a novel IH636 grape seed proanthocyanidin extract and a niacin-bound chromium in a hamster atherosclerosis model. *Mol. Cell Biochem.* **2002**, *240*, 99–103. [CrossRef] [PubMed]
101. Tong, H.; Song, X.; Sun, X.; Sun, G.; Du, F. Immunomodulatory and antitumor activities of grape seed proanthocyanidins. *J. Agric. Food Chem.* **2011**, *59*, 11543–11547. [CrossRef] [PubMed]
102. Ding, Y.; Dai, X.; Jiang, Y.; Zhang, Z.; Bao, L.; Li, Y.; Zhang, F.; Ma, X.; Cai, X.; Jing, L.; *et al.* Grape seed proanthocyanidin extracts alleviate oxidative stress and er stress in skeletal muscle of low-dose streptozotocin- and high-carbohydrate/high-fat diet-induced diabetic rats. *Mol. Nutr. Food Res.* **2013**, *57*, 365–369. [CrossRef] [PubMed]
103. Ding, Y.; Dai, X.; Jiang, Y.; Zhang, Z.; Li, Y. Functional and morphological effects of grape seed proanthocyanidins on peripheral neuropathy in rats with type 2 diabetes mellitus. *Phytother. Res.* **2014**, *28*, 1082–1087. [CrossRef] [PubMed]
104. Katiyar, S.K. Dietary proanthocyanidins inhibit uv radiation-induced skin tumor development through functional activation of the immune system. *Mol. Nutr. Food Res.* **2016**. [CrossRef] [PubMed]
105. Milenkovic, D.; Deval, C.; Gouranton, E.; Landrier, J.F.; Scalbert, A.; Morand, C.; Mazur, A. Modulation of miRNA expression by dietary polyphenols in apoE deficient mice: A new mechanism of the action of polyphenols. *PLoS ONE* **2012**, *7*, e29837. [CrossRef] [PubMed]
106. Feng, Y.; Niu, L.L.; Wei, W.; Zhang, W.Y.; Li, X.Y.; Cao, J.H.; Zhao, S.H. A feedback circuit between miR-133 and the ERK1/2 pathway involving an exquisite mechanism for regulating myoblast proliferation and differentiation. *Cell Death Dis.* **2013**, *4*, e934. [CrossRef] [PubMed]
107. Castell-Auvi, A.; Cedo, L.; Movassat, J.; Portha, B.; Sanchez-Cabo, F.; Pallares, V.; Blay, M.; Pinent, M.; Ardevol, A. Procyanidins modulate microRNA expression in pancreatic islets. *J. Agric. Food Chem.* **2013**, *61*, 355–363. [CrossRef] [PubMed]
108. Ma, J.; Fang, B.; Ma, C.; Pang, H.; Zeng, F.; Xia, J. Proanthocyanidins inhibit pancreatic cancer AsPC-1 cell growth and migration through up-regulation of let-7a. *Nan Fang Yi Ke Da Xue Xue Bao* **2015**, *35*, 1110–1115. [PubMed]
109. Prasad, R.; Katiyar, S.K. Down-regulation of miRNA-106b inhibits growth of melanoma cells by promoting G1-phase cell cycle arrest and reactivation of p21/WAF1/Cip1 protein. *Oncotarget* **2014**, *5*, 10636–10649. [CrossRef] [PubMed]
110. Baselga-Escudero, L.; Blade, C.; Ribas-Latre, A.; Casanova, E.; Salvado, M.J.; Arola, L.; Arola-Arnal, A. Grape seed proanthocyanidins repress the hepatic lipid regulators miR-33 and miR-122 in rats. *Mol. Nutr. Food Res.* **2012**, *56*, 1636–1646. [CrossRef] [PubMed]

2

Regulation of Sphingolipid Metabolism by MicroRNAs: A Potential Approach to Alleviate Atherosclerosis

Zainab Jahangir [1], Ahmed Bakillah [2] and Jahangir Iqbal [2,*]

1 John P. Stevens High School, North Edison, NJ 08820, USA; 3006676@edison.k12.nj.us
2 King Abdullah International Medical Research Center, Ministry of National Guard Health Affairs, Al Ahsa 31982, Saudi Arabia; bakillahah@ngha.med.sa
* Correspondence: iqbalja@ngha.med.sa.

Abstract: The rapidly expanding field of bioactive lipids is exemplified by the many sphingolipids, which are structurally and functionally diverse molecules with significant physiologic functions. These sphingolipids are main constituents of cellular membranes and have been found associated with plasma lipoproteins, and their concentrations are altered in several metabolic disorders such as atherosclerosis, obesity, and diabetes. Understanding the mechanisms that regulate their biosynthesis and secretion may provide novel information that might be amenable to therapeutic targeting in the treatment of these diseases. Several sphingolipid synthesis genes have been targeted as potential therapeutics for atherosclerosis. In recent years, significant progress has been made in studying the role of microRNAs (miRNAs) in lipid metabolism. However, little effort has been made to investigate their role in sphingolipid metabolism. Sphingolipid biosynthetic pathways involve various enzymes that lead to the formation of several key molecules implicated in atherosclerosis, and the identification of miRNAs that regulate these enzymes could help us to understand these complex pathways better and may prove beneficial in alleviating atherosclerosis.

Keywords: atherosclerosis; ceramides; lipids; lipoproteins; miRNA; sphingolipids; sphingomyelin

1. Introduction

High plasma lipid levels are major risk factors for several cardiovascular and metabolic disorders such as atherosclerosis, obesity, and diabetes. Some of the most important risk factors for atherosclerosis are the circulating levels of low-density lipoprotein (LDL) and high-density lipoprotein (HDL) cholesterol [1,2]. Besides traditional risk factors, changes in sphingolipids may contribute to the pathogenesis of cardiovascular disease (CVD) [3,4]. Sphingolipids are a class of lipids that contain a sphingoid base, an aliphatic amino alcohol including sphingosine. Sphingoid bases such as dihydrosphingosine and sphingosine are the fundamental building blocks of all sphingolipids. Sphingolipids are biologically active cell components that regulate cellular processes and play an important role in signal transduction and cellular stress responses. The synthesis and degradation of sphingolipids, which serve as both structural lipids as well as signaling molecules, are regulated to maintain homeostasis [3]. Sphingolipids are either derived from other sphingolipids through catabolism via the salvage pathway or synthesized de novo in the endoplasmic reticulum [5]. Ceramide is a simple sphingolipid composed of sphingosine and a fatty acid. Ceramide constitutes the hydrophobic backbone and serves as the key precursor for the de novo synthesis of other biologically active complex sphingolipids (Figure 1) [6]. It represents a nodal point in the sphingolipid de novo pathway. It can be glycosylated or acquire a polar head group to form glycosphingolipids or sphingomyelin (SM), respectively [3]. It can also be reversibly degraded to form sphingosine which in turn can be phosphorylated by sphingosine kinase (SPK) to form sphingosine-1-phospate (S1P). Once synthesized, sphingolipids can be transported to plasma lipoproteins [7] or remain associated with cellular membranes. An imbalance of sphingolipid levels in plasma and tissue is associated with several metabolic diseases including atherosclerosis [4].

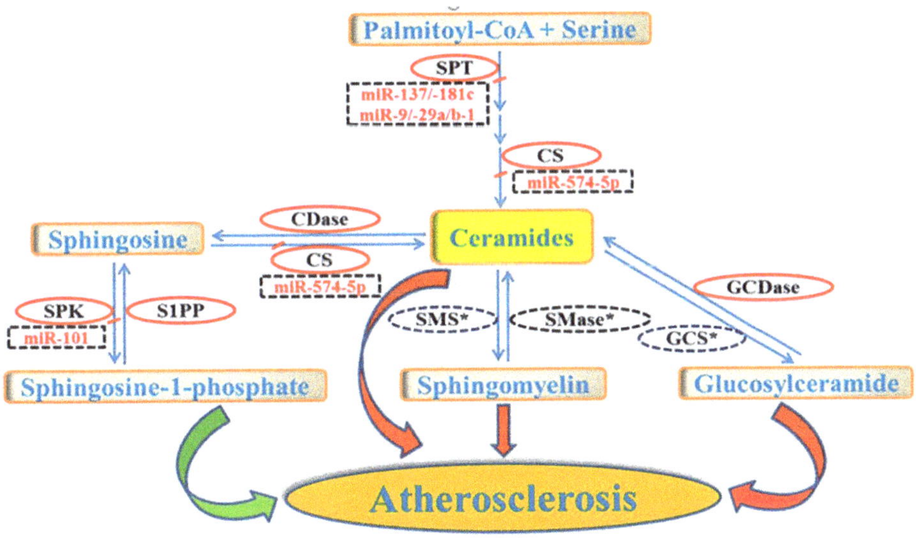

Figure 1. Schematic representation of miRNAs implicated in the synthesis of some key sphingolipid molecules. The figure is representative of some key enzymatic steps involved in sphingolipid biosynthetic pathways that are known to be regulated by various miRNAs (dotted rectangles). Enzymes with asterisks (dotted ovals) may be potential targets for other miRNAs that need to be identified. Increased levels of ceramides, sphingomyelin and glucosylceramide (red arrows) and decreased levels of sphingosine-1-phosphate (green arrow) have been implicated in atherosclerosis. The identification of novel miRNAs that regulate sphingolipid metabolism may be a potential therapeutic target to treat atherosclerosis. Abbreviations: SPT, serine palmitoyl transferase; CDase, ceramidase; CS, ceramide synthase; SPK, sphingosine kinase; S1PP, sphingosine-1-phosphate phosphatase; SMS, sphingomyelin synthase; SMase, sphingomyelinase; GCS, glucosylceramide synthase; GCDase, glucosylceramidase.

An increased plasma SM level has been proposed as an independent risk factor for coronary heart disease and has been shown to be associated with increased atherosclerosis in humans [8]. A reduction in SM levels in sphingomyelin synthase (SMS) knockout mice is known to decrease atherosclerosis [9,10]. Like SM, increased plasma and aortic ceramide levels are also associated with an increased risk of CVD [11]. Genetic deficiency or the inhibition of type 2-neutral sphingomyelinase (nSMase2), a key enzyme in sphingolipid metabolism, has been shown to decrease atherosclerosis in ApoE knockout mice by reducing inflammatory responses due to a decrease in ceramide levels [12]. Furthermore, plasma glycosphingolipid concentrations have been reported to be elevated in patients at increased risk of atherosclerosis [13]. The pharmacological inhibition of glucosylceramide synthase that is responsible for the synthesis of glucosylceramides has been reported to ameliorate atherosclerosis in ApoE knockout mice and rabbits [14]. In contrast to ceramide, glycosphingolipid, and SM, plasma S1P is believed to be cardioprotective [3]. Regulation of the interconvertible sphingolipid metabolites, ceramide and S1P, and their opposing signaling pathways may determine the net biological effect, a concept referred to as the "sphingolipid rheostat". Plasma S1P levels significantly decrease after myocardial infarction [15] and increase in patients after percutaneous coronary intervention [16]. It is believed that S1P bound to HDL may predict the severity of coronary heart disease [17]. Recently, it has been shown that apoM acts as a carrier and modulator of S1P that largely affects its homeostasis [18].

Lipid metabolism is a multi-faceted process that involves synthesis, accumulation, secretion, distribution to various tissues, degradation as well as excretion. Lipid metabolism is regulated by finely modulating a set of rate-limiting enzymes and transporters based on the needs of the cells [19]. A variety of cellular regulators including transcription factors that are involved in its synthesis and degradation are responsible for maintaining lipid homeostasis. The discovery that the cellular lipid and lipoprotein metabolism is targeted by miRNAs has provided new insight into the molecular mechanism of the pathogenesis of atherosclerosis, diabetes and obesity [20]. This has led to the exploitation of miRNAs as regulators of lipid metabolism with subsequent potential use in the treatment of lipid metabolism-related disorders [21,22]. This mini-review provides an overview of what is known about the regulation of lipid metabolism in general and sphingolipid metabolism in particular by miRNAs and emphasize their role in atherosclerosis.

2. Regulation of Lipid and Lipoprotein Metabolism by miRNAs

miRNAs are a class of small ~22-nucleotide long non-coding sequences which post-transcriptionally regulate gene expression to modulate a wide spectrum of biological processes [23,24]. They act as negative post-transcriptional regulators of gene expression [21,22] and have been proposed to regulate the expression of more than half of the human genes via controlling the expression of mRNA targets. miRNAs repress protein expression by mRNA destabilization and/or translational inhibition after binding the complementary sequences in the 3′ untranslated region (3′-UTR) of target mRNAs via seed sequences (Table 1) [25]. Individual miRNAs can target several mRNAs that are often connected in the same metabolic pathway [26]. Moreover, several miRNAs may regulate a single mRNA, thus enabling a fine-tuning of the targeted mRNA expression, to influence the regulation of cellular events [23,26]. Dysregulation of miRNAs has been correlated with disease pathogenesis [27]. Furthermore, miRNAs can be released from cells as exosomes in many body fluids and may serve as noninvasive biomarkers of diseases [28].

Table 1. Predicted seed sequence, target genes and tissues of miRNAs implicated in lipid and sphingolipid metabolism.

miRNAs	Predicted Seed Sequence *	Target Genes	Target Tissues
miR-33	GUUACGU	ABCA1, CROT, CPT1A, HADHB, ACLY, SREBF1, ACACA	Liver
miR-144	UAUGACA	ABCA1	Liver
miR-758	CAGUGUU	ABCA1	Liver
miR-26	AUGAACU	ABCA1	Liver
miR-106b	CGUGAAA	ABCA1	Liver
miR-27	AUUCGAG	SR-B1	Liver
miR-185	AGAGAGG	SR-B1	Liver
miR-96	CACGGUU	SR-B1, ABCA1	Liver
miR-223	UUGACUG	SR-B1	Liver
miR-30c	CAAAUG	MTTP, LPGAT1, ELOVL5, STARD3, MBOAT1	Liver
miR-128-1	GCCGGGG	LDLR	Liver
miR-148	ACGUGAC	LDLR, ABCA1	Liver
miR-122	UGUGAGG	FASN, SCD1, ACLY, ACC2	Liver
miR-155	CGUAAU	LXRα	Liver
miR-574	GUGUGAG	CerS	Multiple human cancer cells
miR-9	UGGUUUC	SPTLC1, SPTLC2	Primary astrocytes
miR-29a	UUUAGUC	SPTLC1, SPTLC2	Primary astrocytes
miR-29b-1	UUUGGUC	SPTLC1, SPTLC2	Primary astrocytes
miR-101	CUAUUGA	SPK	Colorectal cancer cells

* Conserved seed sequence for each miRNA was predicted by using TargetScan Human database (http://www.targetscan.org/vert_72/).

Over the past few decades, the role of miRNAs in controlling the cellular lipid and lipoprotein metabolism has been extensively studied. A large array of miRNAs participate in the lipid and lipoprotein metabolism by targeting enzymes and proteins that are involved in these metabolic pathways. This discovery has provided new insight into the biology and pathophysiology of CVD such as atherosclerosis [20,29]. Higher levels of HDL are inversely correlated with developing CVD. A number of miRNAs, of which miR-33a and miR-33b are the most well-studied, regulate HDL metabolism and control circulating cholesterol levels by regulating HDL synthesis, efflux and clearance [30,31]. miR-33a and miR-33b target multiple genes that regulate cholesterol efflux, which is regarded as an important cholesterol regulatory mechanism [32]. Studies in various animal models and cell lines have confirmed that miR-33 targets ATP-binding cassette (ABC) transporter A1, *ABCA1*, by binding to *3′-UTR* of *ABCA1* via seed sequence to regulate cholesterol metabolism by attenuating circulating HDL levels and thereby decreasing cholesterol efflux to apolipoprotein A1 (apoA1) [30,32]. Conversely, the silencing of miR-33 by using lentiviral expression clone containing anti-sense to miR-33 in experimental animal models generally results in increased hepatic expression of *ABCA1* and enhanced cholesterol efflux to ApoA1, thus increasing circulating HDL levels [32].

In vivo delivery of other miRNAs such as miR-144, miR-758, miR-26, and miR-106b by using either adenoviral or lentiviral expression clones has been shown to have similar results of decreased ABCA1 expression and reduced levels of circulating HDL-cholesterol [33]. The post-transcriptional repression of scavenger receptor class B type 1 (SR-B1) by miR-27, miR-185, miR-96, and miR-223 has also been shown to reduce selective HDL-cholesterol uptake [33]. Reverse cholesterol transport decreases cholesterol levels in peripheral macrophages and in atherosclerotic plaques to increase plaque stability and inhibit atherosclerosis progression [34]. These studies show that HDL metabolism provides a potential therapeutic target to treat atherosclerosis by regulating *ABCA1* via miRNAs.

Besides the HDL metabolism, reports have shown that miRNAs such as miR-30c, miR128-1, or miR-148a are also involved in controlling plasma LDL-cholesterol levels by regulating genes involved in cholesterol biosynthesis, very low-density lipoprotein (VLDL) secretion, and hepatic LDL receptor expression [30]. Recently, it was shown that miR-30c targets microsomal triglyceride transfer protein (MTTP) to regulate VLDL biogenesis and modulate lipid substrate availability for VLDL assembly by targeting genes involved in lipid biosynthesis such as lysophosphatidylglycerol acyltransferase 1 (LPGAT1), ELOVL fatty acid elongase 5 (ELOVL5), stAR related lipid transfer domain containing 3 (STARD3), and membrane bound *O*-acyltransferase domain containing 1 (MBOAT1) [35]. Inhibition of miR-122 has been shown to increase fatty acid oxidation and inhibit lipid synthesis genes such as fatty acid synthase (FASN), steroyl-coA desaturase 1 (SCD1), ATP citrate lyase (ACLY) and acetyl-coA carboxylase 2 (ACC2) and thereby reduce the availability of lipid substrates for VLDL biogenesis [36]. Similarly, inhibition of miR-33a and miR-33b expression in the liver is shown to increase fatty acid oxidation by increasing the expression of carnitine *O*-octanoyltransferase (CROT), carnitine palmitoyltransferase 1A (CPT1A) and hydroxyacyl-CoA dehydrogenase trifunctional multienzyme complex subunit beta (HADHB), and decrease its synthesis by reducing the synthesis of FASN, ACLY, sterol regulatory element binding factor 1 (SREBF1) and acetyl-CoA carboxylase alpha (ACACA) to reduce VLDL secretion in the circulation [37]. Furthermore, reports have suggested that increased expression of miR-155 in liver macrophages regulates fatty acid metabolism by directly targeting liver X receptor alpha (LXRα) and thereby reducing hepatic lipid accumulation [38]. These studies indicate that miRNAs exert their regulatory impacts at different levels of lipoprotein biosynthesis.

3. Regulation of Sphingolipid Metabolism by miRNAs

Sphingolipids have been recognized to regulate distinct biological functions beyond their role as structural membrane components. Within the past few decades, significant progress has been made toward understanding the role of sphingolipid pathways for atherosclerosis. The role of miRNAs to regulate sphingolipid metabolism has not been widely studied. There are only a few limited studies, done mostly in cancer cell lines, that have investigated the role of miRNAs in the regulation of enzymes involved in sphingolipid biosynthetic pathways (Figure 1) [39–41]. Ceramide is a central molecule in sphingolipid metabolism [6] which is generated either through de novo pathway or salvage pathway [5]. Ceramide acts as a signaling molecule, regulating many cellular responses and functions that may be involved in molecular mechanisms of CVD. The generation of ceramide can be significantly enhanced in certain inflammatory conditions such as atherosclerosis [12]. Ceramide is synthesized by a family of six ceramide synthases (CerS), each of which synthesizes ceramide with distinct acyl chain lengths. An alternatively spliced variant of ceramide synthase, CerS1-2, which is responsible for synthesizing C18-ceramide, is a target for miR-574-5p [41]. The knockdown of miR-574-5p expression has been shown to increase C18-ceramide levels in multiple human cancer cell lines [41].

Serine-palmitoyl transferase (SPT), which converts L-serine and palmitoyl-CoA to 3-ketosphinganine, is the first and rate-limiting enzyme of the de novo biosynthetic pathway of ceramide and SM. SPT has been implicated in the pathogenesis of atherosclerosis and its modulation has been suggested to be a novel therapeutic target in atherosclerosis [42]. Several studies have shown that the pharmacological targeting of SPT through the inhibition with myriocin protects from atherosclerosis in ApoE knockout mice [42–45]. Myriocin has been shown to reduce the levels of not only ceramide but other sphingolipids downstream of ceramide, such as SM, glucosylceramide and S1P as well [42,44,45]. Similar to pharmacological intervention, targeting the enzymes involved in sphingolipid biosynthesis through miRNAs may be a potential therapeutic intervention for alleviating atherosclerosis. Serine-palmitoyl transferase long chain base subunit 1 (SPTLC1) and 2 (SPTLC2) are the subunits of SPT and one study found a negative correlation between the expression levels of miR-137/-181c and SPTLC1 [40]. Transfection of primary rat astrocytes with miR-137 and miR-181c showed a significant suppression of the endogenous SPTLC1 expression and cellular ceramide levels. On the other hand, anti-miR-137 and anti-miR-181c significantly increased the endogenous SPTLC1 expression and cellular ceramide levels. In the same study, a negative correlation was also found between miR-9/-29a/b-1 and SPTLC2 protein expression in sporadic Alzheimer's disease brains. Again, the transient transfection of primary rat astrocytes with miR-9, miR-29a, and miR-29b-1 significantly suppressed the endogenous SPTLC2 and cellular ceramide levels and their antagomirs significantly enhanced the expression of SPTLC2 and cellular ceramide levels.

Finally, plasma S1P levels have been reported to be lower in CVD patients, suggesting its involvement in the pathogenesis of atherosclerosis [15]. Sphingosine kinase is involved in the conversion of sphingosine to S1P that have distinct intracellular and extracellular functions [46,47]. Exogenously expressed miR-101 has been shown to down-regulate SPK mRNA and protein expression in colorectal cancer cells [39]. The downregulation of SPK has been shown to result in increased ceramide levels in miR-101 expressed cells. On the other hand, the expression of SPK was enhanced by treating HT-29 human colorectal adenocarcinoma cells with the antagomiR-101 that resulted in lower levels of ceramide [39].

4. Conclusions

Significant advances have been made in studying the role of miRNAs in lipid metabolism. However, there is a paucity of information on their role in sphingolipid metabolism. Several sphingolipid synthesis genes have been targeted as potential therapeutics for various metabolic disorders, such as atherosclerosis. Sphingomyelin synthase, sphingomyelinase, and glucosylceramide synthase are some of the key enzymes that have been implicated in atherosclerosis [9,10,12,14] (Figure 1). It is likely that studying the panel of miRNAs that regulate sphingolipid metabolism could help us to understand these pathways better, and the modulation of these pathways may prove to be a potential therapeutic strategy. There are only a few studies that have reported the regulation of enzymes involved in sphingolipid biosynthetic pathways by certain miRNAs [39–41], and most of these studies are targeted towards ceramide synthesis in various cancer cell lines [39,41]. However, there is a lack of information on whether miRNAs regulate sphingolipid metabolism in vivo and thereby influence atherosclerosis. Based on the genetic deficiency and pharmacological inhibition studies [9,10,12,14,42], we predict that the regulation of some of the key enzymes in sphingolipid biosynthetic pathways by miRNAs may be useful and a potential therapeutic strategy to alleviate atherosclerosis. Therefore, more focused studies are needed to identify and understand the role of various miRNAs in regulating sphingolipid metabolism.

Author Contributions: Z.J. collected materials and edited the manuscript. A.B. and J.I. conceived and wrote the manuscript. All the authors read and approved the manuscript.

Abbreviations

ABCA1	ATP-binding cassette transporter A1
ACACA	acetyl-CoA carboxylase alpha
ACC2	acetyl-coA carboxylase 2
ACLY	ATP citrate lyase
ApoA1	apolipoprotein A1
CerS	ceramide synthase
CPT1A	carnitine palmitoyltransferase 1A
CROT	carnitine *O*-octanoyltransferase
CVD	cardiovascular disease
ELOVL5	ELOVL fatty acid elongase 5
FASN	fatty acid synthase
HADHB	hydroxyacyl-CoA dehydrogenase trifunctional multienzyme complex subunit beta
HDL	high-density lipoprotein
LDL	low-density lipoprotein
LPGAT1	lysophosphatidylglycerol acyltransferase 1
LXRα	liver X receptor alpha
MBOAT1	membrane bound *O*-acyltransferase domain containing 1
miRNAs	microRNAs
MTTP	microsomal triglyceride transfer protein
SCD1	stearoyl-coA desaturase 1
SM	sphingomyelin
SMS	sphingomyelin synthase
S1P	sphingosine-1-phosphate
SPK	sphingosine kinase
SPT	serine-palmitoyl transferase
SPTLC1	serine-palmitoyl transferase long chain base subunit 1
SPTLC2	serine-palmitoyl transferase long chain base subunit 2
SR-B1	scavenger receptor class B type 1
nSMase2	type 2-neutral sphingomyelinase
SREBF1	sterol regulatory element binding factor 1
STARD3	stAR related lipid transfer domain containing 3
3′-UTR	3′ untranslated region
VLDL	very low-density lipoprotein

References

1. Glass, C.K.; Witztum, J.L. Atherosclerosis. The road ahead. *Cell* **2001**, *104*, 503–516. [CrossRef]
2. Lusis, A.J. Atherosclerosis. *Nature* **2000**, *407*, 233–241. [CrossRef] [PubMed]
3. Borodzicz, S.; Czarzasta, K.; Kuch, M.; Cudnoch-Jedrzejewska, A. Sphingolipids in cardiovascular diseases and metabolic disorders. *Lipids Health Dis.* **2015**, *14*, 55. [CrossRef] [PubMed]
4. Iqbal, J.; Walsh, M.T.; Hammad, S.M.; Hussain, M.M. Sphingolipids and Lipoproteins in Health and Metabolic Disorders. *Trends Endocrinol. Metab.* **2017**, *28*, 506–518. [CrossRef] [PubMed]

5. Pewzner-Jung, Y.; Ben-Dor, S.; Futerman, A.H. When do Lasses (longevity assurance genes) become CerS (ceramide synthases)? Insights into the regulation of ceramide synthesis. *J. Biol. Chem.* **2006**, *281*, 25001–25005. [CrossRef] [PubMed]
6. Ogretmen, B.; Hannun, Y.A. Biologically active sphingolipids in cancer pathogenesis and treatment. *Nat. Rev. Cancer* **2004**, *4*, 604–616. [CrossRef] [PubMed]
7. Iqbal, J.; Walsh, M.T.; Hammad, S.M.; Cuchel, M.; Tarugi, P.; Hegele, R.A.; Davidson, N.O.; Rader, D.J.; Klein, R.L.; Hussain, M.M. Microsomal triglyceride transfer protein transfers and determines plasma concentrations of ceramide and sphingomyelin but not glycosylceramide. *J. Biol. Chem.* **2015**, *290*, 25863–25875. [CrossRef] [PubMed]
8. Jiang, X.C.; Paultre, F.; Pearson, T.A.; Reed, R.G.; Francis, C.K.; Lin, M.; Berglund, L.; Tall, A.R. Plasma sphingomyelin level as a risk factor for coronary artery disease. *Arterioscler. Thromb. Vasc. Biol.* **2000**, *20*, 2614–2618. [CrossRef] [PubMed]
9. Li, Z.; Fan, Y.; Liu, J.; Li, Y.; Huan, C.; Bui, H.H.; Kuo, M.S.; Park, T.S.; Cao, G.; Jiang, X.C. Impact of sphingomyelin synthase 1 deficiency on sphingolipid metabolism and atherosclerosis in mice. *Arterioscler. Thromb. Vasc. Biol.* **2012**, *32*, 1577–1584. [CrossRef] [PubMed]
10. Liu, J.; Huan, C.; Chakraborty, M.; Zhang, H.; Lu, D.; Kuo, M.S.; Cao, G.; Jiang, X.C. Macrophage sphingomyelin synthase 2 deficiency decreases atherosclerosis in mice. *Circ. Res.* **2009**, *105*, 295–303. [CrossRef] [PubMed]
11. Kasumov, T.; Li, L.; Li, M.; Gulshan, K.; Kirwan, J.P.; Liu, X.; Previs, S.; Willard, B.; Smith, J.D.; McCullough, A. Ceramide as a mediator of non-alcoholic Fatty liver disease and associated atherosclerosis. *PLoS. ONE* **2015**, *10*, e0126910. [CrossRef] [PubMed]
12. Lallemand, T.; Rouahi, M.; Swiader, A.; Grazide, M.H.; Geoffre, N.; Alayrac, P.; Recazens, E.; Coste, A.; Salvayre, R.; Negre-Salvayre, A.; et al. nSMase2 (Type 2-Neutral Sphingomyelinase) Deficiency or Inhibition by GW4869 Reduces Inflammation and Atherosclerosis in Apoe (−/−) Mice. *Arterioscler. Thromb. Vasc. Biol.* **2018**, *38*, 1479–1492. [CrossRef] [PubMed]
13. Dawson, G.; Kruski, A.W.; Scanu, A.M. Distribution of glycosphingolipids in the serum lipoproteins of normal human subjects and patients with hypo- and hyperlipidemias. *J. Lipid Res.* **1976**, *17*, 125–131. [PubMed]
14. Chatterjee, S.; Bedja, D.; Mishra, S.; Amuzie, C.; Avolio, A.; Kass, D.A.; Berkowitz, D.; Renehan, M. Inhibition of glycosphingolipid synthesis ameliorates atherosclerosis and arterial stiffness in apolipoprotein E-/- mice and rabbits fed a high-fat and -cholesterol diet. *Circulation* **2014**, *129*, 2403–2413. [CrossRef] [PubMed]
15. Knapp, M.; Lisowska, A.; Zabielski, P.; Musial, W.; Baranowski, M. Sustained decrease in plasma sphingosine-1-phosphate concentration and its accumulation in blood cells in acute myocardial infarction. *Prostag. Other Lipid Mediat.* **2013**, *106*, 53–61. [CrossRef] [PubMed]
16. Egom, E.E.; Mamas, M.A.; Chacko, S.; Stringer, S.E.; Charlton-Menys, V.; El-Omar, M.; Chirico, D.; Clarke, B.; Neyses, L.; Cruickshank, J.K.; et al. Serum sphingolipids level as a novel potential marker for early detection of human myocardial ischaemic injury. *Front. Physiol.* **2013**, *4*, 130. [CrossRef] [PubMed]
17. Sattler, K.; Lehmann, I.; Graler, M.; Brocker-Preuss, M.; Erbel, R.; Heusch, G.; Levkau, B. HDL-bound sphingosine 1-phosphate (S1P) predicts the severity of coronary artery atherosclerosis. *Cell. Physiol. Biochem.* **2014**, *34*, 172–184. [CrossRef] [PubMed]
18. Kurano, M.; Yatomi, Y. Sphingosine 1-Phosphate and Atherosclerosis. *J. Atheroscler. Thromb.* **2018**, *25*, 16–26. [CrossRef] [PubMed]
19. Santulli, G. Effects of low-carbohydrate and low-fat diets. *Ann. Intern. Med.* **2015**, *162*, 392. [CrossRef] [PubMed]
20. Christian, P.; Su, Q. MicroRNA regulation of mitochondrial and ER stress signaling pathways: Implications for lipoprotein metabolism in metabolic syndrome. *Am. J. Physiol. Endocrinol. Metab.* **2014**, *307*, E729–E737. [CrossRef] [PubMed]
21. Novak, J.; Bienertova-Vasku, J.; Kara, T.; Novak, M. MicroRNAs involved in the lipid metabolism and their possible implications for atherosclerosis development and treatment. *Mediat. Inflamm.* **2014**, *2014*, 275867. [CrossRef] [PubMed]
22. Wronska, A.; Kurkowska-Jastrzebska, I.; Santulli, G. Application of microRNAs in diagnosis and treatment of cardiovascular disease. *Acta Physiol. (Oxf.)* **2015**, *213*, 60–83. [CrossRef] [PubMed]

23. Bartel, D.P. MicroRNAs: Target recognition and regulatory functions. *Cell* **2009**, *136*, 215–233. [CrossRef] [PubMed]
24. Filipowicz, W.; Bhattacharyya, S.N.; Sonenberg, N. Mechanisms of post-transcriptional regulation by microRNAs: Are the answers in sight? *Nat. Rev. Genet.* **2008**, *9*, 102–114. [CrossRef] [PubMed]
25. Wilczynska, A.; Bushell, M. The complexity of miRNA-mediated repression. *Cell Death Differ.* **2015**, *22*, 22–33. [CrossRef] [PubMed]
26. Lim, L.P.; Lau, N.C.; Garrett-Engele, P.; Grimson, A.; Schelter, J.M.; Castle, J.; Bartel, D.P.; Linsley, P.S.; Johnson, J.M. Microarray analysis shows that some microRNAs downregulate large numbers of target mRNAs. *Nature* **2005**, *433*, 769–773. [CrossRef] [PubMed]
27. Ha, T.Y. MicroRNAs in Human Diseases: From Cancer to Cardiovascular Disease. *Immune Netw.* **2011**, *11*, 135–154. [CrossRef] [PubMed]
28. Scholer, N.; Langer, C.; Dohner, H.; Buske, C.; Kuchenbauer, F. Serum microRNAs as a novel class of biomarkers: A comprehensive review of the literature. *Exp. Hematol.* **2010**, *38*, 1126–1130. [CrossRef] [PubMed]
29. Olson, E.N. MicroRNAs as therapeutic targets and biomarkers of cardiovascular disease. *Sci. Transl. Med.* **2014**, *6*, 239ps3. [CrossRef] [PubMed]
30. Aryal, B.; Singh, A.K.; Rotllan, N.; Price, N.; Fernandez-Hernando, C. MicroRNAs and lipid metabolism. *Curr. Opin. Lipidol.* **2017**, *28*, 273–280. [CrossRef] [PubMed]
31. Zaiou, M.; Bakillah, A. Epigenetic Regulation of ATP-Binding Cassette Protein A1 (ABCA1) Gene Expression: A New Era to Alleviate Atherosclerotic Cardiovascular Disease. *Diseases* **2018**, *6*, 34. [CrossRef] [PubMed]
32. Rayner, K.J.; Suarez, Y.; Davalos, A.; Parathath, S.; Fitzgerald, M.L.; Tamehiro, N.; Fisher, E.A.; Moore, K.J.; Fernandez-Hernando, C. MiR-33 contributes to the regulation of cholesterol homeostasis. *Science* **2010**, *328*, 1570–1573. [CrossRef] [PubMed]
33. Sud, N.; Taher, J.; Su, Q. MicroRNAs and Noncoding RNAs in Hepatic Lipid and Lipoprotein Metabolism: Potential Therapeutic Targets of Metabolic Disorders. *Drug Dev. Res.* **2015**, *76*, 318–327. [CrossRef] [PubMed]
34. Novak, J.; Olejnickova, V.; Tkacova, N.; Santulli, G. Mechanistic Role of MicroRNAs in Coupling Lipid Metabolism and Atherosclerosis. *Adv. Exp. Med. Biol.* **2015**, *887*, 79–100. [PubMed]
35. Soh, J.; Iqbal, J.; Queiroz, J.; Fernandez-Hernando, C.; Hussain, M.M. MicroRNA-30c reduces hyperlipidemia and atherosclerosis in mice by decreasing lipid synthesis and lipoprotein secretion. *Nat. Med.* **2013**, *19*, 892–900. [CrossRef] [PubMed]
36. Esau, C.; Davis, S.; Murray, S.F.; Yu, X.X.; Pandey, S.K.; Pear, M.; Watts, L.; Booten, S.L.; Graham, M.; McKay, R.; et al. miR-122 regulation of lipid metabolism revealed by in vivo antisense targeting. *Cell Metab.* **2006**, *3*, 87–98. [CrossRef] [PubMed]
37. Rayner, K.J.; Esau, C.C.; Hussain, F.N.; McDaniel, A.L.; Marshall, S.M.; van Gils, J.M.; Ray, T.D.; Sheedy, F.J.; Goedeke, L.; Liu, X.; et al. Inhibition of miR-33a/b in non-human primates raises plasma HDL and lowers VLDL triglycerides. *Nature* **2011**, *478*, 404–407. [CrossRef] [PubMed]
38. Miller, A.M.; Gilchrist, D.S.; Nijjar, J.; Araldi, E.; Ramirez, C.M.; Lavery, C.A.; Fernandez-Hernando, C.; McInnes, I.B.; Kurowska-Stolarska, M. MiR-155 has a protective role in the development of non-alcoholic hepatosteatosis in mice. *PLoS ONE* **2013**, *8*, e72324. [CrossRef] [PubMed]
39. Chen, M.B.; Yang, L.; Lu, P.H.; Fu, X.L.; Zhang, Y.; Zhu, Y.Q.; Tian, Y. MicroRNA-101 down-regulates sphingosine kinase 1 in colorectal cancer cells. *Biochem. Biophys. Res. Commun.* **2015**, *463*, 954–960. [CrossRef] [PubMed]
40. Geekiyanage, H.; Chan, C. MicroRNA-137/181c regulates serine palmitoyltransferase and in turn amyloid beta, novel targets in sporadic Alzheimer's disease. *J. Neurosci.* **2011**, *31*, 14820–14830. [CrossRef] [PubMed]
41. Meyers-Needham, M.; Ponnusamy, S.; Gencer, S.; Jiang, W.; Thomas, R.J.; Senkal, C.E.; Ogretmen, B. Concerted functions of HDAC1 and microRNA-574-5p repress alternatively spliced ceramide synthase 1 expression in human cancer cells. *EMBO Mol. Med.* **2012**, *4*, 78–92. [CrossRef] [PubMed]
42. Hojjati, M.R.; Li, Z.; Zhou, H.; Tang, S.; Huan, C.; Ooi, E.; Lu, S.; Jiang, X.C. Effect of myriocin on plasma sphingolipid metabolism and atherosclerosis in apoE-deficient mice. *J. Biol. Chem.* **2005**, *280*, 10284–10289. [CrossRef] [PubMed]
43. Park, T.S.; Panek, R.L.; Mueller, S.B.; Hanselman, J.C.; Rosebury, W.S.; Robertson, A.W.; Kindt, E.K.; Homan, R.; Karathanasis, S.K.; Rekhter, M.D. Inhibition of sphingomyelin synthesis reduces atherogenesis in apolipoprotein E-knockout mice. *Circulation* **2004**, *110*, 3465–3471. [CrossRef] [PubMed]

44. Glaros, E.N.; Kim, W.S.; Wu, B.J.; Suarna, C.; Quinn, C.M.; Rye, K.A.; Stocker, R.; Jessup, W.; Garner, B. Inhibition of atherosclerosis by the serine palmitoyl transferase inhibitor myriocin is associated with reduced plasma glycosphingolipid concentration. *Biochem. Pharmacol.* **2007**, *73*, 1340–1346. [CrossRef] [PubMed]
45. Glaros, E.N.; Kim, W.S.; Quinn, C.M.; Jessup, W.; Rye, K.A.; Garner, B. Myriocin slows the progression of established atherosclerotic lesions in apolipoprotein E gene knockout mice. *J. Lipid Res.* **2008**, *49*, 324–331. [CrossRef] [PubMed]
46. Shida, D.; Takabe, K.; Kapitonov, D.; Milstien, S.; Spiegel, S. Targeting SphK1 as a new strategy against cancer. *Curr. Drug Targets* **2008**, *9*, 662–673. [CrossRef] [PubMed]
47. Vadas, M.; Xia, P.; McCaughan, G.; Gamble, J. The role of sphingosine kinase 1 in cancer: Oncogene or non-oncogene addiction? *Biochim. Biophys. Acta* **2008**, *1781*, 442–447. [CrossRef] [PubMed]

3

Hsa-miR-375/RASD1 Signaling may Predict Local Control in Early Breast Cancer

Barbara Zellinger [1,2], Ulrich Bodenhofer [3,4], Immanuela A. Engländer [1,5], Cornelia Kronberger [2], Peter Strasser [6], Brane Grambozov [5], Gerd Fastner [5], Markus Stana [5], Roland Reitsamer [7], Karl Sotlar [2], Felix Sedlmayer [1,5] and Franz Zehentmayr [1,5,*]

1 radART—Institute for Research and Development on Advanced Radiation Technologies, Paracelsus Medical University, Müllner Hauptstrasse 48, 5020 Salzburg, Austria; b.zellinger@salk.at (B.Z.); ila.engl@protonmail.com (I.A.E.); f.sedlmayer@salk.at (F.S.)

2 Department of Pathology, Paracelsus Medical University, SALK, Müllner Hauptstrasse 48, 5020 Salzburg, Austria; c.kronberger@salk.at (C.K.); k.sotlar@salk.at (K.S.)

3 School of Informatics, Communications and Media, University of Applied Sciences Upper Austria, Softwarepark 11, 4232 Hagenberg, Austria; ulrich.bodenhofer@fh-hagenberg.at

4 Institute for Machine Learning, Campus Science Park 3, Johannes Kepler University, Altenbergerstrasse 69, 4040 Linz, Austria

5 Department of Radiation Oncology, Paracelsus Medical University, SALK, Müllner Hauptstrasse 48, 5020 Salzburg, Austria; b.grambozov@salk.at (B.G.); g.fastner@salk.at (G.F.); m.stana@salk.at (M.S.)

6 Department of Laboratory Medicine, Paracelsus Medical University, SALK, Müllner Hauptstrasse 48, 5020 Salzburg, Austria; p.strasser@salk.at

7 Department of Gynecology and Obstetrics, Paracelsus Medical University, SALK, Müllner Hauptstrasse 48, 5020 Salzburg, Austria; r.reitsamer@salk.at

* Correspondence: f.zehentmayr@salk.at.

Abstract: Background: In order to characterize the various subtypes of breast cancer more precisely and improve patients selection for breast conserving therapy (BCT), molecular profiling has gained importance over the past two decades. MicroRNAs, which are small non-coding RNAs, can potentially regulate numerous downstream target molecules and thereby interfere in carcinogenesis and treatment response via multiple pathways. The aim of the current two-phase study was to investigate whether hsa-miR-375-signaling through RASD1 could predict local control (LC) in early breast cancer. Results: The patient and treatment characteristics of 81 individuals were similarly distributed between relapse ($n = 27$) and control groups ($n = 54$). In the pilot phase, the primary tumors of 28 patients were analyzed with microarray technology. Of the more than 70,000 genes on the chip, 104 potential hsa-miR-375 target molecules were found to have a lower expression level in relapse patients compared to controls (p-value < 0.2). For RASD1, a hsa-miR-375 binding site was predicted by an in silico search in five mRNA-miRNA databases and mechanistically proven in previous pre-clinical studies. Its expression levels were markedly lower in relapse patients than in controls (p-value of 0.058). In a second phase, this finding could be validated in an independent set of 53 patients using ddPCR. Patients with enhanced levels of hsa-miR-375 compared to RASD1 had a higher probability of local relapse than those with the inverse expression pattern of the two markers (log-rank test, p-value = 0.069). Conclusion: This two-phase study demonstrates that hsa-miR-375/RASD1 signaling is able to predict local control in early breast cancer patients, which—to our knowledge—is the first clinical report on a miR combined with one of its downstream target proteins predicting LC in breast cancer.

Keywords: hsa-miR-375; RASD1; early stage breast cancer; local control; predictive markers

---

1. Introduction

Breast cancer is the most common malignant tumor in women, with 93,300 deaths predicted for 2020 in Europe [1]. Carioli estimated that, if the trend between 1989 and 2020 had continued, 32,500 women more would be prone to die in 2020 [1,2]. The epidemiological numbers reflect advances in all related disciplines including early diagnosis by screening programs as well as improved local and systemic treatment. A study by the Early Breast Cancer Trialist Group (EBCTG) revealed that radical treatment of the primary tumor is a prerequisite for long-term cure [3,4]. The standard of care for these patients is breast conserving therapy (BCT) combined with systemic treatment, which—depending on subtype—achieves local control (LC) rates of 95% or more at five years [5–7].

Over the past decade, molecular markers were integrated in the staging system for breast cancer [8], paving the way for new prognosticators and predictors such as microRNAs (miRs). These short non-coding RNAs, whose role in breast cancer was first described by Iorio in 2005 [9], consist of 21–25 nucleotides. Based on a comprehensive in vitro and in silico analysis, Tang et al. concluded that the up-regulation of a specific microRNA, i.e., hsa-miR-375, is essential for tumor progression in early breast cancer [10]. In line with this finding, we could demonstrate in a previous publication that the enhanced expression levels of hsa-miR-375 were associated with a higher probability of local relapse [11].

From a translational as well as a clinical point of view, it is important to describe how miRs interfere with cancer pathways. In theory, one miR can regulate the expression of numerous downstream molecules. Hence, in order to select appropriate target proteins that directly influence known cancer pathways, the current analysis builds—first and foremost—on the preclinical data published by Souza et al., who demonstrated that hsa-miR-375 has a functional binding site on the mRNA of the dexamethasone-induced ras-related protein (RASD1) [12]. Both molecules were shown to constitute a positive feedback loop together with the estrogen receptor α (ER-α) [12], which was substantiated by interventional studies with phytoestrogens [13–15]. In addition to this, Gao et al. showed that RASD1 overexpression in glioma cells resulted in reduced activity of the Akt/mTOR pathway [16].

Apart from RASD1 and ER-α, Sec23A was considered as a potential target for the current study based on preclinical data in breast [17] and prostate cancer [18]. Sec23A is involved in protein transport from the endoplasmic reticulum to the Golgi apparatus and seems to exert an inhibitory function in cell proliferation. Szczyrba et al. could not only define a binding site for hsa-miR-375 on the mRNA of Sec23A, but also found the protein to be down-regulated in prostate cancer cell lines and tissue specimens [18], which made it worth being included in the current analysis.

It is noteworthy that—apart from the above mentioned analysis by our group [11]—only the investigation by Zhou et al. focused on the role of miRs in LC prediction of early breast cancer [19], while most other studies used progression free survival (PFS) and/or overall survival (OS) as an endpoint [20–25]. As an extension of this work [19], the preclinical analysis by de Souza on hsa-miR-375/RASD1 [12] and our own data [11], the aim of the current two-phase study was to answer the question whether hsa-miR-375-signaling through one of the above mentioned targets, i.e., RASD1, ER-α and Sec23A, could predict LC in early breast cancer.

2. Methods

2.1. Patients

The present analysis is a continuation of a previous publication on the predictive value of hsa-miR-375 in breast cancer [11]. Of the initial 147 patients selected from the institutional database of more than 5000 patients, 81 individuals diagnosed with early breast cancer remained. Due to the age of the pathologic specimen that precluded proper signal detection in some cases and forensic restrictions in others, the cohort had to be reduced to 28 (14 relapses/14 controls) in the pilot phase and 53 (13 relapses/40 controls) in the validation phase. Controls are patients with early breast cancer who did no experience local relapse. Like the previous analysis, this is also a matched pair analysis with the same matching criteria [11]: year of diagnosis, surgery (mastectomy or lumpectomy), radiotherapy (whole breast irradiation with percutaneous or intraoperative boost), age, tumor size, lymph node involvement, grading, histology, hormonal receptor status, her2 status, menopause, and Ki67. In the current report however, due to the reduction in patient numbers for the above-mentioned restrictions, a strict patient-to-patient match was only possible in the pilot phase. The patients of the pilot cohort were not included in the subsequent validation phase.

The endpoint of the current study was LC. This marks a difference between this paper and the majority of publications in the field, which investigated PFS and/or OS. Local relapse was defined as the re-appearance of cancer in the same breast, regardless of whether the former index quadrant was affected or not [26].

All patients gave their informed consent to surgery, radiotherapy and systemic treatment. The study was approved by the local ethics committee in Salzburg (Ethikkommission für das Bundesland Salzburg 415-EP/73/582-2015).

2.2. Experimental Design

The current investigation was a two-phase analysis. The first step of the pilot phase was an in silico search for potential targets in five different databases: TargetScan, miRDB, PITA, DIANA, DIANA Cancer (accessed in May 2015). The second step comprised a screen for potential targets in 28 patients (14 with local relapse and 14 matched controls) by means of microarray technology (Affymetrix GeneChip Human Transcriptome Array 2.0®). The combined results formed the basis for the validation phase in an independent set of 53 patients performed with droplet digital PCR (ddPCR). The clinical data acquisition and the molecular analyses were performed at the Departments of Radiation Oncology and Pathology at the Paracelsus Medical University Clinics Salzburg, respectively. Blinded to all outcomes, data processing including quality check (Figures A3 and A4), and biostatistics were carried out externally by the Institute for Machine Learning, Johannes Kepler University, Linz. Figure A2 summarizes the experimental design.

2.3. Tissue Samples

The current analyses were carried out in the same samples that were used for a previous study published by our group [11]. Immediately after surgery, they were formalin fixed paraffin embedded (FFPE) and archived in the tissue bank of the Department of Pathology without any additional processing. It is noteworthy that the samples were not microdissected to enhance the relative proportion of tumor tissue. Similar to the previous investigation, whole tissue sections were selected, which resulted in a tumor content range of 10% to 90% (see Table 1).

Table 1. Patient and treatment characteristics.

Patient Characteristics and Treatment			Pilot Phase $n = 28$			Validation Phase $n = 53$		
	Parameters		Relapse $n = 14$	Control $n = 14$	p-Value	Relapse $n = 13$	Control $n = 40$	p-Value
Patient characteristics	Age at diagnosis (years)	Median	53.5	56	0.84	57	54	0.06
		Range	40–71	41–74		40–79	41–78	
	Menopause (n)	No	3 (21%)	3 (21%)	0.75	2 (15%)	16 (40%)	0.05
		Yes	7 (50%)	8 (57%)		10 (77%)	18 (45%)	
		Unclear	4 (29%)	3 (22%)		1 (8%)	6 (15%)	
	T (n)	T1	9 (64%)	8 (57%)	1	11 (85%)	33 (83%)	0.86
		T2	5 (36%)	6 (43%)		2 (15%)	7 (17%)	
	n (n)	N0	13 (93%)	12 (86%)	0.77	9 (69%)	30 (75%)	0.55
		N1	1 (7%)	1 (7%)		4 (31%)	10 (25%)	
		N2	0 (0%)	1 (7%)		0 (0%)	0 (0%)	
	M (n)	M0	14 (100%)	14 (100%)	1	13 (100%)	40 (100%)	1

Table 1. *Cont.*

Patient Characteristics and Treatment			Pilot Phase n = 28			Validation Phase n = 53		
Parameters			Relapse n = 14	Control n = 14	p-Value	Relapse n = 13	Control n = 40	p-Value
Patient characteristics	Grading (n)	G1	0 (0%)	0 (0%)	0.87	0 (0%)	4 (10%)	0.56
		G2	7 (50%)	8 (57%)		8 (62%)	22 (55%)	
		G3	7 (50%)	6 (43%)		5 (38%)	14 (35%)	
	Histology (n)	IDC [1]	12 (86%)	13 (93%)	0.77	8 (62%)	33 (83%)	0.39
		ILC [2]	2 (14%)	1 (7%)		3 (23%)	2 (5%)	
		Other	0 (0%)	0 (0%)		2 (15%)	5 (13%)	
	In situ component (n)	Yes	12 (86%)	8 (57%)	0.11	7 (54%)	20 (50%)	0.81
		No	2 (14%)	6 (43%)		6 (46%)	20 (50%)	
	Receptors (n)	ER positive [3]	6 (43%)	8 (57%)	0.54	10 (77%)	30 (75%)	0.89
		ER negative	8 (57%)	6 (43%)		3 (23%)	10 (25%)	
		PR positive [4]	5 (36%)	7 (50%)	0.54	8 (62%)	18 (70%)	0.57
		PR negative	9 (64%)	7 (50%)		5 (38%)	12 (30%)	
	her2neu (n)	Positive	8 (57%)	2 (14%)	0.06	4 (31%)	21 (53%)	0.56
		Negative	4 (29%)	9 (64%)		9 (69%)	13 (32%)	
		Not assessable	2 (14%)	3 (21%)		0 (0%)	6 (15%)	
	Proliferation index (n)	ki67 < 20%	9 (64%)	6 (43%)	0.54	4 (31%)	24 (60%)	0.91
		ki67 > 20%	5 (36%)	7 (50%)		7 (54%)	13 (32%)	
		Not assessable	0 (0%)	1 (7%)		2 (15%)	3 (8%)	
Treatment	Boost	Intraoperative (n)	8 (57%)	7 (50%)	0.8	7 (54%)	21 (53%)	0.22
		Percutaneous (n)	6 (43%)	7 (50%)		4 (31%)	19 (47%)	
		None	0 (0%)	0 (0%)		2 (15%)	0 (0%)	
		Intraoperative dose (Gy)	10	10		10	10	
		Percutaneous dose (Gy)	12	12		12	12	
	WBRT [5] dose (Gy)	Median	54	54	1	54	54	0.2
		Range	52.5–61.2	51.0–57.8		51–57.8	50–54	
	Surgery (n)	BCT [6]	14 (100%)	14 (100%)	1	13 (100%)	40 (100%)	1
		Mastectomy	0 (0%)	0 (0%)		0 (0%)	0 (0%)	
	Re-excision (n)	Yes	6 (43%)	3 (21%)	0.54	5 (38%)	18 (45%)	0.69
		No	8 (57%)	11 (79%)		8 (62%)	22 (55%)	
	Year of surgery (n)	Before 1998	3 (21%)	4 (29%)	1	5 (38%)	15 (38%)	0.434
		Since 1998	11 (79%)	10 (61%)		8 (62%)	25 (62%)	
	Chemotherapy (n)	Yes	7 (50%)	5 (36%)	0.21	4 (31%)	11 (28%)	0.86
		No	7 (50%)	9 (64%)		9 (69%)	29 (72%)	
	Antihormonal treatment (n)	Yes	7 (50%)	6 (43%)	1	9 (69%)	22 (55%)	0.46
		No	7 (50%)	8 (57%)		4 (31%)	16 (40%)	
		Unclear	0 (0%)	0 (0%)		0 (0%)	2 (5%)	
	Tumor burden in biopsy (%)	Median	70	60	0.4	70	70	0.76
		Range	10–90	10–90		10–90	30–90	

[1] Invasive ductal carcinoma, [2] Invasive lobular carcinoma, [3] Estrogen receptor, [4] Progesteron receptor, [5] Whole breast radiotherapy, [6] Breast conserving therapy.

2.4. Pilot Phase: Microarray

High through-put RNA expression analysis was performed with the Affymetrix GeneChip Human Transcriptome Array 2.0®, which contains a panel of 70,523 RNA transcripts. Five nanogram total RNA from each individual sample was processed with the Affymetrix GeneChip WT Pico Reagent Kit® to obtain 5,5 µg ss-cDNA, which was fragmented, labelled and hybridized to the GeneChip® according to the manufacturer's instructions. The technical equipment for chip processing was Gene

Chip Hybridization Oven 645, Gene Chip Scanner and Gene Chip Fluidics Station 450 Dx (Affymetrix®, Thermo Fisher Scientific®, Waltham, MA, USA). Microarray data were processed using the RMA method [27,28] from the *oligo* package [29].

2.5. Validation Phase: Droplet Digital PCR

2.5.1. Tissue Processing

After routine processing of the tumor specimen, total RNA was isolated from tissue sections with the Maxwell® RSC RNA FFPE Kit (Promega®, Madison, WI, USA) according to the manufacturer's instructions. RNA was eluted in water and stored at −80 °C until further use. RNA was quantified with the QuantiFluor RNA System on Quantus® Fluorimeter (Promega®) following the "High Standard Calibration" protocol.

2.5.2. Droplet Digital PCR

Reverse transcription was performed with the iScript® Advanced cDNA Synthesis Kit for RT-qPCR (Bio-Rad®, Hercules, CA, USA). A total of 420 ng of each RNA sample was transcribed into cDNA in 20 µL reactions by incubation at 46 °C for 20 min, followed by incubation at 95 °C for 1 min. Then, 1 µL of the cDNA was directly used for gene expression analyses by QX200® ddPCR® EvaGreen system (Bio-Rad®) including the following assays: EEF2 (assay ID: dHsaEG5017938) and RASD1 (assay ID: dHsaEG5004980) and self-designed intron-spanning primers for the amplification of RASD1 (F: TCTCCATCCTCACAGGAGAC; R: GTTCTTGAGGCAAGACTTGG). Each 20 µL reaction contained 10 µL 2 × QX200 ddPCR EvaGreen supermix, 1 µL 20 × primers, 1 µL cDNA template and 8 µL RNase/DNase free water. Each sample was prepared in technical duplicates. The cycling conditions were as follows: 95 °C, 5 min; 40 cycles, 96 °C, 30 s; 58 °C, 60 s; 4 °C, 5 min; 90 °C, 5 min; 4 °C, storage; the ramp time was set to 2 °C/s.

2.5.3. Data Analysis

The raw ddPCR data were analyzed with QuantaSoft® (version 1.7.4.0917) supplied by Bio-Rad® in accordance with the MIQE guidelines (minimum information for publication of quantitative digital PCR experiments) [30]. Copies/µL of RASD1 were normalized to the housekeeping gene eucaryotic elongation factor 2 (EEF2) for each sample (https://doi.org/10.1016/j.ymeth.2012.09.012). The normalized values were the basis for further statistical analyses.

2.6. Statistics

Potentially prognostic and predictive patient and treatment related parameters were compared between groups by means of the Mann–Whitney U test.

In the pilot phase, de-regulated mRNA transcripts were detected using linear models for microarray analysis (LIMMA) [31]. In accordance with other explorative studies, the *p*-value threshold (α) for the microarray was set at 0.2 [32]. An overly restrictive threshold might have led to losing potential target molecules. To make up for this rather permissive limit for first-order errors (α), the stringency filter to narrow down the vast range of potential target molecules for further validation was the mechanistic proof of a binding site between miRNA and mRNA on the one hand and the result of the in silico search on the other. The statistical data analysis to detect the mRNAs with the highest predictive potential was performed with the potential support vector machine (PSVM), a regularized linear classifier with built-in feature selection [33].

In the validation phase, the levels of hsa-miR-375 and RASD1 were correlated with each other by means of the one-sided Spearman test since an inverse correlation between the two molecules could be assumed based on their biological interaction. LC was estimated with the Kaplan–Meier method. In order to show how far the expression levels of the two molecules are associated with LC, a combined hsa-miR-375/RASD1 marker was generated. Each value was mapped to an index of 1 to

4 representing the quartile which it belonged to. For each sample, the indices were subtracted from each other; while positive results mean higher expression of the microRNA than RASD1, negative values represent the opposite. Zero means "equal" expression levels of hsa-miR-375 and RASD1. The log-rank test was used to compare groups stratified by this combined marker. In order to estimate the predictive value of the combined marker, receiver–operating curve (ROC) analyses were performed for the microRNA/RASD1 ratio. Similar to the microarray, we assumed an α of 0.2 as first-order error, which is the threshold proposed for screening trials [34,35]. Hence, the p-value of reference was 0.20.

3. Results

3.1. Patients

The patient and treatment characteristics of the 81 individuals were similarly distributed between relapse and control groups (Table 1). No significant differences could be detected using the Mann–Whitney U test. The median follow-up in the pilot and validation phase was 130.5 months (range 40–200 months) and 116 months (range 44–214 months), respectively (Table 2). The median time to local relapse was 44 months (range 15–123 months) in the pilot cohort and 77 months (28–140 months) in the validation cohort (Table 2).

Table 2. Clinical outcome.

Patient Characteristics and Treatment		Pilot Phase $n = 28$		Validation Phase $n = 54$	
Parameters		Relapse $n = 14$	Control $n = 14$	Relapse $n = 13$	Control $n = 41$
Time to local failure (months)	Median	44	x	77	x
	Range	15–123	x	28–140	x
Time to distant metastasis (months)	Median	61	x	71	109
	Range	28–94	x	21–78	x
Follow-up (months)	Median	121.5	130.5	95	122
	Range	40–186	72–200	44–214	45–194
Cancer related deaths (n)		3	0	3	1

3.2. Pilot Phase

In order to assess the biological context of hsa-miR-375, potential target genes were identified by two independent approaches: a computational analysis combined with a microarray experiment. Therefore, five different miRNA–target mRNA prediction databases were accessed [36–38]. Fifteen targets were predicted both in Target Scan (Release 6.2) and PITA (Venn diagram in Figure A1) including RASD1, which is a validated functional target of hsa-miR-375 in breast cancer cells [12]. The fact that a potential target is listed in at least two databases with different miRNA-mRNA correlation algorithms adds to the reliability of the in silico search results. On top of that, RASD1 was proven in a mechanistic study to have a binding site for hsa-miR-375 [12]. In the very same study, ER-α was shown to be involved in a positive feedback loop between hsa-miR-375 via RASD1 [12]. Sec23A was predicted by DIANA and validated in previous studies in breast [17] and prostate cancer [18].

From the 70,523 probes on the chip, we selected 1433 genes that were predicted to have a binding site with hsa-miR-375 in their 3′UTR using the five databases mentioned in the methods section. The heatmap in Figure 1a summarizes 104 of these potential target molecules that were found to have a lower expression level in relapse patients compared to controls (p-value < 0.2). Figure 1b visualizes the proteins whose hsa-miR-375 binding site was predicted by an in silico algorithm and mechanistically proven on a cellular level, i.e., RASD1, Sec23A and ER-α. While RASD1 expression levels were lower in relapse patients than in controls, which reached borderline significance with a p-value of 0.058, no difference could be detected for Sec23A (p-value = 0.628) and ER-α (p-values between 0.546 and 0.902; Figure 2).The microarray data were deposited in the NCBI Gene Expression

Omnibus. The microRNA data and the target protein data were deposited under the accession numbers GSE69951 and GSE 156873, respectively; they can be downloaded via the following links:

http://www.ncbi.nlm.nih.gov/geo/query/acc.cgi?acc=GSE69951

http://www.ncbi.nlm.nih.gov/geo/query/acc.cgi?acc=GSE156873.

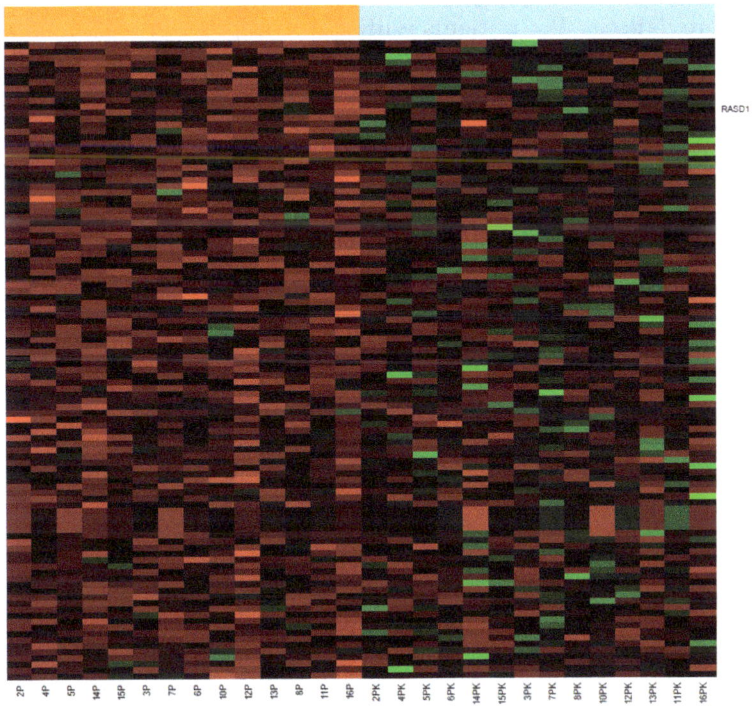

(a)

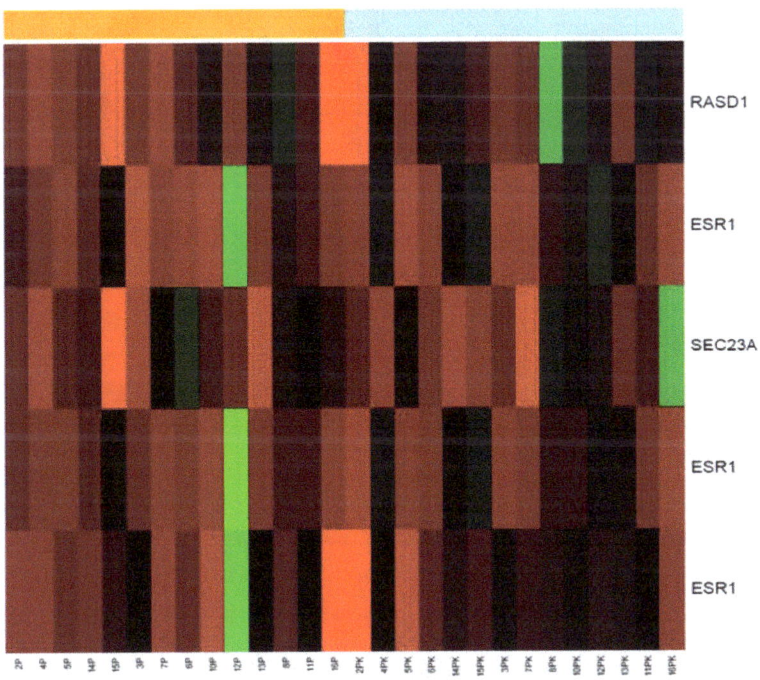

(b)

Figure 1. (**a**) The orange bar on top symbolizes the group of patients with local relapse, who were coded with the extension "P" after the number (bottom line), while the blue bar represents the controls (extension "PK"). The microarray (Affymetrix GeneChip Human Transcriptome Array 2.0®) contained 70,523 transcripts. According to the in silico analysis, 1433 were known targets of hsa-miR-375, of which 104 were down-regulated in the relapse group (patients beneath the orange bar on top of the heat map) compared to controls (patients beneath the blue bar on top of the heat map): raw p-value < 0.20. The transcripts are ordered according to their p-values with the lowest on top. RASD1 (highlighted on the right) is number 12 in this list. High and low expressions of a molecule are depicted in green and red, respectively. Abbreviations: P = patients with local relapse, PK = matched control. (**b**) Shows the differential expression of RASD1, ER-α and Sec23A between relapse patients (beneath the orange bar) and controls (beneath the blue bar). High and low expression levels of a molecule are depicted in green and red, respectively. The differential expression of RASD1 reached borderline significance (p-value = 0.058), while for the two other targets, the p-value was > 0.2. The Affymetrix GeneChip Human Transcriptome Array 2.0® contained three transcripts for ER-α (ESR 1), whereas RASD1 and Sec23A were represented by one transcript each. Abbreviations: P = patients with local relapse, PK = matched control.

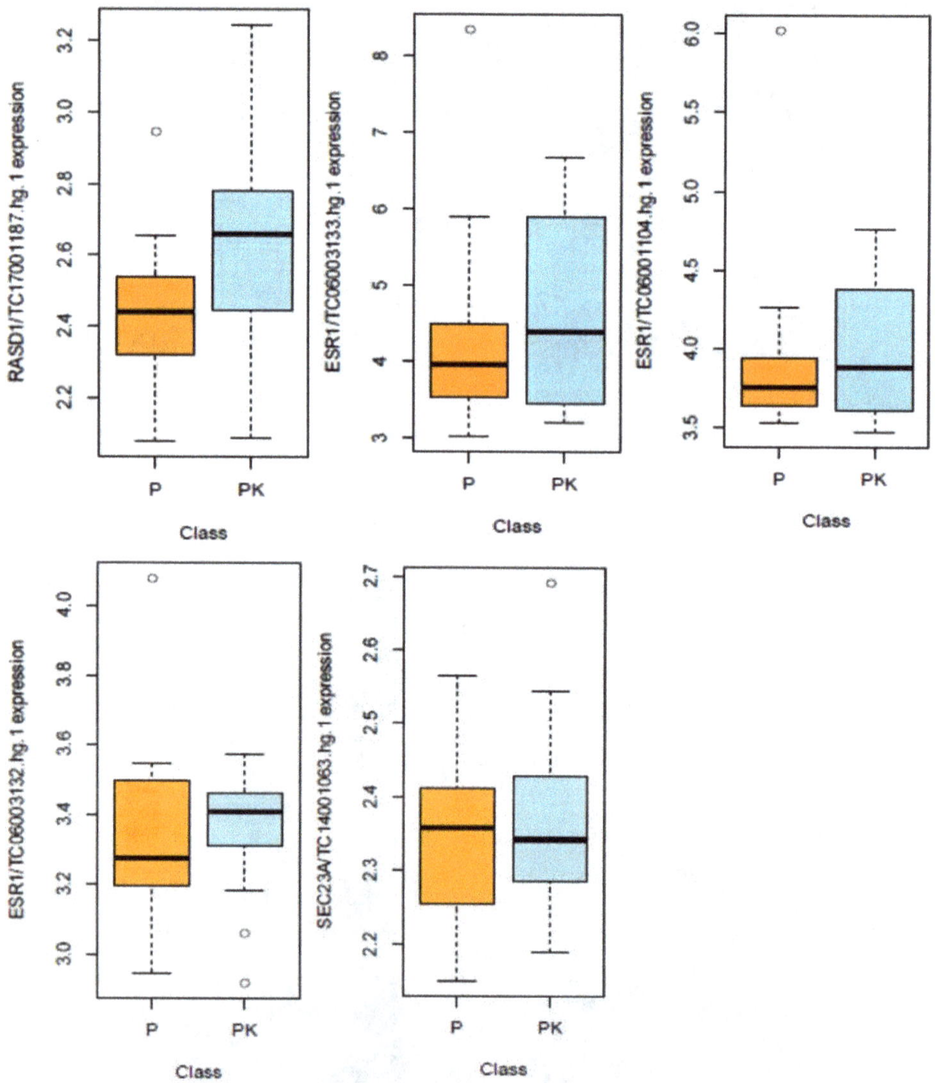

Figure 2. Differential expression between patients and controls for each of the putative target proteins: RASD1, ERα (three transcripts) and Sec23A. The difference in RASD1 expression reached borderline significance (p-value = 0.058), while for all other targets, the p-value was > 0.2. Abbreviations: P = patients with local relapse, PK = matched control, TC = transcript with the respective number on the Affymetrix GeneChip Human Transcriptome Array 2.0®.

3.3. Validation Phase

As described in the previously mentioned previous publication by our group [11], hsa-miR-375 could significantly predict the probability of local control. In the current validation cohort of 53 patients, the same effect could be observed, although with a higher p-value of 0.014 (Figure 3). From the target proteins shown in Figure 2, RASD1 was selected for further validation since the differential expression between relapse, and controls almost reached significance in the microarray (p-value = 0.058, Figures 1 and 2). In order to validate this preliminary finding in an independent set of samples, digital droplet PCR (ddPCR), a highly sensitive method to quantify gene expression levels, was used.

This is in coherence with the physiological functioning of a microRNA. Generally, there are two mechanisms at work in the miRNA/mRNA interaction, which both result in decreased mRNA concentrations: a perfect miRNA/mRNA match leads to RISC-mediated dissection of the mRNA molecule, while a less perfect match stops translation with delayed degradation. Hence, changes in the amount of mRNA are an indirect measure for protein concentrations. For lack of suitable RASD1 antibodies, we chose this approach. In the validation cohort, the expression levels of RASD1 and hsa-miR-375 were inversely correlated (one-sided Spearman test, p-value = 0.021, Figure A5).

Figure 4 shows the correlation of the combined biomarker hsa-miR-375 and RASD1 to LC. Patients with enhanced levels of hsa-miR-375 compared to RASD1 have a higher probability of local relapse than those with the inverse expression pattern of the two markers. This difference was below the limit of 0.2 for α (log-rank test, p-value = 0.069) in the 39/53 (74%) patients who had a clearly higher expression level of one marker (Figure 4a). When the same analysis was performed for the whole cohort including the 14/53 (26%) patients who had an equal expression level of both markers, this difference still persisted (log-rank test, overall p-value = 0.177; Figure 4b). The ROC analyses revealed an area under the curve(AUC) of 0.615 (p-value = 0.158) and 0.552 (p-value = 0.294) for the two cohorts, respectively.

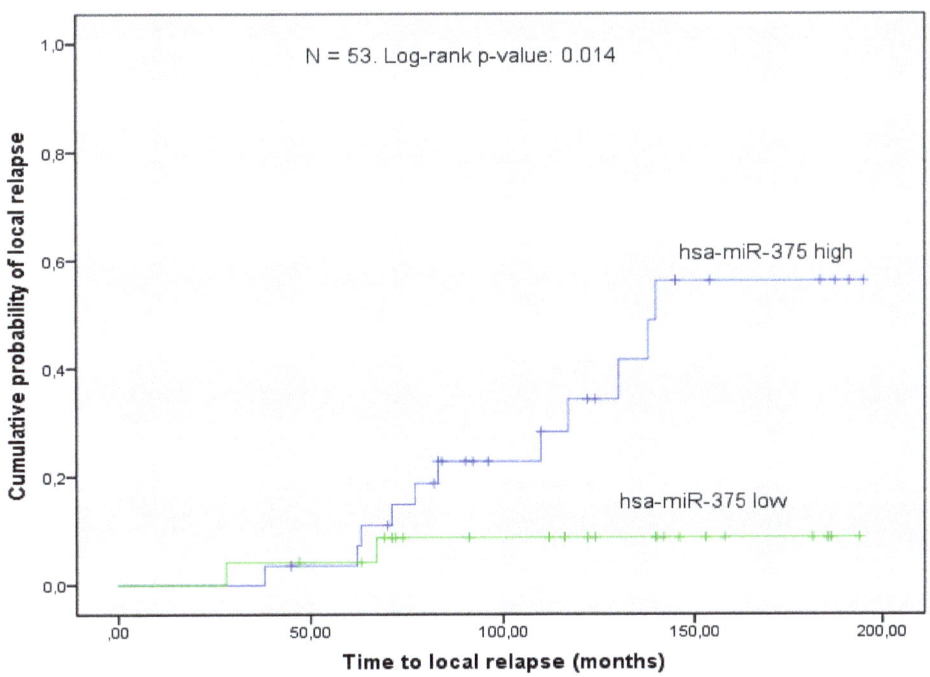

Figure 3. Comparison between patients stratified by the levels of hsa-miR-375 revealed that low concentrations (i.e., below the median) of the micro-RNA are associated with a lower risk of local relapse.

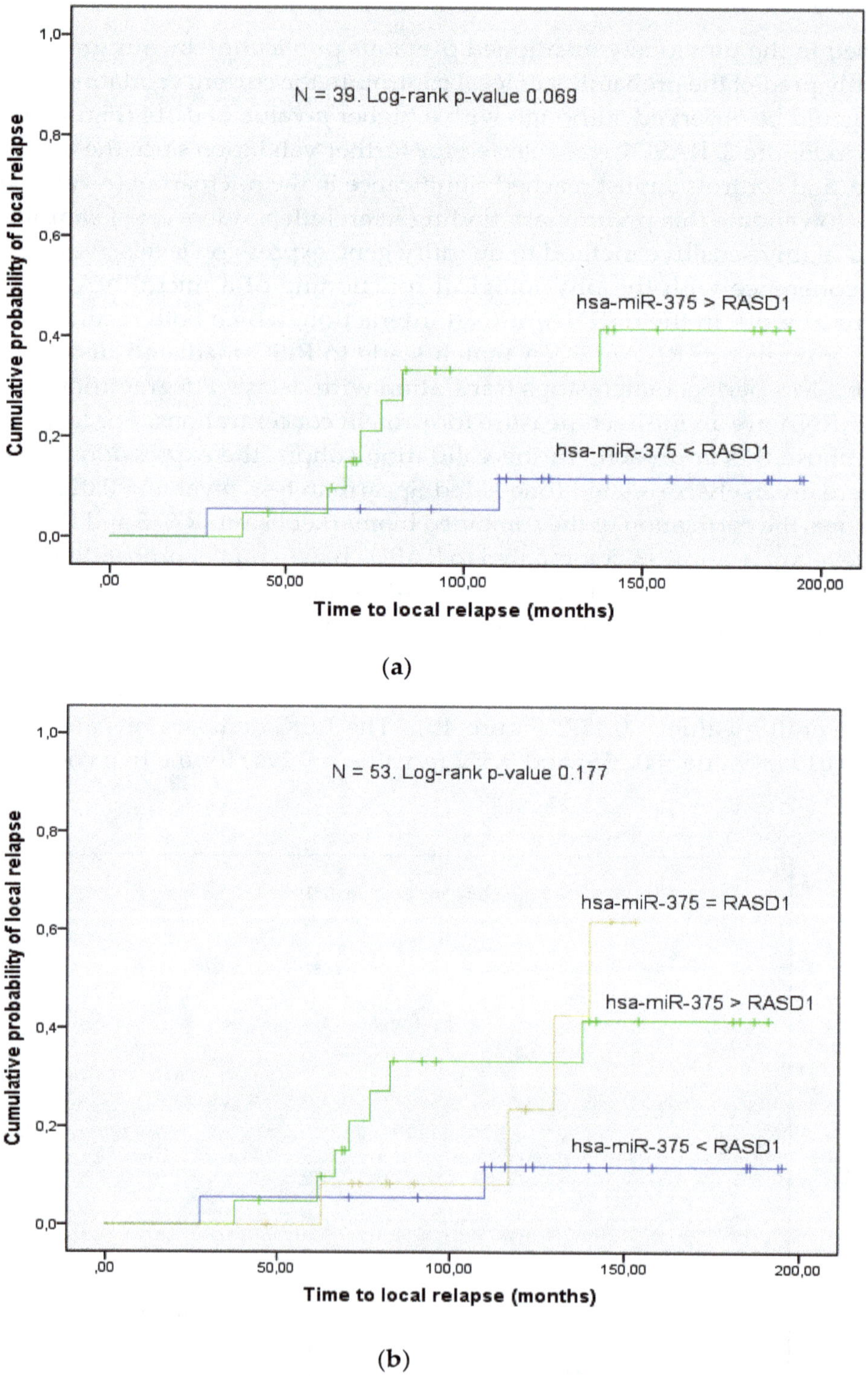

Figure 4. High levels of hsa-miR-375 compared to RASD1 correlate to a higher probability of local relapse. (**a**) The standardized comparison of hsa-miR-375 and RASD1 for each individual patient revealed inverse expression patterns in 39/53 patients. The combined marker hsa-miR-375/RASD1 in these cases was able to predict local control (log-rank *p*-value = 0.069). Hsa-miR-375 > RASD1 means that the expression level of the microRNA is higher than that of the target protein, whereas hsa-miR-375 < RASD1 is the opposite. Hsa-miR-375 = RASD1 signifies that the concentration of both markers is equal. (**b**) As depicted by the clearly separating survival curves, this difference persisted in the whole cohort of 53 patients (log-rank *p*-value = 0.177).

4. Discussion

The current two-phase study demonstrates that hsa-miR-375/RASD1 signaling is able to predict local control in early breast cancer. To our knowledge, this is the first clinical report on a miRNA combined with one of its downstream target proteins showing a strong correlation with in-breast tumor control. Since biomarkers have been added to the panel of predictive and prognostic markers in cancer, our work may help to better select patients for BCT.

The first analysis that established the importance of miRs in breast cancer was published 15 years ago by Iorio and co-workers [9]. In this preclinical study, the authors showed that the de-regulated expression levels of 15 miRNAs could discriminate breast cancer from normal tissue. In order to elucidate the functional role of miRs, it is important to know which pathways they interfere in. The most appropriate search process for potential target proteins from the huge number of possible downstream molecules is a miR-mRNA database search combined with the results of mechanistic preclinical studies in cells and rodents. For the current analysis, this approach yielded three potential targets, including RASD1, ER-α and Sec23A, of which only the first was selected for further validation.

Souza et al. provided the mechanistic proof that hsa-miR-375 has a binding site on the UTR-3′-end of the RASD1 mRNA molecule [12]. By gain and loss of function experiments, the authors showed that down-regulated RASD1, which belongs to the Ras superfamily of small G-proteins, resulted in increased cell growth via enhanced ER-α activity [12]. The potential function of RASD1 was investigated in breast and lung cancer cells as well as a xenograft model [39]. High expression levels of this small G-protein stop dysregulated cell growth and may therefore exert anti-cancer activity. Its locus on chromosome 17p11.2 is characterized by frequent loss of heterozygosity and deletions [40], which leads to pro-proliferative signaling in cancer cells [16,39,41–43].

The fact that RASD1 is a direct functional target of hsa-miR-375 [12] harbors the potential of a therapeutic approach. This was investigated in two preclinical studies using phytoestrogens [13,14]. Chen showed that the up-regulation of ER-α by the phytoestrogen formononetin and hsa-miR-375 is a universal phenomenon, which is also present in pluripotent endothelial cells (HUVECs) and not only in ER-α positive breast cancer cells [13]. In their study in rodents, Wang et al. demonstrated that calycosin, another phytoestrogen, induces hsa-miR-375 and ER-α, while down regulating RASD1 at the same time [14].

Taken together, the current analysis is the first clinical verification of experimental data on hsa-miR-375 combined with RASD1 in breast cancer. Our results allow us to assume that hsa-miR-375/RASD1 signaling helps to predict local failure, which corroborates the above mentioned preclinical findings [9,12–15]. The clinical endpoint of the current investigation was LC, while the majority of the studies in the field use PFS and/or OS instead. To date, only one study in a smaller cohort than ours investigated miRs and LC in breast cancer [19], whereas none considered the role of RASD1, let alone the hsa-miR-375/RASD1 combination. Discrepancies with the results of other clinical studies may be related to sample processing. First, in the current analysis, only tumor specimens were compared to each other, while previous studies, e.g., that by Iorio [9], compared tumor and normal tissue leading to a more pronounced differential expression of miRs and their down-stream molecules. Secondly, in the current analysis, tissue sections including tumor stroma were used as opposed to micro-dissected specimen in comparable studies [19,20]. Third, whole tissue sections were analyzed since we believe that a given signaling cascade should retain its discriminative power even if the tumor content is low.

In this context, the question arises how hsa-miR-375/RASD1 interferes in the cancer pathway system. An extensive in silico and in vitro analysis comparing normal versus tumor tissue in more than 3700 cases revealed that up-regulated hsa-miR-375 accelerated cell proliferation by regulating the PI3K/Akt/mTOR pathway [10]. Gao et al. demonstrated the connection between RASD1 and the Akt/mTOR pathway in glioma cells and a xenograft model by means of an array of 18 intracellular signaling molecules [16]. The overexpression of RASD1 reduced the phosphorylation of p-Akt, p-S6 and p-GSK3β, which finally inactivated the Akt/mTOR pathway. Enhanced RASD1 levels led to rearrangement of the cytoskeleton and reduced cell migration [16]. In addition, the clinical data published in the very same study imply that high levels of RASD1 were associated with improved OS [16]. The authors concluded that RASD1 affected cell migration and invasion presumably through Akt-mediated epithelial-mesenchymal transition [16,44], which may translate into clinically detectable differences. In accordance with Gao, the current study substantiates that high levels of RASD1 combined with a low expression of hsa-miR-375 are beneficial in terms of LC. As the report by Gao and our own computational analysis [45] demonstrated, RASD1 interferes with the PI3K/Akt pathway [16] (Figure A7). Therefore, targeting the hsa-miR-375/RASD1 signaling axis could be a rewarding approach for therapeutic studies aimed at inhibiting aberrant cell growth.

To date, clinicaltrials.gov lists a total of 29 studies involving microRNAs in breast cancer, nine of which are completed (accessed 2020/07). These studies cover the following aspects of cancer diagnostics and treatment: response to systemic treatment (10), response to RT (1), response to physical activity and diet (5), including one that focuses on hsa-miR-375, monitoring toxicity (6), and diagnostics (6). As our analysis corroborates preclinical findings, it underlines the usefulness and necessity for inclusion into prospective clinical trials on a larger scale. Thus far, none of these 29 studies on miRs listed in clinicaltrials.gov was conducted with LC as an oncological endpoint.

An obvious weakness of our analysis is the rather permissive threshold for first-order errors (α). However, this is not unusual in explorative studies with the aim of extracting as much as possible of the potentially important information from a huge amount of data [32]. The initial microarray included more than 70,000 target molecules. Nevertheless, RASD1 with a raw *p*-value of 0.058 almost fulfilled the conventional limit of $\alpha < 0.05$. For reasons described in the methods section, the number of patients had to be reduced to 81 in the current investigation. This may explain why the clinical characteristics such as age, menopause and her2neu were not as evenly distributed as in the previous study [11] (Table 1). Additionally, this rather small number of cases could also be the reason for the moderate predictive power of the combined marker hsa-miR-375/RASD1 in the ROC analyses. While the *p*-value in the subcohort of 39 patients with a clearly inverse correlation between the two markers was still within the limit of 0.2 for first-order errors, in the whole cohort, it was not (Figure A6a,b). Furthermore, the background noise of the multi-level regulatory network, which RASD1 is embedded in, may also play a role. Hence, as the current analysis was designed as a hypothesis-generating study, no definite conclusions can be drawn with respect to sensitivity and specificity.

As for methodological deficiencies, we have to admit that a direct measurement of RASD1 protein concentrations would have certainly strengthened our results. While appropriate antibodies were unavailable, the implementation and validation of a manual staining procedure or the in-house designing of a suitable antibody for RASD1 were beyond the scope of the current project. Using a luciferase reporter construct, de Souza and colleagues provided reliable evidence that RASD1 is a direct target of hsa-miR-375 [12]. These findings were confirmed independently by another group [13]. Similar to our study, neither of these publications included analyses on RASD1 protein levels.

In contrast, these apparent shortcomings are counterbalanced by several strengths. This is the first study that corroborates preclinical data [12] on hsa-miR-375/RASD1 signaling for LC in breast cancer patients. Although the current analysis is retrospective, its two-phase design provides reliable results to constitute a sound basis for prospective investigations. The methodological disadvantage of using whole tissue sections strengthens our results since thereby we keep the specimen processing as close to daily routine as possible, which enhances the transferability of our findings to clinical practice.

5. Conclusions

In summary, the current study demonstrates that hsa-miR-375/RASD1 signaling may contribute to predict local control after BCT for early stage breast cancer. Although prospective studies in larger cohorts are warranted, our results enlarge the panel of potential molecular markers for adequate patient selection in the context of tailored therapies.

Author Contributions: Conceptualization, B.Z., F.S. and F.Z.; data curation, U.B., I.A.E., C.K., P.S., B.G., G.F., M.S., R.R. and F.Z.; formal analysis, U.B., I.A.E., P.S., B.G. and M.S.; funding acquisition, F.S. and F.Z.; investigation, B.Z., I.A.E., C.K., G.F., M.S. and F.Z.; methodology, B.Z., U.B., I.A.E., P.S., B.G., G.F., R.R. and F.S.; project administration, B.G. and F.Z.; resources, C.K., P.S. and K.S.; supervision, R.R., K.S. and F.S.; writing—original draft, B.Z., U.B., F.S. and F.Z.; writing—review and editing, C.K., B.G., G.F., M.S., R.R. and K.S. All authors have read and agreed to the published version of the manuscript.

Abbreviations

AKT	protein kinase B
AUC	area under the curve
BCT	breast conserving therapy
ddPCR	droplet digital PCR
EBCTB	early breast cancer trialist group
EEF	eukaryotic elongation factor 2
ER-α	estrogen receptor α
FFPE	formalin fixed paraffin embedded
HUVEC	human umbilical vein endothelial cells
LC	local control
LIMMA	linear models for microarray analysis
MIQE	minimum information for publication of quantitative digital PCR experiments
miR	microRNA, miRNA
OS	overall survival
p-S6	phosphorylated ribosomal protein S6
p-GSK-3β	phosphorylated glycogen synthase kinase 3 β
PFS	progression free survival
PI3K	phosphatidylinositol 3 kinase
PSVM	potential support vector machine
RASD1	dexamethasone induced ras-related protein 1
RISC	RNA induced silencing complex
ROC	receiver operating curve
RT	radiotherapy
Sec23A	saccharomyces cerevisiae 23 homolog A

Appendix A

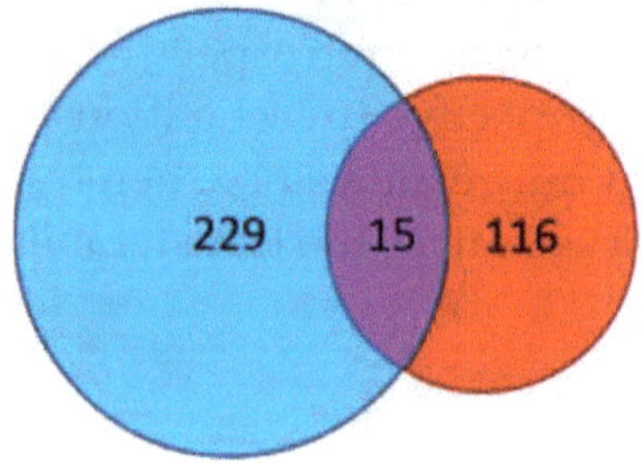

Figure A1. Target prediction for hsa-miR-375. The Venn diagram shows the number of predicted targets from TargetScan (blue) http://www.targetscan.org/ and PITA (red) http://genie.weizmann.ac.il/pubs/mir07/mir07_data.html; 15 genes were suggested by both algorithms.

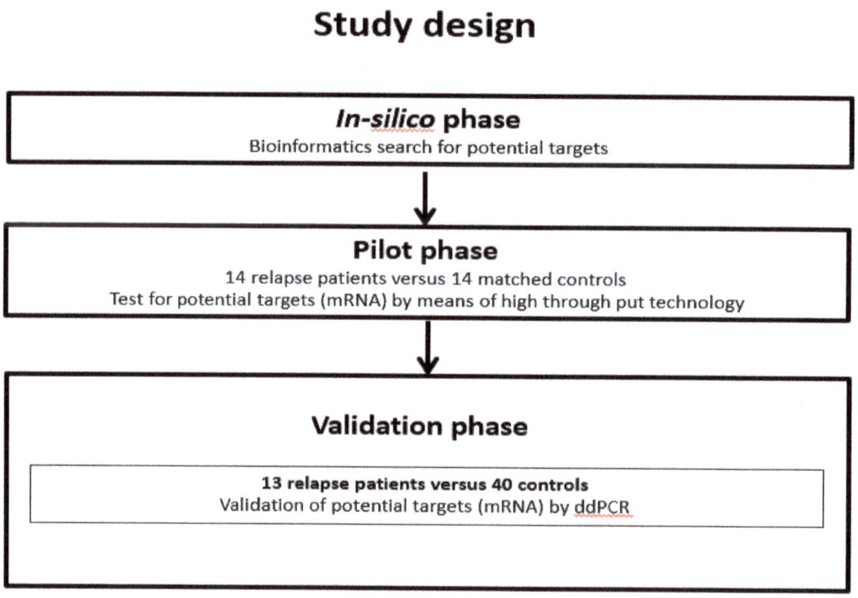

Figure A2. Study design.

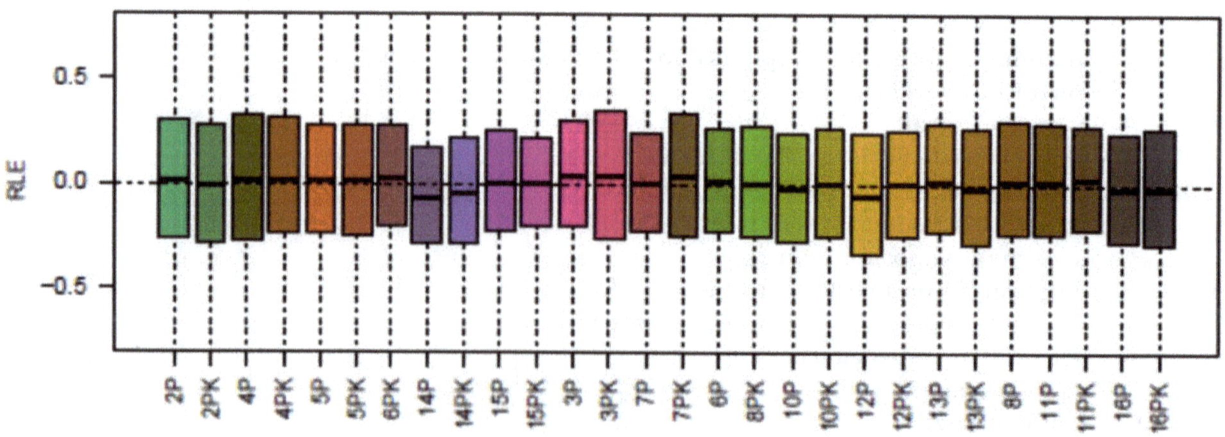

Figure A3. Data quality check.

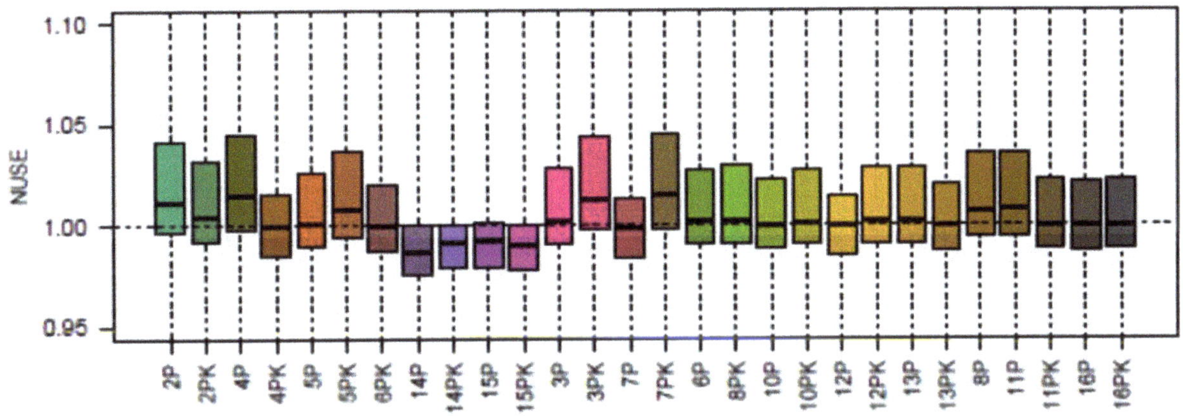

Figure A4. Data quality check.

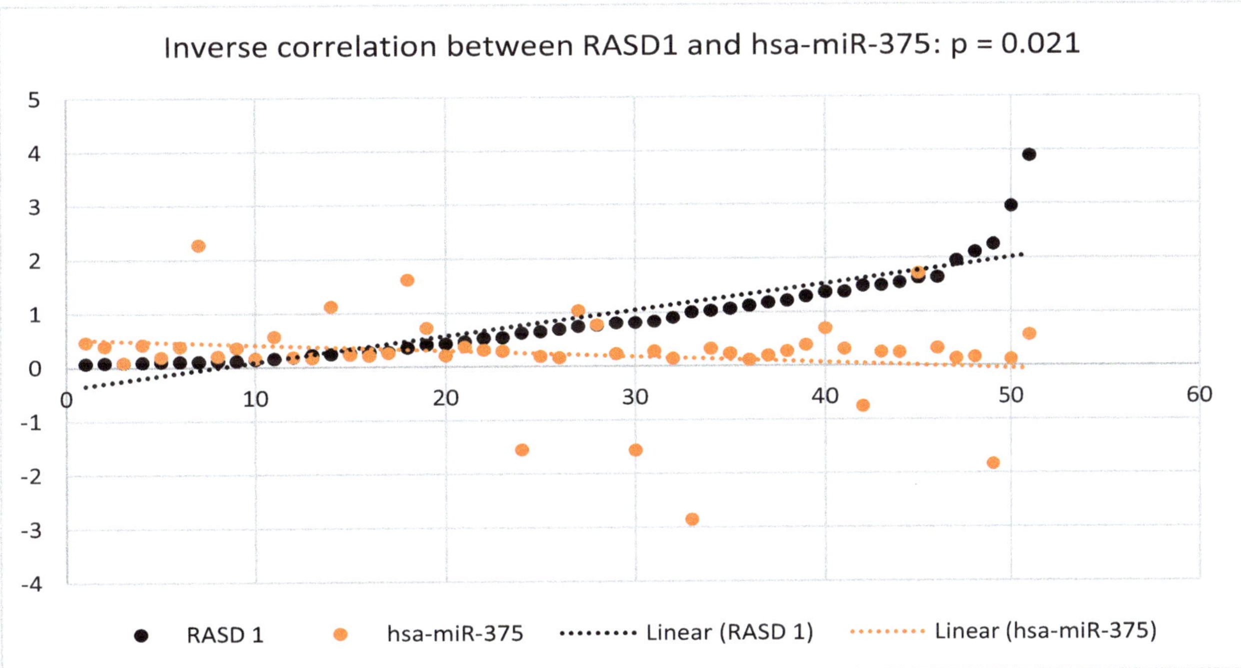

Figure A5. RASD1 and hsa-miR-375 were inversely correlated to each other (one-sided Spearman test, *p*-value: 0.021).

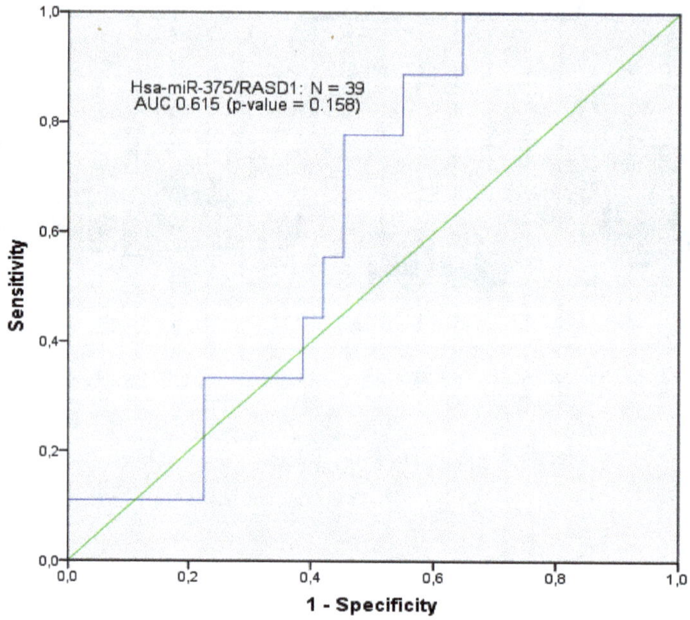

(a)

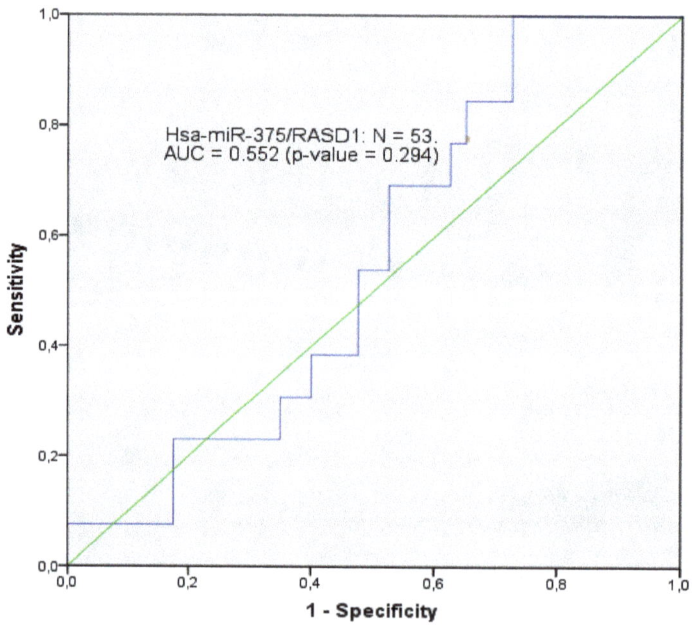

(b)

Figure A6. (**a**) In the 39 patients with a clearly inverse correlation of hsa-miR-375 and RASD1, the receiver–operating curve (ROC) analysis revealed an AUC of 0.615 (p-value 0.158). (**b**) In the whole cohort of 53 patients, the ROC analysis showed an AUC of 0.552 (p-value = 0.294).

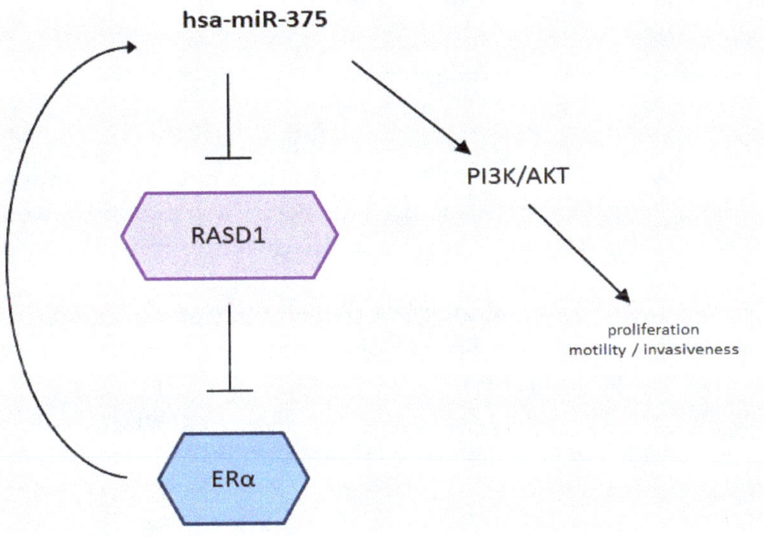

Figure A7. Working model of a putative hsa-miR-375 pathway: Hsa-miR-375 is involved in a positive feedback loop with ERα, thereby blocking RASD1, which in itself interferes with the PI3K/Akt pathway.

References

1. Carioli, G.; Malvezzi, M.; Rodriguez, T.; Bertuccio, P.; Negri, E.; La Vecchia, C. Trends and predictions to 2020 in breast cancer mortality in Europe. *Breast* **2017**, *36*, 89–95. [CrossRef]
2. Malvezzi, M.; Carioli, G.; Bertuccio, P.; Boffetta, P.; Levi, F.; La Vecchia, C.; Negri, E. European cancer mortality predictions for the year 2019 with focus on breast cancer. *Ann. Oncol.* **2019**, *30*, 781–787. [CrossRef]
3. Darby, S.; McGale, P.; Correa, C.; Taylor, C.; Arriagada, R.; Clarke, M.; Cutter, D.; Davies, C.; Ewertz, M.; Godwin, J.; et al. Effect of radiotherapy after breast-conserving surgery on 10-year recurrence and 15-year breast cancer death: Meta-analysis of individual patient data for 10,801 women in 17 randomised trials. *Lancet* **2011**, *378*, 1707–1716. [PubMed]
4. Sautter-Bihl, M.L.; Sedlmayer, F.; Budach, W.; Dunst, J.; Feyer, P.; Fietkau, R.; Haase, W.; Harms, W.; Rodel, C.; Souchon, R.; et al. One life saved by four prevented recurrences? Update of the Early Breast Cancer Trialists confirms: Postoperative radiotherapy improves survival after breast conserving surgery. *Strahlenther Onkol.* **2012**, *188*, 461–463. [CrossRef] [PubMed]
5. Bartelink, H.; Horiot, J.C.; Poortmans, P.; Struikmans, H.; Van den Bogaert, W.; Barillot, I.; Fourquet, A.; Borger, J.; Jager, J.; Hoogenraad, W.; et al. Recurrence rates after treatment of breast cancer with standard radiotherapy with or without additional radiation. *N. Engl. J. Med.* **2001**, *345*, 1378–1387. [CrossRef] [PubMed]
6. Whelan, T.; MacKenzie, R.; Julian, J.; Levine, M.; Shelley, W.; Grimard, L.; Lada, B.; Lukka, H.; Perera, F.; Fyles, A.; et al. Randomized trial of breast irradiation schedules after lumpectomy for women with lymph node-negative breast cancer. *J. Natl. Cancer Inst.* **2002**, *94*, 1143–1150. [CrossRef]
7. Bentzen, S.M.; Agrawal, R.K.; Aird, E.G.; Barrett, J.M.; Barrett-Lee, P.J.; Bliss, J.M.; Brown, J.; Dewar, J.A.; Dobbs, H.J.; Haviland, J.S.; et al. The UK Standardisation of Breast Radiotherapy (START) Trial A of radiotherapy hypofractionation for treatment of early breast cancer: A randomised trial. *Lancet Oncol.* **2008**, *9*, 331–341.
8. Goldhirsch, A.; Wood, W.C.; Coates, A.S.; Gelber, R.D.; Thurlimann, B.; Senn, H.J. Strategies for subtypes–dealing with the diversity of breast cancer: Highlights of the St. Gallen International Expert Consensus on the Primary Therapy of Early Breast Cancer 2011. *Ann. Oncol.* **2011**, *22*, 1736–1747. [CrossRef]
9. Iorio, M.V.; Ferracin, M.; Liu, C.G.; Veronese, A.; Spizzo, R.; Sabbioni, S.; Magri, E.; Pedriali, M.; Fabbri, M.; Campiglio, M.; et al. MicroRNA gene expression deregulation in human breast cancer. *Cancer Res.* **2005**, *65*, 7065–7070. [CrossRef]

10. Tang, W.; Li, G.S.; Li, J.D.; Pan, W.Y.; Shi, Q.; Xiong, D.D.; Mo, C.H.; Zeng, J.J.; Chen, G.; Feng, Z.B.; et al. The role of upregulated miR-375 expression in breast cancer: An in vitro and in silico study. *Pathol. Res. Pract.* **2020**, *216*, 152754. [CrossRef]
11. Zehentmayr, F.; Hauser-Kronberger, C.; Zellinger, B.; Hlubek, F.; Schuster, C.; Bodenhofer, U.; Fastner, G.; Deutschmann, H.; Steininger, P.; Reitsamer, R.; et al. Hsa-miR-375 is a predictor of local control in early stage breast cancer. *Clin. Epigenet.* **2016**, *8*, 28. [CrossRef] [PubMed]
12. Simonini, P.D.S.R.; Breiling, A.; Gupta, N.; Malekpour, M.; Youns, M.; Omranipour, R.; Malekpour, F.; Volinia, S.; Croce, C.M.; Najmabadi, H.; et al. Epigenetically deregulated microRNA-375 is involved in a positive feedback loop with estrogen receptor α in breast cancer cells. *Cancer Res.* **2010**, *70*, 9175–9184. [CrossRef] [PubMed]
13. Chen, J.; Zhang, X.; Wang, Y.; Ye, Y.; Huang, Z.Q. Differential ability of formononetin to stimulate proliferation of endothelial cells and breast cancer cells via a feedback loop involving MicroRNA-375, RASD1, and ER α. *Mol. Carcinog.* **2018**, *57*, 817–830. [CrossRef]
14. Wang, Y.; Dong, X.H.; Li, Z.Z.; Wang, W.B.; Tian, J.; Chen, J. Downregulated RASD1 and upregulated miR-375 are involved in protective effects of calycosin on cerebral ischemia/reperfusion rats. *J. Neurol. Sci.* **2014**, *339*, 144–148. [CrossRef] [PubMed]
15. Munagala, R.; Aqil, F.; Vadhanam, M.V.; Gupta, R.C. MicroRNA 'signature' during estrogen-mediated mammary carcinogenesis and its reversal by ellagic acid intervention. *Cancer Lett.* **2013**, *339*, 175–184. [CrossRef] [PubMed]
16. Gao, S.F.; Jin, L.; Liu, G.P.; Wang, P.; Sun, Z.H.; Cao, Y.J.; Shi, H.L.; Liu, X.J.; Shi, Q.; Zhou, X.P.; et al. Overexpression of RASD1 inhibits glioma cell migration/invasion and inactivates the AKT/mTOR signaling pathway. *Sci. Rep.* **2017**, *7*, 3202. [CrossRef] [PubMed]
17. Luo, D.; Wilson, J.M.; Harvel, N.; Liu, J.; Pei, L.; Huang, S.; Hawthorn, L.; Shi, H. A systematic evaluation of miRNA:mRNA interactions involved in the migration and invasion of breast cancer cells. *J. Transl. Med.* **2013**, *11*, 57. [CrossRef]
18. Szczyrba, J.; Nolte, E.; Wach, S.; Kremmer, E.; Stohr, R.; Hartmann, A.; Wieland, W.; Wullich, B.; Grasser, F.A. Downregulation of Sec23A protein by miRNA-375 in prostate carcinoma. *Mol. Cancer Res.* **2011**, *9*, 791–800. [CrossRef]
19. Zhou, X.; Marian, C.; Makambi, K.H.; Kosti, O.; Kallakury, B.V.; Loffredo, C.A.; Zheng, Y.L. MicroRNA-9 as potential biomarker for breast cancer local recurrence and tumor estrogen receptor status. *PLoS ONE* **2012**, *7*, e39011. [CrossRef]
20. Jonsdottir, K.; Janssen, S.R.; Da Rosa, F.C.; Gudlaugsson, E.; Skaland, I.; Baak, J.P.; Janssen, E.A. Validation of expression patterns for nine miRNAs in 204 lymph-node negative breast cancers. *PLoS ONE* **2012**, *7*, e48692. [CrossRef]
21. Hoppe, R.; Achinger-Kawecka, J.; Winter, S.; Fritz, P.; Lo, W.Y.; Schroth, W.; Brauch, H. Increased expression of miR-126 and miR-10a predict prolonged relapse-free time of primary oestrogen receptor-positive breast cancer following tamoxifen treatment. *Eur. J. Cancer* **2013**, *49*, 3598–3608. [CrossRef] [PubMed]
22. Lyng, M.B.; Laenkholm, A.V.; Sokilde, R.; Gravgaard, K.H.; Litman, T.; Ditzel, H.J. Global microRNA expression profiling of high-risk ER+ breast cancers from patients receiving adjuvant tamoxifen mono-therapy: A DBCG study. *PLoS ONE* **2012**, *7*, e36170. [CrossRef] [PubMed]
23. Perez-Rivas, L.G.; Jerez, J.M.; Carmona, R.; de Luque, V.; Vicioso, L.; Claros, M.G.; Viguera, E.; Pajares, B.; Sanchez, A.; Ribelles, N.; et al. A microRNA signature associated with early recurrence in breast cancer. *PLoS ONE* **2014**, *9*, e91884. [CrossRef] [PubMed]
24. Tuomarila, M.; Luostari, K.; Soini, Y.; Kataja, V.; Kosma, V.M.; Mannermaa, A. Overexpression of microRNA-200c predicts poor outcome in patients with PR-negative breast cancer. *PLoS ONE* **2014**, *9*, e109508. [CrossRef] [PubMed]
25. Svoboda, M.; Sana, J.; Redova, M.; Navratil, J.; Palacova, M.; Fabian, P.; Slaby, O.; Vyzula, R. MiR-34b is associated with clinical outcome in triple-negative breast cancer patients. *Diagn. Pathol.* **2012**, *7*, 31. [CrossRef]
26. Huang, E.; Buchholz, T.A.; Meric, F.; Krishnamurthy, S.; Mirza, N.Q.; Ames, F.C.; Feig, B.W.; Kuerer, H.M.; Ross, M.I.; Singletary, S.E.; et al. Classifying local disease recurrences after breast conservation therapy

based on location and histology: New primary tumors have more favorable outcomes than true local disease recurrences. *Cancer* **2002**, *95*, 2059–2067. [CrossRef]
27. Irizarry, R.A.; Bolstad, B.M.; Collin, F.; Cope, L.M.; Hobbs, B.; Speed, T.P. Summaries of affymetrix GeneChip probe level data. *Nucleic Acids Res.* **2003**, *31*, e15. [CrossRef]
28. Irizarry, R.A.; Hobbs, B.; Collin, F.; Beazer-Barclay, Y.D.; Antonellis, K.J.; Scherf, U.; Speed, T.P. Exploration, normalization, and summaries of high density oligonucleotide array probe level data. *Biostatistics* **2003**, *4*, 249–264. [CrossRef]
29. Carvalho, B.S.; Irizarry, R.A. A framework for oligonucleotide microarray preprocessing. *Bioinformatics* **2010**, *26*, 2363–2367. [CrossRef]
30. Huggett, J.F.; Foy, C.A.; Benes, V.; Emslie, K.; Garson, J.A.; Haynes, R.; Hellemans, J.; Kubista, M.; Nolan, R.D.M.T.; Pfaffl, M.W.; et al. The Digital MIQE Guidelines: Minimum Information for Publication of Quantitative Digital PCR Experiments. *Clin. Chem.* **2013**, *59*, 892–902. [CrossRef]
31. Smyth, G.K. Linear models and empirical bayes methods for assessing differential expression in microarray experiments. *Stat. Appl. Genet. Mol. Biol.* **2004**, *3*. [CrossRef] [PubMed]
32. Dhawan, A.; Scott, J.G.; Harris, A.L.; Buffa, F.M. Pan-cancer characterisation of microRNA across cancer hallmarks reveals microRNA-mediated downregulation of tumour suppressors. *Nat. Commun.* **2018**, *9*, 5228. [CrossRef]
33. Hochreiter, S.; Obermayer, K. Support vector machines for dyadic data. *Neural. Comput.* **2006**, *18*, 1472–1510. [CrossRef] [PubMed]
34. Palma, D.A.; Olson, R.; Harrow, S.; Gaede, S.; Louie, A.V.; Haasbeek, C.; Mulroy, L.; Lock, M.; Rodrigues, G.B.; Yaremko, B.P.; et al. Stereotactic ablative radiotherapy versus standard of care palliative treatment in patients with oligometastatic cancers (SABR-COMET): A randomised, phase 2, open-label trial. *Lancet* **2019**, *393*, 2051–2058. [CrossRef]
35. Rubinstein, L.V.; Korn, E.L.; Freidlin, B.; Hunsberger, S.; Ivy, S.P.; Smith, M.A. Design issues of randomized phase II trials and a proposal for phase II screening trials. *J. Clin. Oncol.* **2005**, *23*, 7199–7206. [CrossRef] [PubMed]
36. Witkos, T.M.; Koscianska, E.; Krzyzosiak, W.J. Practical Aspects of microRNA Target Prediction. *Curr. Mol. Med.* **2011**, *11*, 93–109. [CrossRef] [PubMed]
37. Lewis, B.P.; Burge, C.B.; Bartel, D.P. Conserved seed pairing, often flanked by adenosines, indicates that thousands of human genes are microRNA targets. *Cell* **2005**, *120*, 15–20. [CrossRef]
38. Kertesz, M.; Iovino, N.; Unnerstall, U.; Gaul, U.; Segal, E. The role of site accessibility in microRNA target recognition. *Nat. Genet.* **2007**, *39*, 1278–1284. [CrossRef]
39. Vaidyanathan, G.; Cismowski, M.J.; Wang, G.; Vincent, T.S.; Brown, K.D.; Lanier, S.M. The Ras-related protein AGS1/RASD1 suppresses cell growth. *Oncogene* **2004**, *23*, 5858–5863. [CrossRef]
40. Koga, T.; Iwasaki, H.; Ishiguro, M.; Matsuzaki, A.; Kikuchi, M. Losses in chromosomes 17, 19, and 22q in neurofibromatosis type 1 and sporadic neurofibromas: A comparative genomic hybridization analysis. *Cancer Genet. Cytogenet.* **2002**, *136*, 113–120. [CrossRef]
41. Tang, Y.J.; Huang, J.; Tsushima, H.; Ban, G.I.; Zhang, H.; Oristian, K.M.; Puviindran, V.; Williams, N.; Ding, X.; Ou, J.; et al. Tracing Tumor Evolution in Sarcoma Reveals Clonal Origin of Advanced Metastasis. *Cell Rep.* **2019**, *28*, 2837–2850.e5. [CrossRef] [PubMed]
42. Both, J.; Wu, T.; Ten Asbroek, A.L.; Baas, F.; Hulsebos, T.J. Oncogenic Properties of Candidate Oncogenes in Chromosome Region 17p11.2p12 in Human Osteosarcoma. *Cytogenet. Genome. Res.* **2016**, *150*, 52–59. [CrossRef] [PubMed]
43. Wong, K.Y.; Yao, Q.; Yuan, L.Q.; Li, Z.; Ma, E.S.K.; Chim, C.S. Frequent functional activation of RAS signalling not explained by RAS/RAF mutations in relapsed/refractory multiple myeloma. *Sci. Rep.* **2018**, *8*, 13522. [CrossRef] [PubMed]
44. Ward, A.; Balwierz, A.; Zhang, J.D.; Kublbeck, M.; Pawitan, Y.; Hielscher, T.; Wiemann, S.; Sahin, O. Re-expression of microRNA-375 reverses both tamoxifen resistance and accompanying EMT-like properties in breast cancer. *Oncogene* **2013**, *32*, 1173–1182. [CrossRef]
45. Tabas-Madrid, D.; Nogales-Cadenas, R.; Pascual-Montano, A. GeneCodis3: A non-redundant and modular enrichment analysis tool for functional genomics. *Nucleic Acids Res.* **2012**, *40*, W478–W483. [CrossRef]

4

Epigenetic Regulation of ATP-Binding Cassette Protein A1 (*ABCA1*) Gene Expression: A New Era to Alleviate Atherosclerotic Cardiovascular Disease

Mohamed Zaiou [1,*] and Ahmed Bakillah [2]

1 School of Pharmacy, University of Lorraine, 5 rue Albert Lebrun, 54000 Nancy, France
2 Department of Medicine, Downstate Medical Center, State University of New York, 450 Clarkson Ave, Brooklyn, NY 11203, USA; Ahmed.Bakillah@downstate.edu
* Correspondence: mohamed.zaiou@univ-lorraine.fr.

Abstract: The most important function of high density lipoprotein (HDL) is its ability to remove cholesterol from cells and tissues involved in the early stages of atherosclerosis back to the liver for excretion. The ATP-binding cassette transporters ABCA1 and ABCG1 are responsible for the major part of cholesterol efflux to HDL in macrophage foam cells. Thus, promoting the process of reverse cholesterol transport (RCT) by upregulating mainly ABCA1 remains one of the potential targets for the development of new therapeutic agents against atherosclerosis. Growing evidence suggests that posttranscriptional regulation of HDL biogenesis as well as modulation of ABCA1 expression are under the control of several genetic and epigenetic factors such as transcription factor (TFs), microRNAs (miRNAs) and RNA-binding proteins (RBPs).These factors may act either individually or in combination to orchestrate ABCA1 expression. Complementary to our recent work, we propose an exploratory model for the potential molecular mechanism(s) underlying epigenetic signature of ABCA1 gene regulation. Such a model may hopefully provide the basic framework for understanding the epigenetic regulation of RCT and contribute to the development of novel therapeutic strategies to alleviate the burden of cardiovascular diseases (CVD).

Keywords: ABCA1; HDL; miRNA; circular RNA; gene expression; RNA-binding proteins; reverse cholesterol transport; cardiovascular diseases

1. Introduction

Based on the general consensus that HDL protects against atherosclerotic CVD, several attempts have been made to design drugs that raise HDL cholesterol (HDL-C) levels or enhance its cardioprotective function. However, several prospective HDL-C-raising trials have failed to demonstrate improved efficacy for major adverse cardiovascular outcomes. In this prospect, there are continuous efforts to increase HDL functionality to enhance RCT and potentially achieve further cardiovascular event reduction.

Over the last decades, genetic breakthroughs have revolutionized cardiovascular research by providing opportunities to elucidate novel molecular mechanisms underlying HDL biogenesis and RCT pathways. Some preclinical and clinical studies have shown that apoA-I Milano, a naturally occurring mutant of ApoA-I (Arg173Cysteine) has beneficial athero-protective and anti-inflammatory effects [1]. Furthermore, research has suggested that the expression of genes involved in the RCT combines several complex regulatory networks that are controlled by at least two types of trans-factors: transcription factors (TFs) in the nucleus and posttranscriptional epigenetic factors that bind to cis-regulatory RNA elements mostly located in the 3′UTR of their target mRNAs. Epigenetic factors including miRNAs and RBPs have been shown to be associated with several pathophysiological conditions. For instance, the dysregulation of miRNAs has been associated with the disruption of multiple gene networks leading to metabolic disorders including diabetes, obesity, metabolic syndrome and atherosclerosis [2,3]. This clearly indicates that beyond the existing therapies, advance in epigenetics mechanisms could offer additional opportunities to develop novel treatment strategies for atherosclerosis. In this regard, attempts have been made to develop Apabetalone (RVX-208) as the first epigenetic approach to treat CVD. In recent clinical trials such as SUSTAIN and ASSURE, RVX-208, an orally active molecule has been shown to increase plasma apoA-I, HDL-C (pre-beta and alpha-HDL), and enhance the ability of serum to efflux cholesterol via ABCA1, ABCG1, and scavenger receptor class B type I (SR-BI)-dependent pathways [4]. RVX-208 increases apoA-I transcription through an epigenetic mechanism by inhibiting the bromodomain and extra-terminal domain (BET) protein 4 (BRD4) [5]. Most importantly, emerging evidence has suggested that miRNAs act as a novel class of epigenetic regulators of RCT and HDL-C from synthesis to clearance, and thus contributing remarkably to the pathogenesis of atherosclerosis [6]. Hence, understanding epigenetic mechanisms that control RCT genes network, mainly *ABCA1* gene that is involved in the initiation of this process, may bring insights into novel therapeutic approaches for treating atherosclerotic vascular disease.

2. Discussion

ABCA1 is a key mediator of cholesterol efflux to lipid-poor apolipoprotein A-1. Several studies have confirmed that compromised ABCA1 activity leads to accelerated and early atherogenesis. As summarized in our recent paper [6], the RCT gene network is a complex process highly controlled at both transcriptional and posttranscriptional levels. Herein, we consider the *ABCA1* gene, for which some experimental knowledge is available both at the transcriptional and posttranscriptional levels, as an example to illustrate potential epigenetic mechanisms driving the regulation of RCT. We propose a hypothetical model for the potential dynamic interplay between different genetic and epigenetic regulators that may serve to regulate *ABCA1* gene enabling the cell to respond to different environment changes (Figure 1). However, all proposed axes of interaction, if they occur, need to be experimentally validated in different models.

At the transcriptional level, several nuclear receptors including peroxisome proliferator-activated receptors (PPARs), liver X-receptor (LXR), and farnesoid X receptor (FXR), have been shown to influence lipid metabolism along with genes involved in RCT pathway including *ABCA1*. Unfortunately, despite the effectiveness of these nuclear receptors in preclinical studies, their translation to human clinical trials is still facing many challenges.

At the posttranscriptional level, the 3′-UTR of *ABCA1* gene has been shown to be directly targeted by multiple miRNAs including miR-33, miR-758, miR-145, miR-27, miR-144, miR-26 and miR-106, which lead to cholesterol efflux and HDL-C levels repression [6]. In addition to miRNAs, RNA-binding proteins (RBPs) are also known to bind to AU-rich elements (AREs) in the 3′UTR of many genes, and thereby modulate their expression by increasing or decreasing mRNAs' translation and/or stability. With respect to ABCA1, human antigen R (HuR), an RBP, has been reported to bind to the 3′-UTR of this transporter and increase its expression by enhancing protein translation [7]. Based on this information, it would be tempting to suggest the existence of a possible regulatory relationship between miRNAs and RBPs that could influence ABCA1 expression. In this context, specific miRNAs and RBPs may act via cooperation/competition to directly or indirectly regulate gene expression. Furthermore, post-translational modifications of RBPs, including their phosphorylation and methylation, provide additional layers of complexity, as they control RNA-binding, function and localization [8]. Therefore, phosphorylation and nuclear transit of RBPs could be another possible mechanism to influence RBP-mediated regulation of *ABCA1* gene expression as suggested in a previous study [9]. In addition to these potential mechanisms, certain miRNA species may also control the expression of other important epigenetic regulators such as DNA methyltransferases and histone deacetylases. Conversely, DNA/RNA methylation and histone modification may contribute to the regulation of these miRNAs.

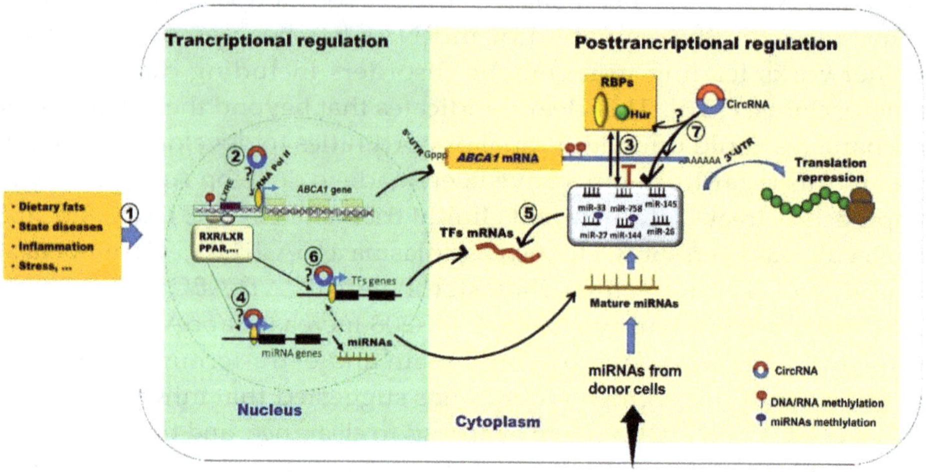

Abbreviations: **TFs**; transcription factors, **RBPs**; RNA-binding proteins, **Hur**; Human antigen R, **CircRNA**; circular RNA, **ABCA1**; ATP-binding cassette transporter A1, **RXR**; Retinoid X receptor, **LXR**; Liver X receptor, **PPAR**; peroxisome proliferator-activated receptor.

Figure 1. Schematic representation of proposed scenarios for the mechanisms by which *ABCA1* gene expression is regulated. Various factors including nutrition, state diseases, stress and inflammation can alter epigenetic *ABCA1* gene regulation in different cells mainly macrophages and hepatocytes (**1**). At the transcriptional level, *ABCA1* gene expression is regulated by key nuclear receptors including LXR family and their heterodimeric partners, retinoic acid receptors (RXR) via functional LXREs (**2**). At the post-transcriptional level, miRNAs and RBPs (Hur) cooperate/compete for target binding regulation (**3**). miRNAs and TFs may cooperate to tune gene expression by forming feedback or feedforward loops (**4**) and (**5**). CircRNAs may also be part of the interplay by their potential interaction with TFs, miRNAs and RBPs (**2**), (**4**), (**6**) and (**7**). Finally, the proposed interplay between TFs-miRNAs-RBPs-CircRNAs could be taken into consideration while elucidating epigenetic mechanisms regulating the first step of RCT gene network.

As is true for protein-coding genes, the expression of miRNAs is also under the control of numerous transcription factors. Aberrant regulation of miRNAs by TFs can cause phenotypic variations and diseases. As an example, miR-26, known to directly target the *ABCA1* 3′-UTR, thereby repressing cholesterol efflux and HDL-C levels, has been shown to be inhibited in cells treated with LXR agonists [10]. Moreover, the activation of LXR has been associated with an increase in miR-144 and ABCA1 expression to fine-tune RCT by macrophages [11]. On the other hand, miRNAs/TFs feedback-loop regulation may occur and represent another possible mechanism involved in the regulation of the RCT gene networks [12]. As an example of the TFs–miRNA regulatory network, RXRα has been reported to be regulated by miRNAs including miR-128-2 [13], while other miRNAs have been shown to regulate the expression and activity of different TFs such PPARs. Hence, it is reasonable to assume the existence of a possible cross-talk between TFs and miRNAs that could modulate *ABCA1* gene and RCT circuit. However, such mechanisms of regulation are still not well defined and could represent an important future area of research.

Circular RNAs (circRNAs), a class of RNAs with the linking of 3′ and 5′ ends, are predicted to function as robust transcriptional and posttranscriptional regulators of gene expression. Evidence is arising that some circRNAs might regulate miRNA function as microRNA sponges. Relevant information in this respect can be provided by the web tool CircInteractome, freely available at http://circinteractome.nia.nih.gov. This new database has been developed for exploring circRNAs and their interacting proteins and microRNAs [14]. Further studies have suggested that circRNA-miRNA-mRNA axes play a prominent role in different pathologies including CVD [15]. Some circRNAs have been predominantly localized in the nucleus where they can directly promote host-gene transcription through interaction with RNA polymerase II (pol II) in the promoter region of genes. Furthermore, these RNAs may interact with TFs to influence various diseases [16].

Because no data are yet available regarding the impact of circRNAs on lipid homeostasis and HDL biogenesis, this has prompted us to propose a model integrating TFs, miRNAs, RBPs and CirRNAs into the *ABCA1* regulatory gene network (Figure 1). However, we are aware that certain circuits of this model are much more difficult to directly translate in vivo mostly due to the complexity of several highly connected pathways with feedback and feedforward loops. Thus, to validate the proposed epigenetic transcriptional and post-transcriptional mechanisms for *ABCA1* gene expression regulation, future well-designed in vivo experimental studies combined with high-throughput sequencing and bioinformatic data are required.

3. Conclusions and Perspectives

Emerging research suggests that deregulated miRNAs can impact RCT gene networks, mainly *ABCA1* gene. These observations have generated a renewed interest in novel targets for epigenetic regulators that may pave the way for establishing novel strategies to raise functional HDL and promote RCT. However, since the field of epigenetic regulation by small RNAs is still in its infancy, there are only a few experimentally validated data associating the dysregulation of miRNAs with genes involved in the HDL biogenesis and RCT. Our proposed model (Figure 1) for potential epigenetic regulation of *ABCA1* is expected to be translated into specific questions and hypothesis which may benefit further research in this area. However, one of the greatest challenges one may confront here is how to fully gain knowledge from the potential interplay and/or cross-talk between various miRNAs and other cellular regulatory factors controlling specific pathway(s) of cholesterol removal from cells and how all these networks can shape a physiological function in both normal and pathological cells in response to a behavior or a stimulus. To answer these questions, more focus should be given to the integration of computational approaches such as systems biology and molecular networks modeling in combination with high-throughput experiments to provide a list of potential miRNA target genes involved in the RCT process. This might help us to better understand the consequences of the complex interplay of miRNAs with other epigenetic regulators on *ABCA1* gene and possibly other relevant genes involved in RCT regulation.

Author Contributions: M.Z. and A.B. conception, design of perspectives, wrote, read and approved the manuscript.

References

1. Chyu, K.Y.; Shah, P.K. HDL/ApoA-1 infusion and ApoA-1 gene therapy in atherosclerosis. *Front. Pharmacol.* **2015**, *6*, 187. [CrossRef] [PubMed]
2. Rottiers, V.; Näär, A.M. MicroRNAs in metabolism and metabolic disorders. *Nat. Rev. Mol. Cell Biol.* **2012**, *13*, 239–250. [CrossRef] [PubMed]
3. Zaiou, M.; El Amri, H.; Bakillah, A. The clinical potential of adipogenesis and obesity-related microRNAs. *Nutr. Metab. Cardiovasc. Dis.* **2018**, *28*, 91–111. [CrossRef] [PubMed]
4. Gilham, D.; Wasiak, S.; Tsujikawa, L.M.; Halliday, C.; Norek, K.; Patel, R.G.; Kulikowski, E.; Johansson, J.; Sweeney, M.; Wong, N.C. RVX-208, a BET-inhibitor for treating atherosclerotic cardiovascular disease, raises ApoA-I/HDL and represses pathways that contribute to cardiovascular disease. *Atherosclerosis* **2016**, *247*, 48–57. [CrossRef] [PubMed]
5. Picaud, S.; Wells, C.; Felletar, I.; Brotherton, D.; Martin, S.; Savitsky, P.; Diez-Dacal, B.; Philpott, M.; Bountra, C.; Lingard, H.; et al. RVX-208, an inhibitor of BET transcriptional regulators with selectivity for the second bromodomain. *Proc. Natl. Acad. Sci. USA* **2013**, *110*, 19754–19759. [CrossRef] [PubMed]
6. Zaiou, M.; Rihn, B.; Bakillah, A. Epigenetic regulation of genes involved in the reverse cholesterol transport through interaction with miRNAs. *Front. Biosci. Landmark* **2018**, *23*, 2090–2105.
7. Ramírez, C.M.; Lin, C.S.; Abdelmohsen, K.; Goedeke, L.; Yoon, J.H.; Madrigal-Matute, J.; Martin-Ventura, J.L.; Vo, D.T.; Uren, P.J.; Penalva, L.O.; et al. RNA binding protein HuR regulates the expression of ABCA1. *J. Lipid Res.* **2014**, *55*, 1066–1076. [CrossRef] [PubMed]

8. Glisovic, T.; Bachorik, J.L.; Yong, J.; Dreyfuss, G. RNA-binding proteins and post-transcriptional gene regulation. *FEBS Lett.* **2008**, *582*, 1977–1986. [CrossRef] [PubMed]
9. Kurischko, C.; Broach, J.R. Phosphorylation and nuclear transit modulate the balance between normal function and terminal aggregation of the yeast RNA-binding protein Ssd1. *Mol. Biol. Cell.* **2017**, *28*, 3057–3069. [CrossRef] [PubMed]
10. Sun, D.; Zhang, J.; Xie, J.; Wei, W.; Chen, M.; Zhao, X. MiR-26 controls LXR-dependent cholesterol efflux by targeting ABCA1 and ARL7. *FEBS Lett.* **2012**, *586*, 1472–1479. [CrossRef] [PubMed]
11. Ramirez, C.M.; Rotllan, N.; Vlassov, A.V.; Dávalos, A.; Li, M.; Goedeke, L.; Aranda, J.F.; Cirera-Salinas, D.; Araldi, E.; Salerno, A.; et al. Control of cholesterol metabolism and plasma high-density lipoprotein levels by microRNA-144. *Circ. Res.* **2013**, *112*, 1592–1601. [CrossRef] [PubMed]
12. Martinez, N.J.; Walhout, A.J. The interplay between transcription factors and microRNAs in genome-scale regulatory networks. *Bioessays* **2009**, *31*, 435–445. [CrossRef] [PubMed]
13. Adlakha, Y.K.; Khanna, S.; Singh, R.; Singh, V.P.; Agrawal, A.; Saini, N. Pro-apoptotic miRNA-128-2 modulates ABCA1, ABCG1 and RXRα expression and cholesterol homeostasis. *Cell Death Dis.* **2013**, *4*, e780. [CrossRef] [PubMed]
14. Dudekula, D.B.; Panda, A.C.; Grammatikakis, I.; De, S.; Abdelmohsen, K.; Gorospe, M. CircInteractome: A web tool for exploring circular RNAs and their interacting proteins and microRNAs. *RNA Biol.* **2016**, *13*, 34–42. [CrossRef] [PubMed]
15. Holdt, L.M.; Stahringer, A.; Sass, K.; Pichler, G.; Kulak, N.A.; Wilfert, W.; Kohlmaier, A.; Herbst, A.; Northoff, B.H.; Nicolaou, A.; et al. Circular non-coding RNA ANRIL modulates ribosomal RNA maturation and atherosclerosis in humans. *Nat. Commun.* **2016**, *7*, 12429. [CrossRef] [PubMed]
16. Yang, Z.G.; Awan, F.M.; Du, W.W.; Zeng, Y.; Lyu, J.; Wu, D.; Gupta, S.; Yang, W.; Yang, B.B. The Circular RNA Interacts with STAT3, Increasing Its Nuclear Translocation and Wound Repair by Modulating Dnmt3a and miR-17 Function. *Mol. Ther.* **2017**, *25*, 2062–2074. [CrossRef] [PubMed]

5

Computational Analysis of Targeting SARS-CoV-2, Viral Entry Proteins ACE2 and TMPRSS2, and Interferon Genes by Host MicroRNAs

Jacob B. Pierce [1,2,†], Viorel Simion [1,†], Basak Icli [1,†], Daniel Pérez-Cremades [1,†], Henry S. Cheng [1,†] and Mark W. Feinberg [1,*]

1 Department of Medicine, Cardiovascular Division, Brigham and Women's Hospital, Harvard Medical School, Boston, MA 02115, USA; Jacob.pierce@northwestern.edu (J.B.P.); simiond.viorel@gmail.com (V.S.); BICLI@BWH.HARVARD.EDU (B.I.); dperez-cremades@bwh.harvard.edu (D.P.-C.); sccheng@bwh.harvard.edu (H.S.C.)

2 Feinberg School of Medicine, Northwestern University, Chicago, IL 60611, USA

* Correspondence: mfeinberg@bwh.harvard.edu.

† These authors contributed equally to this work.

Abstract: Rapid spread of severe acute respiratory syndrome coronavirus 2 (SARS-CoV-2), the virus responsible for coronavirus disease 2019 (COVID-19), has led to a global pandemic, failures of local health care systems, and global economic recession. MicroRNAs (miRNAs) have recently emerged as important regulators of viral pathogenesis, particularly among RNA viruses, but the impact of host miRNAs on SARS-CoV-2 infectivity remains unknown. In this study, we utilize the combination of powerful bioinformatic prediction algorithms and miRNA profiling to predict endogenous host miRNAs that may play important roles in regulating SARS-CoV-2 infectivity. We provide a collection of high-probability miRNA binding sites within the SARS-CoV-2 genome as well as within mRNA transcripts of critical viral entry proteins ACE2 and TMPRSS2 and their upstream modulators, the interferons (IFN). By utilizing miRNA profiling datasets of SARS-CoV-2-resistant and -susceptible cell lines, we verify the biological plausibility of the predicted miRNA–target RNA interactions. Finally, we utilize miRNA profiling of SARS-CoV-2-infected cells to identify predicted miRNAs that are differentially regulated in infected cells. In particular, we identify predicted miRNA binders to SARS-CoV-2 ORFs (miR-23a (1ab), miR-29a, -29c (1ab, N), miR-151a, -151b (S), miR-4707-3p (S), miR-298 (5′-UTR), miR-7851-3p (5′-UTR), miR-8075 (5′-UTR)), ACE2 3′-UTR (miR-9-5p, miR-218-5p), TMPRSS2 3′-UTR (let-7d-5p, -7e-5p, miR-494-3p, miR-382-3p, miR-181c-5p), and IFN-α 3′-UTR (miR-361-5p, miR-410-3p). Overall, this study provides insight into potential novel regulatory mechanisms of SARS-CoV-2 by host miRNAs and lays the foundation for future investigation of these miRNAs as potential therapeutic targets or biomarkers.

Keywords: SARS-CoV-2; COVID-19; microRNA; ACE2; TMPRSS2

1. Introduction

Since December 2019, coronavirus disease 2019 (COVID-19) has rapidly spread from Wuhan, China across the globe, with nearly four million confirmed cases and two-hundred and fifty thousand deaths as of May, 2020, receiving designation as a global pandemic by the World Health Organization [1,2]. A novel betacoronavirus termed severe acute respiratory syndrome coronavirus 2 (SARS-CoV-2) has been identified as the infectious agent responsible for COVID-19 [3]. Infection with SARS-CoV-2 may lead to a wide variety of clinical syndromes primarily involving the respiratory tract, ranging from asymptomatic infection to acute respiratory distress syndrome and death [4,5]. While many clinical trials are ongoing, there are currently minimal established therapies to prevent or treat COVID-19, highlighting the urgent need for further mechanistic insights into the cellular mechanisms underlying SARS-CoV-2 infection.

Much like other betacoronaviruses such as SARS-CoV and MERS-CoV, SARS-CoV-2 is a single-strand, positive-sense RNA virus that uses its viral spike glycoprotein (S protein) to bind to specific target-cell surface receptors and initiate viral entry [6]. Similar to SARS-CoV, SARS-CoV-2 binds specifically to host-cell angiotensin-converting enzyme 2 (ACE2), after which host serine protease TMPRSS2 cleaves the S protein to facilitate membrane fusion [7–10]. Several recent analyses of single-cell RNA sequencing datasets have demonstrated that co-expression of both ACE2 and TMPRSS2 occurs in only a minority of cells, primarily secretory epithelial cells in the upper respiratory tract [11–13], suggesting that cell type-specific mechanisms may be important for SARS-CoV-2 infection and propagation. Similarly, commonly used human cell lines have demonstrated varying degrees of infectivity with SARS-CoV-2 [14], although the mechanisms governing cell line-specific phenotypes remain unknown. Identifying host factors that influence ACE2 and TMPRSS2 expression may provide crucial insights into cellular mechanisms that might be therapeutically exploited against SARS-CoV-2 infection.

MicroRNAs (miRNAs) are well-conserved, short (20–22 nt) non-coding RNA molecules that regulate a broad array of cellular functions at the post-transcriptional level. Cellular miRNAs play an important role in the regulation of gene expression in eukaryotes. Canonical gene regulation by miRNAs involves the binding of miRNAs to the 3′ untranslated region (UTR) of target messenger RNA (mRNA) transcripts, promoting translational repression and mRNA degradation [15]. As a post-transcriptional gene expression regulatory molecule, miRNAs act as key effector molecules in the complicated interaction network between virus and host. Both virus and host can encode miRNAs [16]. Virus miRNAs can resist human cell antiviral immune defense systems by changing various host gene expression to control cell growth and development [17,18]. In contrast, cell-encoded miRNAs can directly affect viral entry and replication cycle [19]. In addition, miRNAs can affect viral infectivity indirectly by regulating host factors involved with viral pathogenesis [19]. Therefore, studies on host cell-derived miRNAs can contribute to the further understanding of mechanisms underlying the interactions between virus and host cells and provide a framework for the discovery of novel antiviral agents and strategies. For example, miRNAs have been shown to target and regulate viral gene expression by targeting different regions of the viral genome, including structural (membrane, envelope, nucleocapsid), non-structural proteins, or the non-coding (untranslated) regions [19]. However, the role of miRNAs in dictating SARS-CoV-2 pathogenesis is unknown.

To address these knowledge gaps, we investigated potential interactions between host miRNAs and SARS-CoV-2. The goals of our study were 2-fold: (1) identify host miRNAs that may directly bind to SARS-CoV-2 genomic RNA and thus affect viral infectivity; and (2) identify host miRNAs that may regulate ACE2 and TMPRSS2 expression, or their upstream interferon (IFN) regulators, and thus control the susceptibility of cells to infection by SARS-CoV-2. By utilizing bioinformatic prediction algorithms to predict miRNA binding to viral and host RNA transcripts in combination with RNA-sequencing data in cell lines with known resistance or susceptibility to SARS-CoV-2 infection, we provide an atlas of miRNAs that are predicted to bind to the SARS-CoV-2 genome directly and thus potentially affect genomic stability, translation, or replication. We also describe host miRNAs that may regulate ACE2, TMPRSS2, or IFN-α, -β, or -γ expression. Finally, we use miRNA profiling from SARS-CoV-2-infected cells to investigate changes in expression of predicted miRNAs following infection. These data establish a detailed framework for future mechanistic studies investigating the relationship of these important miRNA host cell–SARS-CoV-2 interactions and provide a foundation for potential new targets for therapy.

2. Materials and Methods

2.1. SARS-CoV-2 Genomic Analysis

National Center for Biotechnology Information (NCBI) Virus was queried for SARS-CoV-2 complete genomes deposited with NCBI through 19 April 2020. Five genomes were excluded due to poor sequence quality, resulting in 820 unique SARS-CoV-2 genomes used in further analyses. Viral genomic sequences were aligned using MUltiple Sequence Comparison by Log-Expectation (MUSCLE) [20], and a rooted phylogenetic tree was constructed by the unweighted pair group method with six base nucleotide k-mers [21] and visualized using iTOL [22]. NCBI Basic Local Alignment Search Tool (BLAST) [23] was used to investigate genomic sequence identity. SARS-CoV-2 isolate Wuhan-Hu-1 (GenBank accession: MN908947.3) was used as a reference sequence similar to previous phylogenetic analyses [24,25].

2.2. MiRNA Binding Site Prediction

MiRNA binding sites within the SARS-CoV-2 genome were predicted using miRDB (http://mirdb.org/index.html) [26]. Similar to previous analyses [27], candidate miRNAs with a miRDB target score ≥70 were retained for further analysis. RNAhybrid (https://bio.tools/rnahybrid) [28] was used to verify miRDB-predicted miRNA–SARS-CoV-2 genomic binding. RNAhybrid queries specified seed region binding (specified as nucleotides 2–8 on candidate miRNAs), a minimum energy threshold of -18 Kcal/mol, and a maximum bulge and internal loop length of 1. The insertion of a polybasic amino acid sequence (12 nt) in the SARS-CoV-2 S protein has been implicated in the unique pathogenic properties of SARS-CoV-2 [25]. To maximize the sensitivity for potential miRNA binding sites, the stringency of the miRDB target score and RNAhybrid minimum energy thresholds were relaxed for a dedicated analysis of this genomic region. Candidate miRNAs meeting the above inclusion criteria of both miRDB and RNAhybrid databases were retained for expression analyses. RNA secondary structure predictions were performed using RNAfold http://rna.tbi.univie.ac.at/cgi-bin/RNAWebSuite/RNAfold.cgi) and presented as the predicted minimum free energy conformation [29]. MiRWalk 2.0 (http://mirwalk.umm.uni-heidelberg.de/) [30] was used to predict miRNA binding sites in the 3′ UTR of viral entry proteins ACE2 and TMPRSS2 as well as IFN-α, -β, and -γ as ACE2 has recently been shown to be an IFN-responsive gene [11].

2.3. MiRNA Expression in Human Tissue and Cell Lines

The expression of candidate miRNAs predicted to bind both the viral genome as well as mRNA transcripts of proteins of interest (ACE2, TMPRSS2, and IFNs) was analyzed in human tissue samples to verify the biologic plausibility of binding interactions. The Human MicroRNA Tissue Atlas (https://ccb-web.cs.uni-saarland.de/tissueatlas/) was used to investigate candidate miRNA expression in human tissue samples [31]. Tissues analyzed in the Human MicroRNA Tissue Atlas project originated from two male subjects. The first was from a 65-year-old male patient who died due to multiple myeloma. The body was stored at 4 °C after arrival at the anatomical institute and dissected 2 days post-mortem. The second was from a 59-year-old male individual, who died a natural death. The body was frozen at −20 °C after arrival at the anatomical institute and dissected after 3 weeks of storage. Autopsy showed no signs of cancer. Lung, liver, kidney, and lymph nodes were amongst the 37 types of tissues collected from both. MiRNA expression was also investigated in several commonly used cell lines. Cell lines were chosen based on proven infectivity with SARS-CoV-2 as determined by post-inoculation viral titers, copy number quantification, luciferase reporter assays, and their reported expression of ACE2 and TMPRSS2. Huh7 [8,14,32,33] and Calu3 [8,14,32,34] were evaluated as cell lines with high SARS-CoV-2 infectivity, and A549 [8,14,32,33] and primary human lung fibroblasts (LF) [12,13] were evaluated as cells with low infectivity. Changes in miRNA expression following SARS-CoV-2 infection were based on miRNA expression at 24 h post-inoculation with either SARS-CoV-2 or mock [35].

2.4. Statistical Analysis

Absolute miRNA expression in human lung tissue and cell lines was expressed in copies per million (CPM). MiRNA expression profiles of Huh7, Calu-3, A549 and human primary lung fibroblasts were downloaded from https://www.ncbi.nlm.nih.gov/geoprofiles/. MiRNA profiling of untreated Huh7 and A549 was performed as a part of the same experiment (GSE116179). We therefore directly compared miRNA profiles between these two cell lines with high and low SARS-CoV-2 infectivity, respectively [36]. Relative expression between Huh7 and A549 was expressed as log2-fold change in Huh7 relative to A549. Differential expression analyses were performed using DESeq2 R package [37]. MiRNA profiling of Calu-3 (GSE139516) [38] and LF (GSE125183) [39] was performed as a part of independent experiments, limiting direct statistical comparisons between these cell lines. However, miRNAs that play important roles in the susceptibility or resistance to SARS-CoV-2 infection will likely display similar relative expression across cell lines (i.e., miRNAs important to viral susceptibility will likely be more highly expressed relative to other miRNAs in both Huh7 and Calu-3 and lowly expressed in A549 and LF) [40]. We therefore conducted semi-quantitative analyses in which we divided miRNAs into terciles of relative expression within each cell line, allowing comparison across all four cell lines. Relative expression of RNAseq data from Calu-3 cells infected with SARS-CoV-2 was conducted using DESeq2 R package and calculated as log2-fold change in Calu-3 cells infected for 24 h with SARS-CoV-2 compared to mock-infected cells (GSE148729). Adjusted p values calculated from the DESeq2 R package were used in our analyses. Student t-test was used to determine statistical significance between groups. Two-tailed $p < 0.05$ was considered statistically significant. Statistical analysis and data visualization were performed in STATA/IC software version 15.1 (College Station, TX, USA), GraphPad Prism version 7.0a (La Jolla, CA), and R version 3.6.2 (Vienna, Austria).

3. Results

3.1. Globally Representative SARS-CoV-2 Genomes Demonstrate Minimal Variation in Nucleotide Sequence and Predicted miRNA Binding Sites

First, we sought to determine whether a single SARS-CoV-2 genome sequence (i.e., reference sequence MN908947.3) could be used to predict binding by potential hosts miRNAs to SARS-CoV-2 more broadly. Phylogenetic analysis of the 820 included SARS-CoV-2 genomes demonstrated detectable deviation from the SARS-CoV-2 reference sequence (Figure 1A). However, when compared to the SARS-CoV-2 reference sequence using BLAST, there was remarkably little sequence variation in all included sequences (mean percent sequence identity = 99.97%, standard deviation = 0.02%; Figure 1B). To investigate the impact of the small degree of genetic variation between SARS-CoV-2 genomes on potentially important miRNA binding sites, we selected three genomic regions—the 5′ UTR, S protein, and 3′ UTR—of the five sequences with the lowest sequence identity compared to the reference sequence (Table 1). These sites were selected based on previously described interactions between host miRNAs and 5′ and 3′ UTRs of RNA viral genomes [41–43] as well as the critical importance of the S protein for SARS-CoV-2 pathogenicity [7,9,10]. Overall, the genetic sequences were highly conserved compared to the reference sequence. Four of these samples originated in the United States, three of which were from Washington state. The fifth sample was from Punjab, Pakistan. The most common cause of sequence deviation from the reference sequence was short "N" repeats (indicating unknown nucleotides due to unreliable sequencing). Point mutations (C–>T) were identified in the 5′-UTRs of SARS-CoV-2 genomes from Washington (MT345855.1) and California (MT258382.1), USA, but overall sequence identity remained at 99.85% and 99.87%, respectively (Table 1). Additional point mutations (T–>W, either T or A; or A–>G) occurred in the S spike protein domain or (G–>R, either G or A; or C–>M, either C or A) in the 3′UTR of the SARS-CoV-2 genome from California (MT258382.1) Truncations to the 5′ UTR, 3′ UTR, or both were common among all five sequences sampled. Additionally, one United States sequence (MT293170.1) contained a 9 nt poly-A insertion within the poly-A tail of the 3′ UTR. Despite the minor detected differences in genomic sequence, there was very little change in predicted miRNA binding sites when the full SARS-CoV-2 genomes were queried in miRDB (Figure 1C). Based on these results, the SARS-CoV-2 reference sequence Wuhan-Hu-1 (GenBank accession: MN908947.3) was selected for use in further miRNA prediction analyses to maintain comparability with previously published studies.

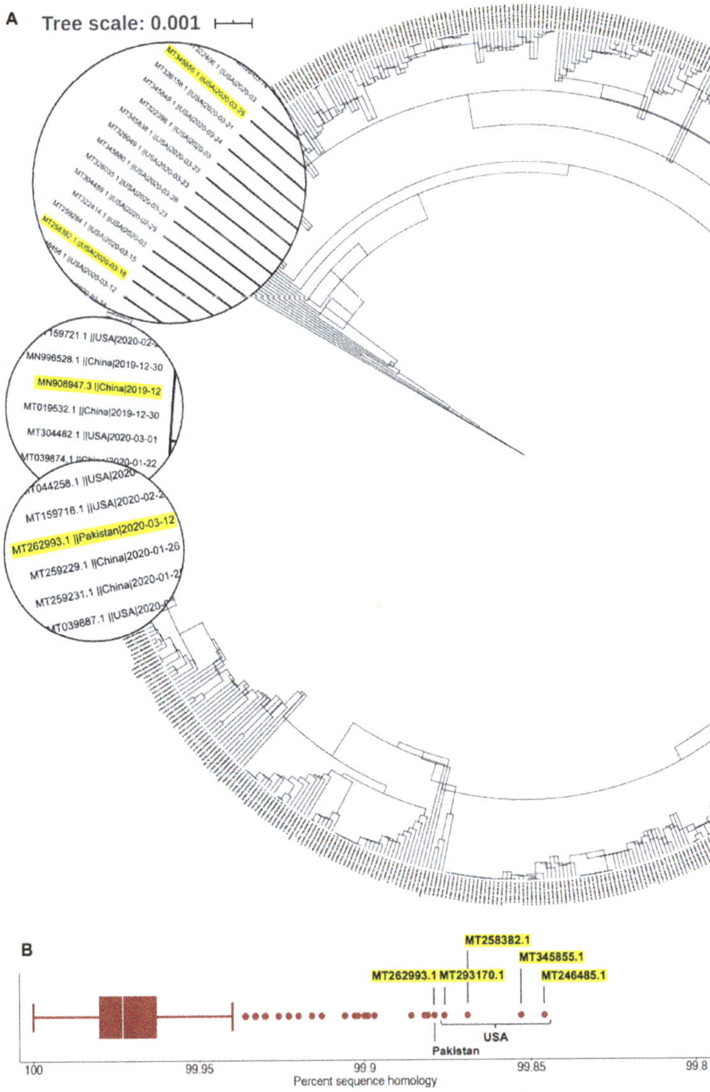

C

	Predicted miRNAs	% similarity
MN908947.3-Wuhan RefSeq	900	(ref.)
MN988669.1	900	100.00
MT262911.1	900	100.00
MT326178.1	906	99.78
MT326117.1	898	99.67
MT050493.1	895	99.33
MT326153.1	896	99.56
MT358705.1	898	99.67
MT106053.1	899	99.89
MT066156.1	899	99.89
MT292575.1	896	99.22
MT262993.1	898	99.88
MT293170.1	894	99.22
MT258382.1	894	98.88
MT345855.1	893	99.11
MT246485.1	893	99.00

Figure 1. Phylogenetic analysis of severe acute respiratory syndrome coronavirus 2 (SARS-CoV-2) genomes demonstrate minimal sequence variation. (**A**) Phylogenetic tree of 820 SARS-CoV-2 complete genomes deposited in the National Center for Biotechnology Information (NCBI) database through 19 April 2020. Highlighted sequences indicate those selected for further sequence analysis shown in Table 1. (**B**) Percent sequence identity of SARS-CoV-2 complete genomes compared with the reference sequence (MN908947.3). (**C**) Changes in miRDB-predicted microRNA binding sites predicted along the full SARS-CoV-2 genomic sequence for genomes with variable identity.

Table 1. Comparison of select genomic regions between severe acute respiratory syndrome coronavirus 2 (SARS-CoV-2) reference sequence and the five largest outliers found during BLAST sequence identity analysis.

Location	Wuhan, China (RefSeq)	Washington, USA	Washington, USA	California, USA	Washington, USA	Punjab, Pakistan
GenBank	MN908947.3	MT246485.1	MT345855.1	MT258382.1	MT293170.1	MT262993.1
Overall						
Bases	29,903	29,775	29,859	29,885	29,798	29,836
Sequence identity [a]	-	99.85% [b]	99.85% [b]	99.87% [b]	99.88% [b]	99.88%
5′ UTR						
Bases	265	137	234	247	151	265
Sequence identity [a]	-	82.5% [b]	91.45% [b]	99.60	92.72% [b]	100%
Mutations	-	-	C→T	C→T	-	-
S protein						
Bases	3822	3822	3822	3822	3822	3822
Sequence identity [a]	-	100%	100%	99.95%	100%	100%
Mutations	-	-	-	T→W (either T or A); A→G	-	-
3′ UTR						
Bases	229	229	198	229	238	198
Sequence identity [a]	-	98.69%	100%	93.45% [b]	97.38% [b]	100%
Mutations	-	TGAC → AAAA	-	G→R (either G or A); C→M (either C or A)	9 nt insertion in poly-A tail	-

[a] Sequence identity was based on BLAST analyses with Wuhan-Hu-1 (GenBank accession: MN908947.3) as a reference sequence. [b] NCBI sequence contains short "N" repeat (unknown sequence) due to unreliable sequencing, accounting for a large proportion of sequence deviation from reference sequence.

3.2. Genomic Regions of SARS-CoV-2 Demonstrate Unique Predicted miRNA Binding Profiles

To predict miRNAs that may bind directly to the SARS-CoV-2 genome, we conducted individual miRDB queries for each open reading frame (ORF) of the SARS-CoV-2 genome as well as the 5′ and 3′ UTRs (1019 total miRNAs; Figure 2A,B). MiRNAs with a miRDB target score ≥70 (459 miRNAs) were then verified using RNAhybrid, resulting in a final collection of 288 miRNAs predicted to bind to the SARS-CoV-2 genome with high probability. Individual miRNAs and their predicted binding sites within the SARS-CoV-2 genome are schematically displayed in Figure 2C. ORF1ab contained the majority of predicted miRNA binding sites (221 of 288 total binding sites) in part due to its longer sequence. The S protein and nucleoprotein (N protein) ORFs contained 24 and 17 miRNA binding sites, respectively. Similarly, the other domains had a reduced number or no miRNA binding sites (5′-UTR (3 sites), ORF3a (7 sites), E (1 site), M (4 sites), ORF 6 (1 site), ORF 7 (8 sites), ORF 8 (2 sites), and ORF 10 (0 sites)). Notably, the 3′ UTR contained no binding sites meeting pre-specified miRDB and RNAhybrid cut offs. The interaction between miRNAs and viral nucleotide secondary structure has previously shown to play important roles in viral pathogenesis, particularly in the viral UTRs [41]. We therefore investigated the binding sites of the three miRNAs predicted to bind the 5′ UTR in the context of its predicted nucleotide secondary structure (Figure 2D). Interestingly, all three miRNAs are predicted to bind either within or adjacent to stem–loop (SL) structures that are highly conserved among other viruses, miR-298, miR-7851-3p, and miR-8075, in the *Coronaviridae* family and known to be important for viral function [44].

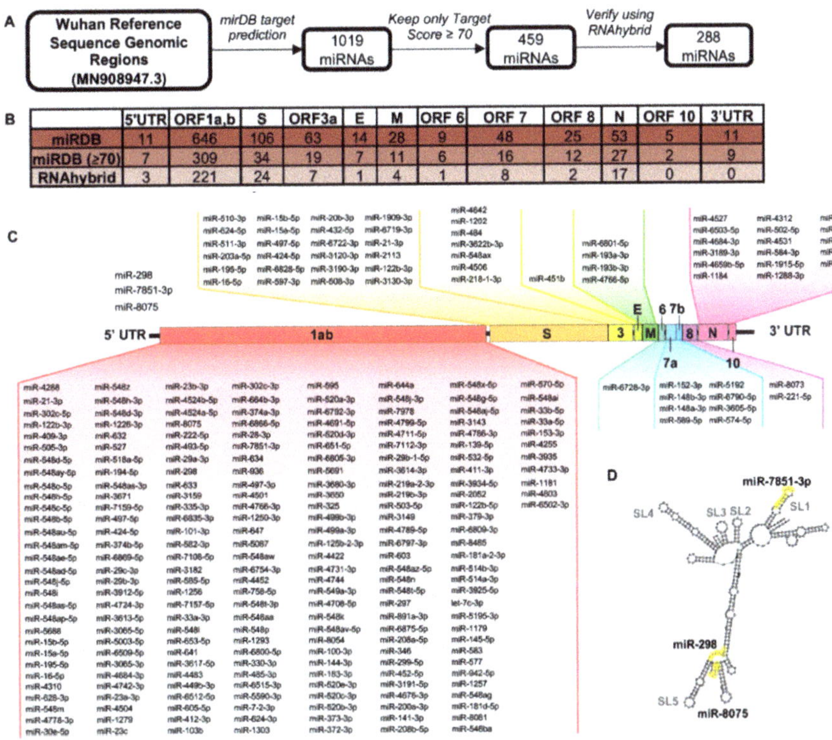

Figure 2. Bioinformatic analysis of the severe acute respiratory syndrome coronavirus 2 (SARS-CoV-2) genome reveals potential microRNA (miRNA) binding sites within distinct genomic regions. (**A**) Open reading frames (ORFs) from the SARS-CoV-2 reference genome (MN908947.3) were probed for potential binding sites of human miRNAs. MiRNAs with a target score ≥70 were validated using RNAhybrid. (**B**) Number of predicted miRNAs binding to individual SARS-CoV-2 genomic regions. Intergenic sequences are included in ORF following intergenic sequence. (**C**) Predicted miRNAs binding to indicated SARS-CoV-2 ORFs. (**D**) Highlighted miRNA binding sites within 5′ UTR. Evolutionarily conserved stem–loop (SL) structures are highlighted.

3.3. Expression of miRNAs in Lungs Predicted to Bind SARS-CoV-2

We further assessed the lung expression of miRNAs predicted to bind to the SARS-CoV-2 genome using the miRNA Tissue ATLAS database. Of the 288 miRNAs interrogated, 53 miRNAs were expressed in the lungs while the rest were undetectable (Figure 3A). The miR-6869-5p targeting ORF1ab showed the highest expression in the lungs. High expression was also observed for miR-16-5p, -15a-5p, -15b-5p, -195-5p and -497-5p targeting both the ORF1ab and gene encoding the S protein while miR-21-3p and 424-5p showed an average expression in the same regions. The miRNA family miR-29a, -29b, -29c (-3p), which are predicted to bind both ORF1ab and ORF9 encoding the non-structural proteins and nucleocapsid protein, respectively, were highly expressed in the lungs. More specifically, within ORF1ab, miR-29a-3p is predicted to bind to the region encoding Nsp14, and the miR-23a-3p binding site resides in the region encoding Nsp3. MiR-15a and -15b and miR-16 have 7, 8, and 6 binding sites within ORF1ab, respectively. The miRNAs miR-23a and -23b (-3p), predicted to bind the ORF1ab, and miR-1202, predicted to bind the ORF3a region, showed similarly high expression in human lungs. In contrast, relatively low lung expression was observed for miR-193a and -193b (-3p), which target the membrane (M) protein coding region, and for miR-148a and -148b (-3p), which target the ORF7a, b region of SARS-CoV-2. (Figure 3A).

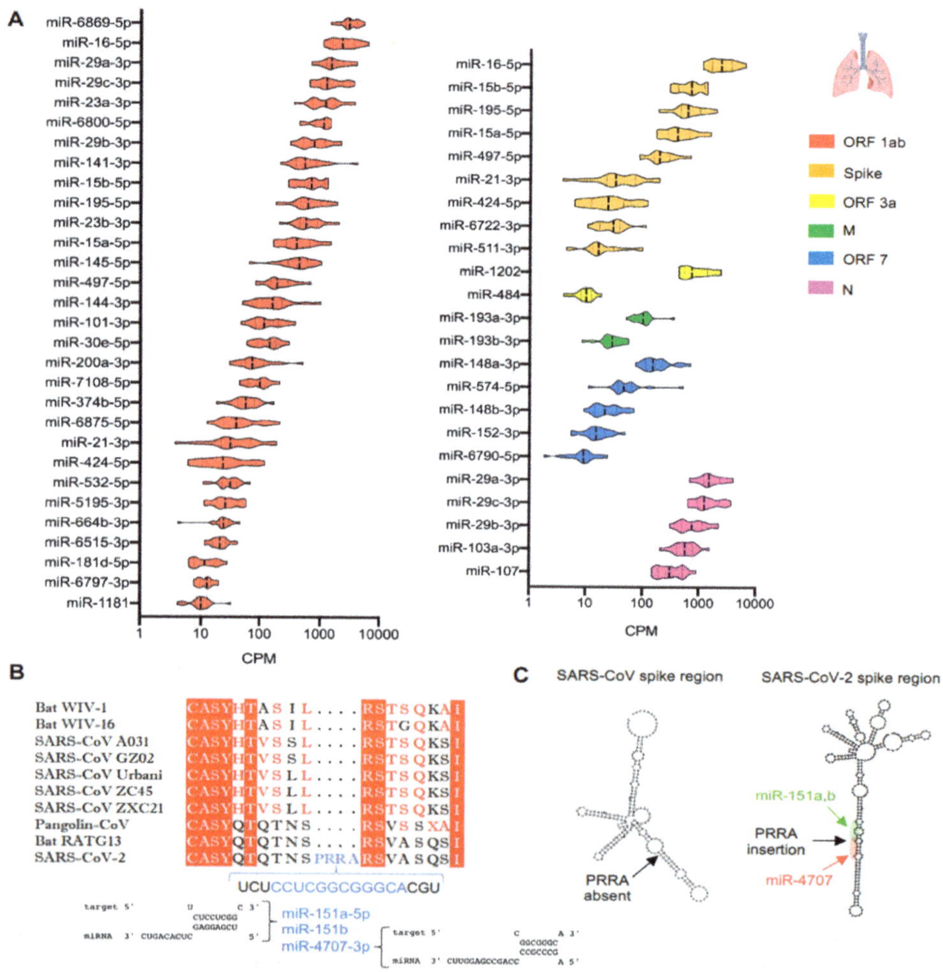

Figure 3. Expression in lungs of microRNAs (miRNAs) targeting the severe acute respiratory syndrome coronavirus 2 (SARS-CoV-2) genome. (**A**) Expression in human healthy lungs of miRNAs predicted to target distinct genomic regions of SARS-CoV-2. (**B**) Amino acid sequence alignment of the spike protein and its phylogenic coronaviruses, and the miRNAs binding in the unique sequence representing the polybasic cleavage site (PRRA). (**C**) Secondary structures of genomic region encoding SARS-CoV and SARS-CoV-2 S proteins in a 200 nt region of the SARS-CoV-2-specific PRRA polybasic cleavage site and highlighted microRNA binding sites within the PRRA region.

One of the most important domains that defines specificity of the SARS-CoV-2 sequence is the acquisition in the spike (S) domain of a polybasic cleavage site (RRAR) with a leading proline (P), a finding absent in SARS-CoV and other bat and pangolin coronaviruses [45]. Using the same prediction algorithms, we found that miR-151a-5p, -151b (RNA hybrid MFE = −17.2 Kcal/mol) and miR-4707-3p (RNA hybrid MFE = −22 Kcal/mol) target this specific PRRA region of the S protein gene (Figure 3B). The nucleotide secondary structure of the PRRA region demonstrated that the miRNAs binding sites are accessibly located in an extended stem–loop structure of the S protein (Figure 3C) and the PRRA region is part of a stem loop in the nucleotide secondary structure of the SARS-CoV-2 S protein (Suppl. Figure S1B). The nucleotide secondary structure of the SARS-CoV genome, where PRRA region is absent, is depicted for comparison (Suppl. Figure S1A). We further assessed the three miRNAs expression in a database from primary lung fibroblasts (a cell type relatively resistant to SARS-CoV-2 infectivity) and observed high enrichment for miR-151a-5p and 151b, while miR-4707-3p was barely detectable (Suppl. Figure S1C). Interrogation of the miRNA issue ATLAS demonstrated similar enrichment of miR-151a-5p and -151b in the lungs compared to other organs, e.g., kidney, liver and lymph node, while miR-4707-3p showed similarly low expression in all organs (Suppl. Figure S1D). Collectively, these findings raise the possibility that miRNAs enriched in either SARS-CoV-2-resistant or -susceptible cell lines may provide further insights for miRNA-mediated control of SARS-CoV-2 infectivity.

3.4. *Identification and Expression of Predicted miRNAs in SARS-CoV-2-Resistant and -Susceptible Cells*

Here we identified several well-documented SARS-CoV-2-resistant and-susceptible cell types (Table 2) that were utilized to further investigate the expression of miRNAs predicted to bind to SARS-CoV-2 (from Figure 2) through in-silico approaches. In these cell types, ACE2/TMPRSS2 mRNA expression roughly correlates with SARS-CoV-2 infectivity. A total of 288 predicted miRNAs binding to the SARS-CoV-2 genome were overlaid on the miRNA expression profile sets of SARS-CoV-2 infection susceptible (Calu-3, Huh7) and resistant (A549, human primary lung fibroblast) cells. As a result, we identified 115 miRNAs binding to the 5′-UTR and ORF regions that were expressed in A549 (resistant) and Huh7 (susceptible) cell lines (Figure 4A,B and Suppl. Figure S2A). Specifically within the region of ORF1ab that encodes the replicase protein [46], miR-320a-3p and miR-320b are predicted to bind the *Nsp8* gene, and miR-3149 is predicted to bind to the *Nsp12* gene (Suppl. Figure S2B). MiRNA datasets for A549 and Huh7 cell lines were acquired from the same study, thereby minimizing any technical issues associated with extraction and processing. Following DESeq2 analyses of A549 and Huh7 raw datasets, 32 miRNAs were identified by differential expression analyses with an arbitrary fold change (FC) cut off of >1.9-fold and $p < 0.05$ (Figure 4C). MiR-23b-3p, miR-452-5p, and miR-181a-2-3p, the most differentially upregulated miRNAs, were enriched in the resistant A549 cells, whereas miR-624-5p, miR-1303 and miR-29c-3p were the most differentially downregulated miRNAs enriched in the susceptible Huh7 cell line (Figure 4C,D). In addition, miR-379-3p, miR-485-3p, and miR-409-3p were among the top miRNAs enriched to the resistant A549 cell line (Figure 4C). To further delineate expression of miRNAs in independent cell datasets, we extended our analyses to additional resistant (human primary lung fibroblasts) and susceptible (Calu-3) cell lines. Interestingly, 19 of the 29 miRNAs identified in A549 and Huh7 datasets were also expressed in both human primary lung fibroblasts (resistant) and Calu-3 (susceptible) cells (Figure 4E). MiR-23a-3p and miR-29a-3p were the most differentially upregulated miRNAs enriched in the resistant human lung fibroblasts, whereas miR-485-3p and miR-299-5p were amongst the top differentially downregulated miRNAs that were enriched in the resistant human lung fibroblasts (Figure 4F). Furthermore, we analyzed a recently deposited RNA-seq dataset of differentially expressed transcripts including miRNAs in Calu-3 cells after infection with SARS-CoV-2 [35]. In line with our findings in permissive and susceptible cell lines, miR-23a-3p and miR-23b-3p were significantly downregulated in Calu-3 cells infected for 24 h with SARS-CoV-2, in comparison to mock-infected cells (Table 3). Additionally, commonly enriched miRNAs between resistant and susceptible cell lines are shown in Table 4. The human tissue expression patterns of the most differentially regulated miRNAs are shown in Suppl. Figure S3.

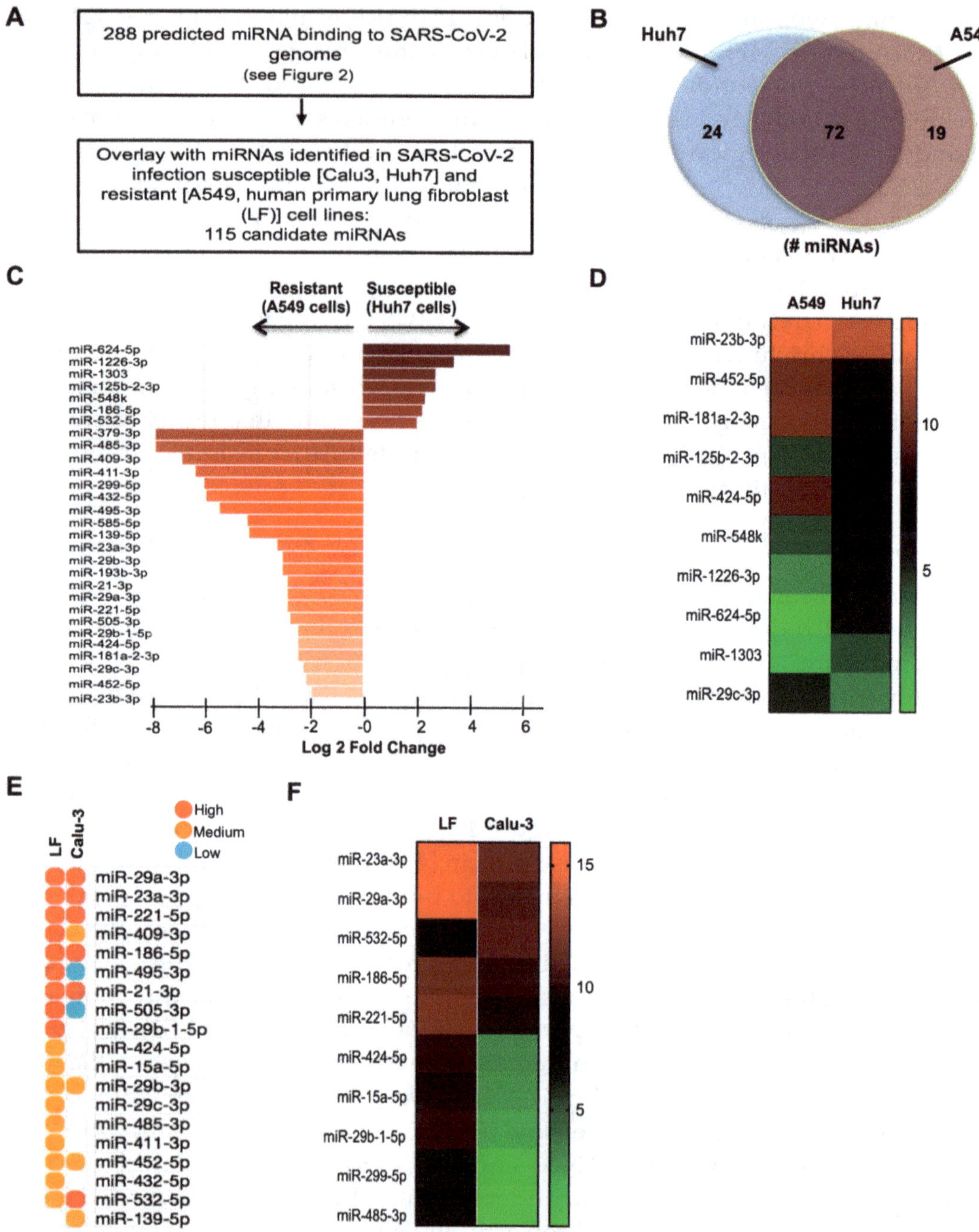

Figure 4. In silico analysis of predicted candidate microRNAs (miRNAs) expressed in severe acute respiratory syndrome coronavirus 2 (SARS-CoV-2)-resistant and -susceptible cell lines. (**A**) Workflow of candidate miRNAs. (**B**) Venn diagram representing the number of differentially regulated miRNAs. (**C**,**D**) Log2-transformed fold changes of most differentially regulated miRNAs (**D**) and heat map showing the top 10 differentially regulated miRNAs in A549 (resistant) and Huh7 (susceptible) cell lines. (**E**) The most differentially regulated miRNAs grouped by high (1st tercile), medium (2nd tercile), and low expression (3rd tercile) (**F**) and heat map showing the top 10 differentially regulated miRNAs expressed as log2-transformed fold changes in human primary lung fibroblasts (LF) and Calu-3 (susceptible) cells.

Table 2. ACE2 and TMPRSS2 expression and reported severe acute respiratory syndrome coronavirus 2 (SARS-CoV-2) infectivity in various liver and lung cells.

	Source	ACE2/TMPRSS2 Expression	SARS-CoV-2 Infectivity
Huh7 [8,14,32,33]	Liver	High [47]	Moderate
A549 [8,14,32,33]	Lung	Variable [48]	Low to Moderate
Calu-3 [8,14,32,34]	Lung	High [49]	High
Primary human lung fibroblasts [12,13]	Lung	None	Low

Predicted miRNAs that target host genes associated with the pathogenesis of SARS-CoV-2.

Table 3. Differential expression of candidate microRNAs (miRNAs) in severe acute respiratory syndrome coronavirus 2 (SARS-CoV-2)-infected (24 h) vs. non-infected Calu-3 cells from Wyler et al. [35].

miRNA	Log2-Fold Change	*p*-value	Favors [a]
ORF1ab			
miR-139-5p	0.945	0.0305	Resistant
miR-664b-3p	0.733	0.0206	-
miR-23a-3p	−0.761	0.0015	Resistant
miR-548d-3p	−0.807	0.0517	-
miR-15b-5p	−0.812	0.0001	-
miR-23c	−0.910	0.0049	-
miR-23b-3p	−0.991	<0.0001	Resistant
miR-374b-3p	−1.069	0.0217	-
miR-374a-3p	−1.582	0.0001	-
Nucleocapsid			
miR-103a-3p	0.904	0.0005	-
miR-107	1.040	0.0003	-
ACE2			
miR-483-3p	4.460	<0.0001	Susceptible
miR-4463	3.048	<0.0001	-
TMPRSS2			
miR-181c-5p	0.892	0.0134	Resistant
miR-664b-3p	0.733	0.0206	-
miR-182-5p	0.704	0.0134	Susceptible
let-7d-5p	0.464	0.0053	Resistant
miR-181a-5p	0.419	0.0422	-
miR-15b-3p	−0.810	0.0008	-
miR-494-3p	−0.961	0.0326	Resistant
IFN-β			
miR-450b-5p	−1.441	0.0020	Resistant
IFN-γ			
miR-664b-3p	0.733	0.0206	-
miR-26b-5p	−0.475	0.0313	-
miR-374c-5p	−1.228	0.0096	-

[a] Susceptible and resistant refers to cell line analyses as per Figures 4, 6, or 7.

Table 4. Concordant microRNA expression between A549 and Huh7 cells.

	log2FC	p_{adj}
miR-140-3p	−0.007953398	0.974817048
miR-32-3p	0.013920456	0.974817048
miR-374a-5p	0.020538776	0.945449302
miR-196b-5p	0.025080141	0.946019864
miR-339-5p	0.029116789	0.949059007
miR-3158-3p	0.029234925	0.97722196

Table 4. *Cont.*

	log2FC	p_{adj}
miR-320c	0.033177006	0.970481801
miR-423-3p	0.039739622	0.860137595
miR-331-5p	0.04015155	0.957965014
miR-1180-3p	0.042520619	0.897231993
miR-1229-3p	0.049456491	0.979826711
miR-6803-3p	0.069110586	0.956513365
miR-339-3p	0.075904611	0.883114045
miR-4326	0.149533463	0.815150225
miR-16-2-3p	0.157983774	0.530911494
miR-6087	0.178211797	0.945394126
miR-25-5p	0.181172206	0.692047229
miR-148b-3p	0.20381122	0.369688028
miR-126-3p	0.20432634	0.517498738
miR-760	0.204547203	0.876649033
miR-200b-5p	0.206829005	0.813533675
miR-1307-5p	0.221466282	0.90711043
miR-454-3p	0.226527458	0.372951464
miR-107	0.232189082	0.645482953
miR-301a-3p	0.234614774	0.88095324
miR-200b-3p	0.241301831	0.267692191
miR-324-3p	0.254720466	0.852842736

Given the importance of viral entry in mediating infectivity and pathogenesis of SARS-CoV-2, we explored the predicted miRNAs that target mRNAs of host receptor proteins ACE2 and TMPRSS2. In addition, we also explored miRNAs that target IFN-α, -β, and -γ as ACE2 was recently identified to be an IFN-stimulated gene [11]. Using five miRNA prediction algorithms (miRWalk, MicroT4, miRMap, RNAhybrid, and TargetScan) we identified 43 miRNAs targeting ACE2, 107 for TMPRSS2, 20 for IFN-α, 29 for IFN-β, and 47 for IFN-γ in their respective 3′UTRs (Figure 5A). Utilizing the Human MicroRNA Tissue Atlas (https://ccb-web.cs.uni-saarland.de/tissueatlas/), which contained miRNA expression profiles of 18 individual human lung tissues, we quantified the expression of these predicted miRNAs in human lungs (Figure 5B). By using a cut off of >10 CPM, the list of predicted miRNAs was reduced by more than 4-fold. There were 7 predicted lung-enriched miRNAs (the top 5 miRNAs were miR-141-3p, miR-4270, miR-331-3p, miR-200a-3p, and miR-218-5p) that targeted the ACE2 mRNA 3′-UTR and 25 predicted lung-enriched miRNAs (the top 5 miRNAs were miR-4763-3p, let-7d-5p, miR-4530, let-7e-5p, and miR-181a-5p) that targeted the TMPRSS2 mRNA 3′-UTR (Figure 5B). In addition, the 3′-UTRs of IFN-α and IFN-β mRNAs contained two (miR-203a-3p and miR-361-5p) and one (miR-145-5p) predicted binding sites, respectively, for lung-enriched miRNAs, whereas IFN-γ harbored nine (miR-128-3p, miR-143-3p, miR-181b-5p, miR-181d-5p, miR-24-3p, miR-26a-5p, miR-26b-5p, miR-340-5p, and miR-664b-3p). Notably, miR-181b-5p, miR-181d-5p, and miR-664b-3p are the only miRNAs to share both target mRNAs (TMPRSS2 and IFN-γ) in human lung tissue.

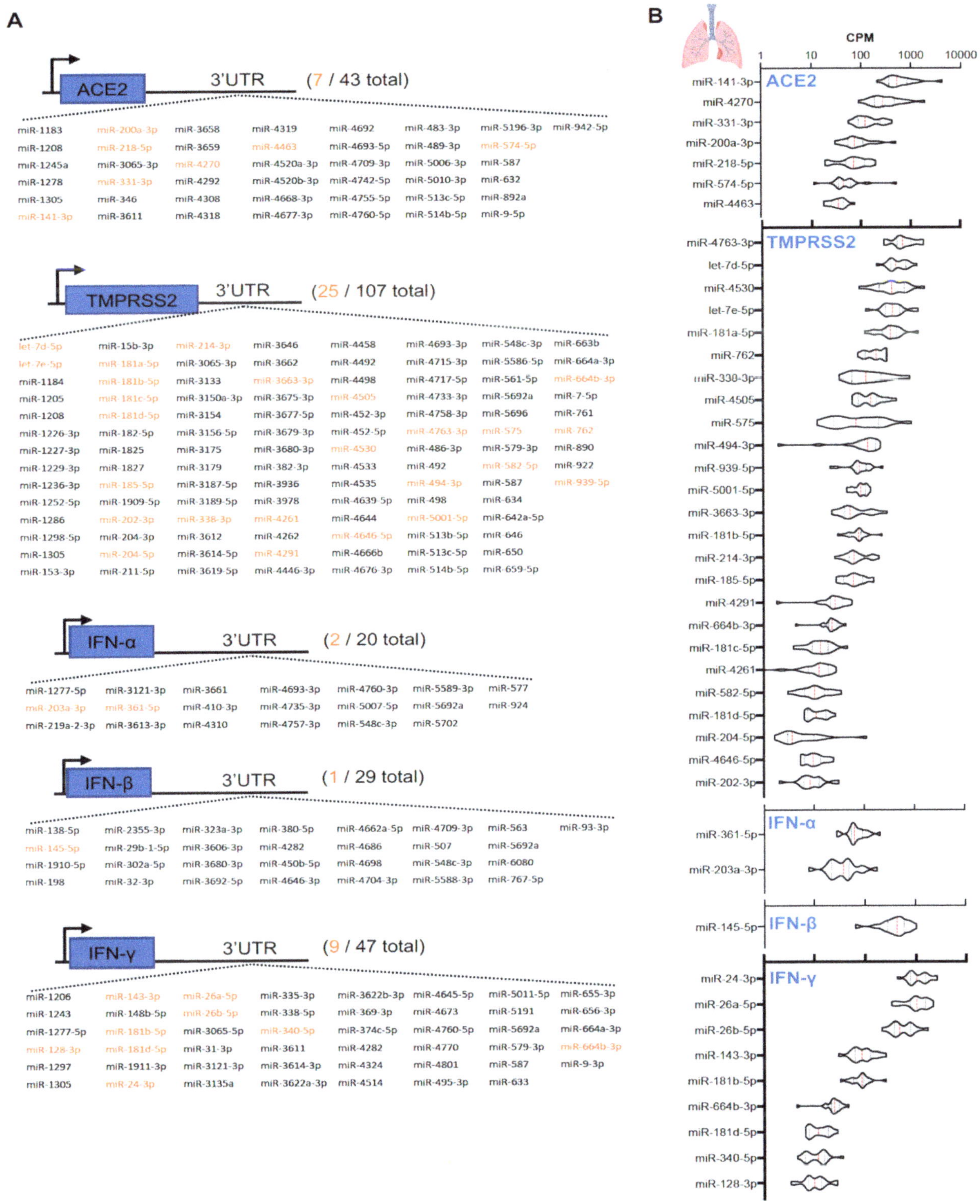

Figure 5. Predicted candidate microRNAs (miRNAs) targeting ACE2, TMPRSS2, interferon (IFN)-α, IFN-β, and IFN-γ mRNAs. (**A**) List of miRNAs targeting the 3′UTRs of ACE2 (total 43), TMPRSS2 (total 107), IFN-α (total 20), IFN-β (total 29), and IFN-γ mRNAs (total 47). Highlighted orange are miRNAs expressed in human lung samples [31]. (**B**) Counts per million (CPM) value of miRNAs predicted to target ACE2, TMPRSS2, IFN-α, IFN-β, and IFN-γ mRNAs found in human lungs (n = 18).

3.5. Identification of Predicted miRNAs to ACE2 and TMPRSS2 in SARS-CoV-2-Resistant and -Susceptible Cells

The human lung is comprised of a plethora of cell types that vary in their ability to resist infectivity of SARS-CoV-2 (Table 1). Utilizing several miRNA expression profiles of 4 different cell types (2 resistant and 2 susceptible), we applied our predicted miRNAs to identify certain miRNAs that are expressed in cells resistant to SARS-CoV-2 infectivity (Figure 6A). Similar to the human lung tissue database, we set a criterion of >10 CPM for each cell type, followed by grouping miRNAs into terciles (highest third, middle third, and lowest third). The list of predicted miRNAs for ACE2 and TMPRSS2 vary slightly between cell types (Figure 6B,C). Similar to analyses of Huh7 and A549 miRNA profiles, we performed differential expression analysis on predicted miRNAs with certain criteria ($p < 0.05$ and >log2-fold change) to identify miRNAs more unique to a particular cell type. MiR-9-5p and miR-218-5p, predicted miRNAs binding to ACE2 mRNA, are expressed higher in resistant A549 cells, while miR-483-3p is expressed higher in susceptible Huh-7 cells (Figure 6D). MiRNAs predicted to bind TMPRSS2 mRNA including let-7d-5p, miR-494-3p, miR-382-3p, let-7e-5p, miR-181c-5p, and miR-452-5p are expressed higher in A549 cells, while only miR-1226-3p is expressed higher in Huh-7 cells. These miRNAs identified to be expressed higher in A549 cells would suggest a potentially important role in providing resistance to SARS-CoV-2 infectivity by reducing ACE2 and TMPRSS2 mRNA. Furthermore, we expanded our query by incorporating Calu-3 (susceptible) and human primary lung fibroblast (LF, resistant) to identify favorable miRNAs to ACE2 and TMPRSS2 (Figure 6E). Because the miRNA expression profiles were generated by different groups, we therefore limited our analysis to semi-quantitative expression comparisons between cell types. MiRNAs that are in the same tercile group between all four cell types were excluded from the analysis (i.e., miR-452-5p and let-7e-5p targeting TMPRSS2 mRNA). Here we confirm that miR-218-5p, let-7d-5p, miR-494-3p, miR-382-3p, and miR-181c-5p are favorably expressed in resistant cells. Furthermore, we also identified that miR-214-3p targeted TMPRSS2 mRNA predominantly in resistant cells. Finally, we conducted tissue expression analyses of miRNAs targeting ACE2 and TMPRSS2 mRNAs in various human tissues including lung, kidney, lymph node, and liver (Suppl. Figure S4A).

3.6. Identification of Predicted miRNAs to IFN Genes in SARS-CoV-2-Resistant and -Susceptible Cells

IFN-α and -γ are upstream drivers of ACE2 and TMPRSS2 expression [11], key proteins for SARS-CoV-2 viral entry. Similar to our efforts in finding candidate miRNAs for targeting ACE2 and TMPRSS2 mRNAs, we applied our workflow to identify miRNAs to target IFN mRNAs in resistant and susceptible cell lines (Figure 7A–C). Notably, predicted miRNAs targeting IFN-α and IFN-β mRNAs were found less in SARS-CoV-2 susceptible cells compared to resistant cell lines, suggesting there might be an overall greater amount of maintenance of these key mediators of ACE2 at the post-transcriptional level. Interestingly, when using differential expression analysis of Huh-7 and A549 cells, all the miRNAs identified are predominantly expressed in the resistant cells A549. Specifically, IFN-γ-targeting miRNAs included miR-655-3p, miR-31-3p, miR-656-3p, miR-495-3p, miR-9-3p, miR-143-3p, and miR-24-3p; IFN-β-targeting miRNAs included miR-138-5p, miR-323a-3p, and miR-9b-1-5p; and IFN-α-targeting miRNAs included miR-361-5p and miR-410-3p (Figure 7D). Furthermore, additional analysis with Calu3 (susceptible) and lung fibroblasts (resistant) cells also confirmed many of these captured miRNAs and revealed several new candidates (Figure 7E). Particularly, miR-1277-5p is shown to target both IFN-α and IFN-γ mRNAs, but was difficult to capture due to it being lowly expressed in Calu3, Huh-7, and A549 cells. Lastly, we highlight expression of IFN-α-targeting miRNAs miR-361-5p and miR-410-3p in different tissues (lung, kidney, lymph node, and liver) (Suppl. Figure S4B). Collectively, these findings highlight key miRNAs predicted to target ACE2, TMPRSS2, or IFN-α, -β, -γ mRNAs and enriched in SARS-CoV-2-resistant or -susceptible cell lines, potentially providing new insights for potentially modulating viral entry. Moreover, some of selected miRNAs targeting ACE2, TMPRSS2 and IFN mRNAs from the above analyses were dysregulated in the recent RNAseq data from lung cells infected with SARS-CoV-2. Specifically, differential expression analysis of SARS-CoV-2-infected

Calu-3 cells showed upregulated expression of miR-483-3p (targeting ACE2 mRNA) and let-7d-5p (targeting TMPRSS2), while miR-450b-5p (targeting IFN-γ mRNA) expression was downregulated (Table 3).

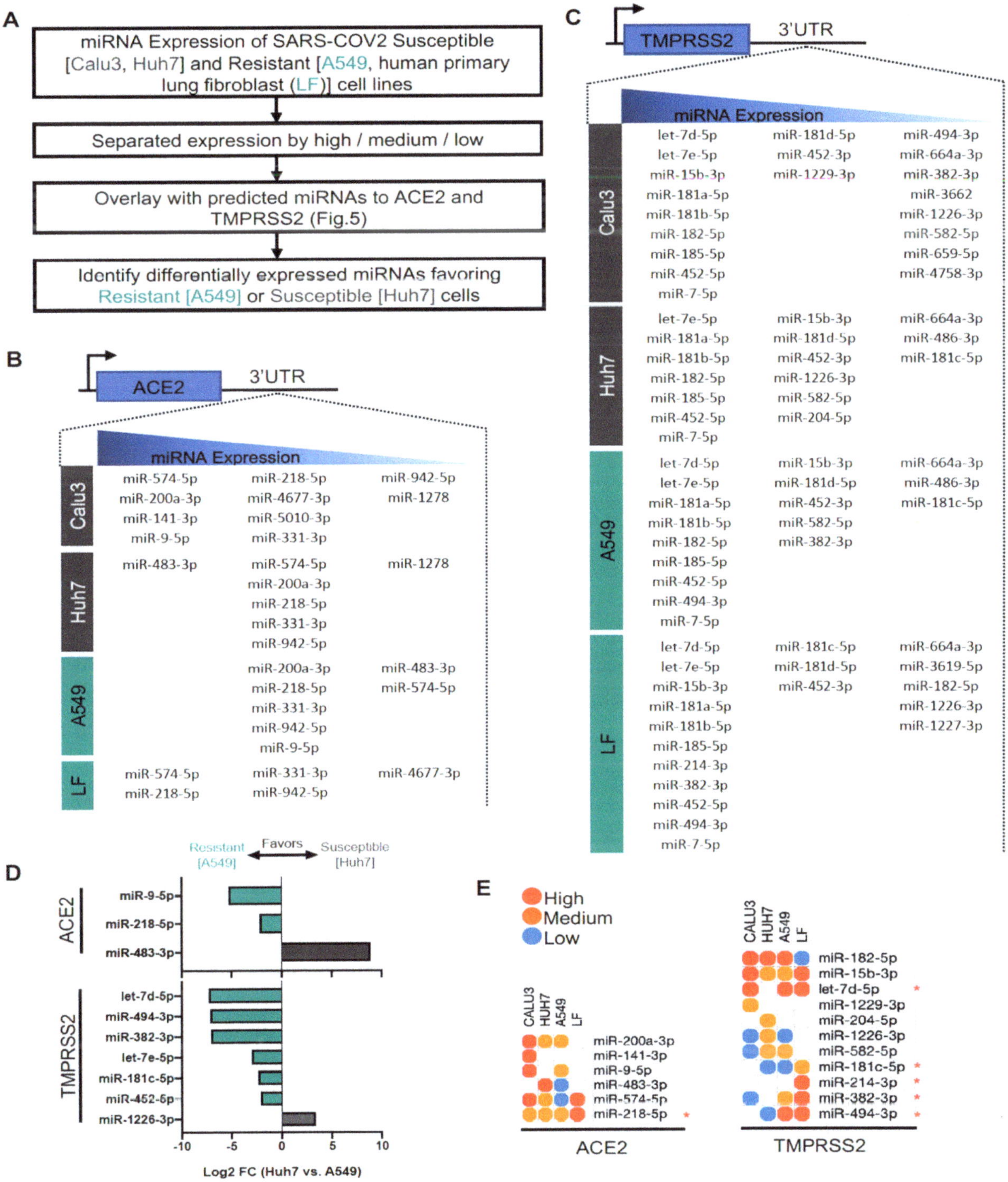

Figure 6. Predicted ACE2 and TMPRSS2 mRNA-targeting microRNA (miRNA) expression from severe acute respiratory syndrome coronavirus 2 (SARS-CoV-2)-resistant and -susceptible cell lines. (**A**) Workflow. (**B**,**C**) List of miRNAs grouped by high (1st tercile), medium (2nd tercile), and low expression (3rd tercile) targeting ACE2 and TMPRSS2 mRNAs, respectively. (**D**) Differential expression from Huh7 and A549 cell lines ($p < 0.05$ and >log2FC). (**E**) Heatmap of shared miRNAs between four cell lines. * identifies miRNAs more favorable to resistant cell types.

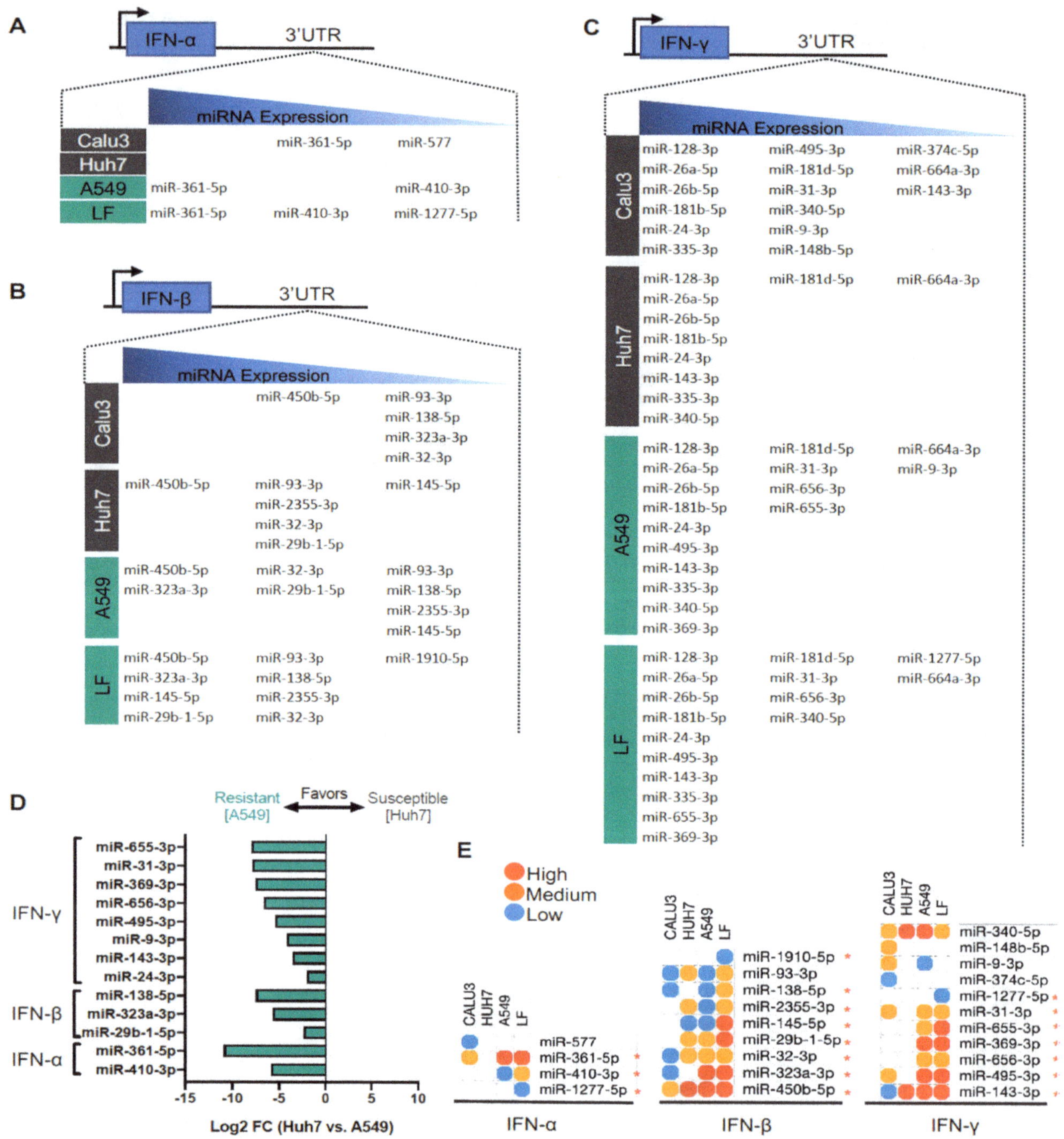

Figure 7. Predicted interferon (IFN)-targeting microRNA (miRNA) expression from severe acute respiratory syndrome coronavirus 2 (SARS-CoV-2)-resistant and -susceptible cell lines. (**A–C**) List of miRNAs grouped by high (1st tercile), medium (2nd tercile), and low expression (3rd tercile) targeting IFN-α, IFN-β, and IFN-γ, respectively. (**D**) Differential expression from Huh7 and A549 cell lines with criteria of $p < 0.05$ and >log2FC. (**E**) Heatmap of shared miRNAs between 4 cell lines. * identifies miRNAs more favorable to resistant cell types.

4. Discussion

In this study, we report a collection of host miRNAs that may regulate SARS-CoV-2 infectivity in humans. We utilized validated prediction algorithms in combination with established miRNA profiling datasets to investigate potential miRNA binding to the SARS-CoV-2 RNA genome, mRNAs of viral entry proteins ACE2 and TMPRSS2, and IFN regulators implicated in driving ACE2 and TMPRSS2 expression (Figure 8). In addition, we examined differential expression of these predicted miRNAs in cells resistant or susceptible to SARS-CoV-2 infection to further deduce potential mechanisms through which cellular host miRNAs may affect SARS-CoV-2 infectivity.

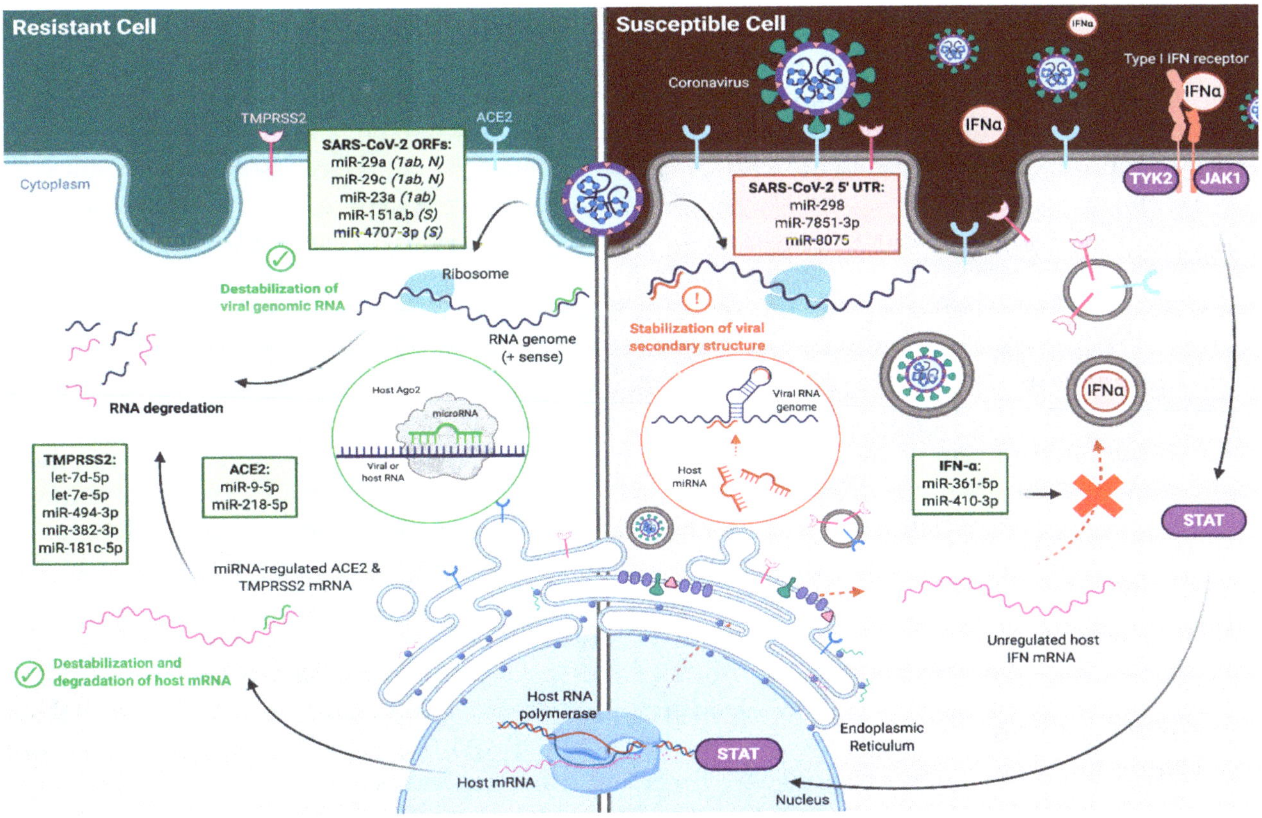

Figure 8. Mechanisms by which microRNAs (miRNAs) may regulate severe acute respiratory syndrome coronavirus 2 (SARS-CoV-2) infectivity. MiRNAs facilitate cellular resistance (**left**) or susceptibility (**right**) to infection. Potential regulatory mechanisms include the destabilization of viral genomic RNA; destabilization and degradation of host ACE2, TMPRSS2, and interferon (IFN) expression; or the stabilization of viral secondary structure. The specific ORFs within SARS-CoV-2 containing binding sites for miRNAs are listed in parentheses where 1ab denotes ORF1ab, N denotes nucleocapsid, and S denotes spike protein. Created using Biorender.com.

4.1. MiRNAs Targeting the SARS-CoV-2 Viral Genome

Certain miRNAs are expressed ubiquitously, whereas others are expressed in a highly tissue-specific manner [31]. One such example is miR-122, specifically expressed in the liver, that regulates HCV replication and infectivity [50]. MiR-122 binds and stabilizes the 5′ UTR of the HCV genome, conferring resistance to enzymatic degradation and thereby contributing to increased protein production. Initial studies in different immortalized cell lines highlighted that miR-122 is highly expressed in cultured human Huh7 and mouse Hepa1-6 liver cells, but not in the human liver-derived HepG2 and cervical carcinoma-derived HeLa cells. In high correlation with miR-122 expression, HCV can only replicate in Huh7 cells, while HepG2 and HeLa are not permissive cell lines [50]. In our study, we have employed a similar approach. Previous studies established that SARS-CoV-2 exhibits varying degrees of infectivity in different commonly used cell lines [14]. To this end, we employed miRNA datasets including cell lines with documented susceptibility (Huh7 and Calu3) and resistance to SARS-CoV-2 infection (A549 and LF) in order to verify the biologic plausibility of these miRNA–target RNA interactions in vitro and predict which miRNAs may be important in regulating SARS-CoV-2 infectivity. In addition, we further analyzed available RNA-seq datasets to uncover miRNA profiling from Calu-3 cells directly infected with SARS-CoV-2 to explore whether infection resulted in altered expression profiles of the miRNAs of interest.

The single-stranded, positive-sense RNA composition of the SARS-CoV-2 genome makes it susceptible to binding and regulation by host miRNAs. Indeed, host miRNAs likely play key roles in the pathogenesis of coronaviruses across species [51]. Additionally, several miRNAs that directly target viral genomes have been described previously. For example, binding of miR-122 to HCV promotes protein translation by specifically stabilizing the secondary structure of the internal ribosomal entry site in the 5′ UTR of HCV [41]. Binding of miR-122 is necessary for viral propagation and has been successfully exploited therapeutically by the pharmaceutical miravirsen, an antisense oligonucleotide that sequesters and inhibits miR-122 to reduce HCV viral load [52,53]. Here, we have identified three miRNAs—miR-298, miR-7851-3p, and miR-8075—that are predicted to bind with high affinity to the 5′ UTR of the SARS-CoV-2 genome. Interestingly, the binding sites of all three miRNAs are within or in close proximity to stem–loop structures that are highly conserved within the *Coronaviridae* family (Figure 2D) [44]. However, it is difficult to predict in silico how endogenous miRNA binding might regulate viral genomes. For example, in contrast to miR-122 and HCV, miR-548g-3p binds to the 5′ UTR and inhibits the translation and replication of dengue virus, another positive-sense, single-strand RNA virus [42]. Similarly, other miRNAs have been shown to bind throughout the viral genome to inhibit the translation and/or promote the degradation of many other disease-causing RNA viruses including human immunodeficiency virus (HIV), hepatitis B virus, enterovirus, human T cell leukemia virus, and eastern equine encephalitis virus (EEEV) [43,54–57]. Future studies investigating the resulting changes to genomic secondary structure following miRNA binding may shed light on the potential regulation of SARS-CoV-2 by these miRNAs.

The polybasic cleavage site (RRAR) is a unique sequence in the S protein of SARS-CoV-2, at the junction of S1 and S2, the two subunits of the spike protein and absent in SARS-CoV and other bat and pangolin coronaviruses. A leading proline results in the addition of O-linked glycans to S673, T678 and S686, which flank the cleavage site (PRRA) (Figure 3B) [25,58]. As previously suggested, this allows effective cleavage by furin and other proteases and has a role in determining viral infectivity [59]. We show that miR-151a-5p, -151b and miR-4707-3p may target the novel PRRA region in the SARS-CoV-2 genome, and assessment of the secondary structure in this region supports accessibility for miRNAs binding. The enriched expression of miR-151-5p and 151b in the lungs and lung primary fibroblasts suggests a potential implication of these miRNAs in the spike protein stability and SARS-CoV-2 RNA stability. Additionally, Nsp7, Nsp8, and Nsp12 have also recently been identified as important proteins in viral replication and are targeted by the investigational therapeutic remdesivir [60,61]. Here, we also show that miRs-320a-3p, -320b and -3149 were predicted to bind to the Nsp8 and Nsp12 genomic regions of the replicase polyprotein 1ab (PP1ab), a region that is important in transcription and replication of viral RNAs [46]. MiR-320 family was implicated in a number of viral infections. MiR-320a-3p was shown to inhibit mink enteritis virus (MEV) replication [62], whereas miR-320b was significantly increased in human plasma samples following Ebola infection [63]. The relevance and applicability of these findings to SARS-CoV-2 infection needs to be further delineated.

RNA viruses can quickly mutate their genomes in response to evolutionary pressure due to the lack of proofreading activity from virally-encoded polymerases [64]. MiRNA binding sites, particularly the 6 to 8 nt seed regions, are therefore likely evolutionarily conserved due to strong positive selection. As noted previously, it is difficult to predict how miRNA binding may regulate viral translation or replication in vivo. Moreover, even miRNAs that negatively regulate genomic stability and limit viral replication may paradoxically result in increased viral infectivity at the organismal level due to cell type-specific miRNA expression. For example, miR-142-3p is exclusively expressed in hematopoietic cells and is necessary for proper hematopoietic cell differentiation [65,66]. Eastern Equine Encephalitis virus (EEEV), another single-strand, positive sense RNA virus, contains three highly-conserved miR-142-3p binding sites in its 3′ UTR, and viral replication is potently inhibited in cells expressing miR-142-3p [43]. Trobaugh et al. demonstrated that repression of EEEV replication in myeloid cells is alleviated upon deletion of miR-142-3p binding sites, and infected myeloid cells contribute to potent type I IFN response [43]. However, despite increased viral replication in myeloid cells and lymph nodes, disease severity was reduced in mice infected with EEEV containing deleted binding sites due in part to the limited capacity for activation of the innate immune system. In this way, there is a strong positive selective pressure for the maintenance of miR-142-3p binding sites in the 3′ UTR of EEEV due to cell-specific miRNA expression. Interestingly, in addition to miR-142-3p, miR-29a has also been identified as a critical regulator of hematopoietic cell differentiation, and loss of hematopoietic cell miR-29a expression has been associated with the development of myelodysplastic syndromes and leukemias [65,67,68]. We also predict that many other potential miRNAs target the 5′ UTR, coding regions, and intergenic non-coding regions of the SARS-CoV-2 genome (Figure 2), and cell-specific expression of these miRNAs may considerably impact the regulatory control over SARS-CoV-2 replication and infectivity on both the cellular and organismal level.

Several independent analyses of single-cell RNA-sequencing datasets demonstrated that ACE2 and TMPRSS2 are co-expressed in only a limited subset of cell types, primarily secretory cells of the upper and middle airways [11–13]. Because SARS-CoV-2 is an enveloped virus, it does not require host-cell lysis for viral release, leading to speculation that SARS-CoV-2 preferentially infects secretory cell types to promote the rapid release of viral particles. In this context, viral mechanisms that promote host-cell survival would likely be advantageous in enhancing viral replication and propagation. MiR-29 has recently emerged as an important regulator of cell survival and apoptosis by targeting several components of pathways critical to cell proliferation and survival including p53 inhibitors *p85α* and *CDC42* as well as *Bcl2*, *CCND2*, and *c-Myc* [69–71]. Similarly, miR-15a -15b, and miR-16 have been shown to play critical roles in these processes by targeting a similar cassette of genes such as *Bcl2*, *CCND1*, *CCNE*, and *Mcl1* [72–74], and miR-23b regulates lungs cells proliferation by targeting Cyclin G1 (*CCNG1*) [75]. Here, we show that miR-29a, -15a, b, -16, and -23a,b have several high-probability predicted binding sites throughout the SARS-CoV-2 genome and exhibited high expression in human lungs (Figures 2 and 3). By binding these miRNAs at multiple high-affinity sites, SARS-CoV-2 may functionally sequester their cytosolic pool [76] and promote the survival of infected cells through the de-repression of cell cycle and apoptosis target genes. Data from SARS-CoV-2-infected Calu-3 cells may support this paradigm, demonstrating a decrease in miR-23a and miR-23b following SARS-CoV-2 infection (Table 3).

We also show that several miRNAs may directly play a role in regulating viral infectivity using established miRNA profiling datasets of both Huh7 (high SARS-CoV-2 infectivity) and A549 (low SARS-CoV-2 infectivity). Our correlative results suggest that several miRNAs may be of potential interest as therapeutic targets. For example, miR-23a, -23b, miR-29a, -29c (-3p) and -29b-1-5p are predicted to bind several regions of the SARS-CoV-2 genome, including the regions encoding ORF1ab (miR-23a and miR-29a, -29-c, -29b-1), the Nucleocapsid (miR-29a, -29c) proteins (Figure 2). At the same time, we observed that miR-29a, -29c and miR-23a, -23b are highly expressed in the cell lines that show low SARS-CoV-2 infectivity (A549 and primary lung fibroblasts), while showing much lower expression in cell lines with high SARS-CoV-2 infectivity (Calu-3 and Huh7; Figure 4). As previously observed, miRNA binding to different genomic regions that code for viral proteins can interfere with the viral replication, translation or infectivity [19]. Whether the increased expression of miR-23a and miR-29a, -29c in non-permissive cell lines may play a role in SARS-CoV-2 infectivity needs to be further assessed by miRNA loss- and gain-of function studies in respective cell lines. Moreover, the two miRNA families (-23 and -29) are highly expressed in healthy lungs (Figure 3). Future analysis of these miRNAs in the lung tissues or plasma of COVID-19 infected patients will be informative for their potential regulation and correlation with increasing severity of disease.

Our study builds upon previous computational analyses that predict interactions between host miRNAs and SARS-CoV-2. Similar to our analytic strategy, Fulzele et al. recently utilized miRDB (a cut off of target score ≥95) to predict miRNAs that bind to the SARS-CoV-2 genome [77]. Because miRNA target prediction algorithms are not specifically designed to predict miRNA-viral genome interactions, we utilized a more liberal cut off of target score ≥70 along with verification by a second prediction tool, RNAhybrid. Using this strategy, we identify many of the same miRNA targets such as miR-15a-5p and miR-15b-5p in addition to a host of other potentially pathogenically important host miRNAs. Chow and Salmena utilized RNA22 to predict miRNAs that may bind the SARS-CoV-2 genome and analyzed their expression in SARS-CoV-2-infected versus uninfected Calu3 cells [78]. By utilizing more robust miRNA-binding prediction methods through two independent prediction tools (miRDB and RNAhybrid), our analysis builds on that of Chow and Salmena and provides additional candidate miRNAs for future in vitro and in vivo investigation.

In addition to potential regulation by host miRNAs, SARS-CoV-2 may encode viral miRNAs or other non-coding RNA transcripts that impact its infectivity [79,80]. For example, SARS-CoV encodes several small viral RNAs 18–22 nt capable of silencing target mRNA expression through non-canonical miRNA regulatory pathways [81]. Treatment of SARS-CoV-infected mice with small viral RNA inhibitors resulted in reduced infection severity, lower levels of inflammation, and higher mRNA expression of antiviral proteins. To date, there have been no confirmed SARS-CoV-2-encoded miRNAs, although several studies predict the existence of such virally-encoded miRNAs [82,83]. It will be important to determine in future studies whether SARS-CoV-2 encodes miRNAs capable of altering viral infectivity and whether they are candidates for therapeutic intervention.

4.2. MiRNAs Targeting Viral Entry Proteins and IFNs

Ziegler et al. [11] recently established ACE2 as an IFN-stimulated gene (ISG), primarily in response to IFN-α stimulation [11]. We therefore investigated potential miRNA regulators of IFN-α, -β, and -γ that may play indirect roles in susceptibility to SARS-CoV-2 infection (Figures 5 and 7). Strikingly, all miRNAs predicted to bind the 3′ UTR of IFN genes are more highly expressed in cell lines resistant to SARS-CoV-2 infection. We found three miRNAs that were co-expressed in both the SARS-CoV-2-resistant cell line A549 and human lung tissue: miR-361-5p targeting IFN-α mRNA and miR-24-3p and miR-143-3p targeting IFN-γ mRNA. In qualitative analyses including additional Calu3 and LF cell lines, miR-361-5p maintained a consistent pattern of high expression in those resistant to SARS-CoV-2 infection and low expression in those permissive to infection, making this an intriguing target for further study. Furthermore, miR-495-3p, which targets the ORF1ab region of the SARS-CoV-2 genome and is highly expressed the resistant cell lines A549 and human primary lung fibroblasts, is also predicted to target IFN-γ 3′-UTR. MiR-495-3p expression was significantly upregulated in mouse lungs infected with H5N2 virus [84] and in human serum following Hepatitis C infection [85]. Whether such upregulation is also valid in response to SARS-CoV-2 infection as a possible protective mechanism to inhibit ACE2 induction via IFN-γ remains to be determined in subsequent in vitro and in vivo studies.

Importantly, dedicated in vitro and in vivo studies demonstrated that ACE2 induction by IFN was considerably weaker in immortalized cell lines and absent in mouse primary tracheal epithelial cells, suggesting a substantial degree of cell and species specificity to this previously unrecognized regulation of SARS-CoV-2 entry factor ACE2 [11]. We have identified several candidate miRNAs that may offer an additional level of regulation of ACE2 with a similar degree of cell specificity. Using several established bioinformatic miRNA target prediction tools, we provide a collection of miRNAs that may target mRNAs of SARS-CoV-2 entry factors ACE2 and TMPRSS2 as well as upstream regulators IFNs with high-probability (Figure 5A). By evaluating the expression of these miRNAs in RNA-sequencing datasets of human lung tissue, we predict 7 miRNAs targeting ACE2 mRNA (e.g., the top 3 are miR-141-3p, miR-4270, miR-331-3p) and 25 miRNAs targeting TMPRSS2 mRNA (e.g., the top 9 are miR-4763-3p, let-7d-5p, miR-4530, let-7e-5p, miR-181a-5p, miR-762, miR-338-3p, miR-4505, miR-575) to be abundant in human lung tissue, establishing biologic plausibility for these miRNA–mRNA interactions (Figure 5B).

We found ten total miRNA—three targeting ACE2 mRNA (miR-9-5p, miR-218-5p, and miR-483-3p) and seven targeting TMPRSS2 mRNA (let-7d-5p, miR-494-3p, miR-382-3p, let-7e-5p, miR-181c-5p, miR-452-5p, and miR-1226-3p)—that demonstrated statistically significant differential expression favoring either resistance (i.e., higher expression in A549) or susceptibility to SARS-CoV-2 infection (i.e., higher expression in Huh7; Figure 6D). We also found that of these eight, miR-218-5p (targeting ACE2 mRNA), miR-382-3p, miR-494-3p, let-7d-5p, let-7e-5p, and miR-181c-5p (targeting TMPRSS2 mRNA) are also expressed in human lung tissue. Importantly, miR 382-3p and miR 494-3p also demonstrated higher expression in SARS-CoV-2-resistant LF cells. Additionally, both miR-218 and let-7d were previously shown to be expressed in bronchoalveolar stem cells, which are susceptible to SARS-CoV expression in part due to surface ACE2 expression [86,87]. Further dedicated miRNA profiling and target gene expression analyses will be insightful regarding the impact of these miRNAs on the relative susceptibility or resistance to infection in cells expressing these miRNAs.

Using RNAseq data from Calu-3 cells infected with SARS-CoV-2, we identified differentially expressed miRNAs targeting the genes involved in virus entry. Interestingly, several miRNAs targeting mRNAs of viral entry proteins were upregulated following infection such as miR-483-3p targeting ACE2 mRNA and miR-181c-5p and let-7d-5p targeting TMPRS2 mRNA. Although we found that miR-483-3p is expressed higher in cells susceptible to SARS-CoV-2 infection, its upregulation (4.46-fold) along with the upregulation of miR-181c-5p and let-7d-5p after infection in Calu-3 cells may represent compensatory responses of host cells to inhibit SARS-CoV-2 entry by targeting ACE2 and TMPRSS2 mRNA. Indeed, ACE2 expression levels were found downregulated in mouse lungs after SARS-CoV infection [88].

5. Conclusions

Overall, in this study, we establish a framework for future investigation of miRNAs regulating SARS-CoV-2 infectivity in human cells. Leveraging miRNA profiling from cells resistant or susceptible to SARS-CoV-2 infection, we provide a preliminary resource of high-probability miRNAs governing SARS-CoV-2 infectivity through two potential mechanisms: (1) the direct binding of host miRNAs to the SARS-CoV-2 genome, which have the potential to both promote and prevent SARS-CoV-2 infection; and (2) regulation of the mRNAs of viral entry proteins ACE2 and TMPRSS2 including regulation via their upstream IFN modulators. Further studies are needed to validate potential miRNA binding sites in both SARS-CoV-2 and mRNAs of entry proteins. Similar to other host miRNA–virus interactions, these miRNAs may represent targets for potential therapeutic intervention in COVID-19 patients.

Supplementary Materials: Figure S1: Secondary structure of SARS-CoV and SARS-CoV-2 spike proteins and lung expression of miRNA binding in the PRRA region of the SARS-CoV-2 spike protein, Figure S2: Differential expression of miRNAs binding the SARS-CoV-2 genome between resistant versus susceptible cell lines. Figure S3: Expression of the most differentially regulated miRNAs in relevant human tissues, Figure S4: Predicted miRNAs of host genes expression across tissues.

Author Contributions: Conceptualization, M.W.F.; Methodology, J.B.P., V.S., B.I., D.P.-C., and H.S.C.; Software, validation, and formal analysis, J.B.P., V.S., B.I., D.P.-C., and H.S.C.; Investigation, J.B.P., V.S., B.I., D.P.-C., and H.S.C.; Data curation and interpretation, J.B.P., V.S., B.I., D.P.-C., H.S.C. and M.W.F.; Writing—original draft preparation, J.B.P., V.S., B.I., D.P.-C., H.S.C., and M.W.F.; Writing—review and editing, J.B.P., V.S., B.I., D.P.-C., H.S.C., and M.W.F.; Visualization, J.B.P., V.S., B.I., D.P.-C., and H.S.C.; Supervision, M.W.F.; Funding acquisition, M.W.F. All authors have read and agreed to the published version of the manuscript.

References

1. World Health Organization. Coronavirus Disease (COVID-19) Pandemic. Available online: https://www.who.int/emergencies/diseases/novel-coronavirus-2019 (accessed on 1 May 2020).
2. Wang, C.; Horby, P.W.; Hayden, F.G.; Gao, G.F. A novel coronavirus outbreak of global health concern. *Lancet* **2020**, *395*, 470–473. [CrossRef]

3. Zhu, N.; Zhang, D.; Wang, W.; Li, X.; Yang, B.; Song, J.; Zhao, X.; Huang, B.; Shi, W.; Lu, R. A novel coronavirus from patients with pneumonia in China, 2019. *N. Engl. J. Med.* **2020**, *382*, 727–733. [CrossRef]
4. Huang, C.; Wang, Y.; Li, X.; Ren, L.; Zhao, J.; Hu, Y.; Zhang, L.; Fan, G.; Xu, J.; Gu, X. Clinical features of patients infected with 2019 novel coronavirus in Wuhan, China. *Lancet* **2020**, *395*, 497–506. [CrossRef]
5. Guan, W.-J.; Ni, Z.-Y.; Hu, Y.; Liang, W.-H.; Ou, C.-Q.; He, J.-X.; Liu, L.; Shan, H.; Lei, C.-l.; Hui, D.S. Clinical characteristics of coronavirus disease 2019 in China. *N. Engl. J. Med.* **2020**, *382*, 1708–1720. [CrossRef]
6. Lu, R.; Zhao, X.; Li, J.; Niu, P.; Yang, B.; Wu, H.; Wang, W.; Song, H.; Huang, B.; Zhu, N. Genomic characterisation and epidemiology of 2019 novel coronavirus: Implications for virus origins and receptor binding. *Lancet* **2020**, *395*, 565–574. [CrossRef]
7. Yan, R.; Zhang, Y.; Li, Y.; Xia, L.; Guo, Y.; Zhou, Q. Structural basis for the recognition of SARS-CoV-2 by full-length human ACE2. *Science* **2020**, *367*, 1444–1448. [CrossRef]
8. Hoffmann, M.; Kleine-Weber, H.; Schroeder, S.; Krüger, N.; Herrler, T.; Erichsen, S.; Schiergens, T.S.; Herrler, G.; Wu, N.-H.; Nitsche, A. SARS-CoV-2 cell entry depends on ACE2 and TMPRSS2 and is blocked by a clinically proven protease inhibitor. *Cell* **2020**, *181*, 271–280.e8. [CrossRef]
9. Wang, Q.; Zhang, Y.; Wu, L.; Niu, S.; Song, C.; Zhang, Z.; Lu, G.; Qiao, C.; Hu, Y.; Yuen, K.-Y. Structural and functional basis of SARS-CoV-2 entry by using human ACE2. *Cell* **2020**, *181*, 894–904.e9. [CrossRef]
10. Shang, J.; Ye, G.; Shi, K.; Wan, Y.; Luo, C.; Aihara, H.; Geng, Q.; Auerbach, A.; Li, F. Structural basis of receptor recognition by SARS-CoV-2. *Nature* **2020**, *581*, 221–224. [CrossRef]
11. Ziegler, C.; Allon, S.J.; Nyquist, S.K.; Mbano, I.; Miao, V.N.; Cao, Y.; Yousif, A.S.; Bals, J.; Hauser, B.M.; Feldman, J.; et al. SARS-CoV-2 receptor ACE2 is an interferon-stimulated gene in human airway epithelial cells and is enriched in specific cell subsets across tissues. *Cell* **2020**, *181*, 1016–1035.e19. [CrossRef]
12. Sungnak, W.; Huang, N.; Bécavin, C.; Berg, M.; Queen, R.; Litvinukova, M.; Talavera-López, C.; Maatz, H.; Reichart, D.; Sampaziotis, F. SARS-CoV-2 entry factors are highly expressed in nasal epithelial cells together with innate immune genes. *Nat. Med.* **2020**, *26*, 681–687. [CrossRef]
13. Lukassen, S.; Chua, R.L.; Trefzer, T.; Kahn, N.C.; Schneider, M.A.; Muley, T.; Winter, H.; Meister, M.; Veith, C.; Boots, A.W. SARS-CoV-2 receptor ACE2 and TMPRSS2 are primarily expressed in bronchial transient secretory cells. *EMBO J.* **2020**, *39*, e105114. [CrossRef] [PubMed]
14. Chu, H.; Chan, J.; Yuen, T.; Shuai, H.; Yuan, S.; Wang, Y.; Hu, B.; Yip, C.; Tsang, J.; Huang, X. Comparative tropism, replication kinetics, and cell damage profiling of SARS-CoV-2 and SARS-CoV with implications for clinical manifestations, transmissibility, and laboratory studies of COVID-19: An observational study. *Lancet Microbe* **2020**, *1*, E14–E23. [CrossRef]
15. Gebert, L.F.; MacRae, I.J. Regulation of microRNA function in animals. *Nat. Rev. Mol. Cell Biol.* **2019**, *20*, 21–37. [CrossRef] [PubMed]
16. Kincaid, R.P.; Sullivan, C.S. Virus-encoded microRNAs: An overview and a look to the future. *PLoS Pathog.* **2012**, *8*, e1003018. [CrossRef]
17. Dölken, L.; Krmpotic, A.; Kothe, S.; Tuddenham, L.; Tanguy, M.; Marcinowski, L.; Ruzsics, Z.; Elefant, N.; Altuvia, Y.; Margalit, H. Cytomegalovirus microRNAs facilitate persistent virus infection in salivary glands. *PLoS Pathog.* **2010**, *6*, e1001150. [CrossRef]
18. Bauman, Y.; Nachmani, D.; Vitenshtein, A.; Tsukerman, P.; Drayman, N.; Stern-Ginossar, N.; Lankry, D.; Gruda, R.; Mandelboim, O. An identical miRNA of the human JC and BK polyoma viruses targets the stress-induced ligand ULBP3 to escape immune elimination. *Cell Host Microbe* **2011**, *9*, 93–102. [CrossRef]
19. Trobaugh, D.W.; Klimstra, W.B. MicroRNA regulation of RNA virus replication and pathogenesis. *Trends Mol. Med.* **2017**, *23*, 80–93. [CrossRef]
20. Edgar, R.C. MUSCLE: Multiple sequence alignment with high accuracy and high throughput. *Nucleic Acids Res.* **2004**, *32*, 1792–1797. [CrossRef]
21. Hatcher, E.L.; Zhdanov, S.A.; Bao, Y.; Blinkova, O.; Nawrocki, E.P.; Ostapchuck, Y.; Schäffer, A.A.; Brister, J.R. Virus Variation Resource–improved response to emergent viral outbreaks. *Nucleic Acids Res.* **2017**, *45*, D482–D490. [CrossRef]
22. Letunic, I.; Bork, P. Interactive Tree Of Life (iTOL) v4: Recent updates and new developments. *Nucleic Acids Res.* **2019**, *47*, W256–W259. [CrossRef]
23. Altschul, S.F.; Gish, W.; Miller, W.; Myers, E.W.; Lipman, D.J. Basic local alignment search tool. *J. Mol. Biol.* **1990**, *215*, 403–410. [CrossRef]

24. Zhang, T.; Wu, Q.; Zhang, Z. Probable pangolin origin of SARS-CoV-2 associated with the COVID-19 outbreak. *Curr. Biol.* **2020**, *30*, 1346–1351. [CrossRef] [PubMed]
25. Andersen, K.G.; Rambaut, A.; Lipkin, W.I.; Holmes, E.C.; Garry, R.F. The proximal origin of SARS-CoV-2. *Nat. Med.* **2020**, *26*, 450–452. [CrossRef] [PubMed]
26. Chen, Y.; Wang, X. miRDB: An online database for prediction of functional microRNA targets. *Nucleic Acids Res.* **2020**, *48*, D127–D131. [CrossRef]
27. Gupta, A.; Nagilla, P.; Le, H.-S.; Bunney, C.; Zych, C.; Thalamuthu, A.; Bar-Joseph, Z.; Mathavan, S.; Ayyavoo, V. Comparative expression profile of miRNA and mRNA in primary peripheral blood mononuclear cells infected with human immunodeficiency virus (HIV-1). *PLoS ONE* **2011**, *6*, e22730. [CrossRef]
28. Krüger, J.; Rehmsmeier, M. RNAhybrid: microRNA target prediction easy, fast and flexible. *Nucleic Acids Res.* **2006**, *34*, W451–W454. [CrossRef]
29. Zuker, M.; Stiegler, P. Optimal computer folding of large RNA sequences using thermodynamics and auxiliary information. *Nucleic Acids Res.* **1981**, *9*, 133–148. [CrossRef]
30. Sticht, C.; De La Torre, C.; Parveen, A.; Gretz, N. miRWalk: An online resource for prediction of microRNA binding sites. *PLoS ONE* **2018**, *13*, e0206239. [CrossRef]
31. Ludwig, N.; Leidinger, P.; Becker, K.; Backes, C.; Fehlmann, T.; Pallasch, C.; Rheinheimer, S.; Meder, B.; Stähler, C.; Meese, E. Distribution of miRNA expression across human tissues. *Nucleic Acids Res.* **2016**, *44*, 3865–3877. [CrossRef]
32. Ou, X.; Liu, Y.; Lei, X.; Li, P.; Mi, D.; Ren, L.; Guo, L.; Guo, R.; Chen, T.; Hu, J. Characterization of spike glycoprotein of SARS-CoV-2 on virus entry and its immune cross-reactivity with SARS-CoV. *Nat. Commun.* **2020**, *11*, 1–12. [CrossRef]
33. Harcourt, J.; Tamin, A.; Lu, X.; Kamili, S.; Sakthivel, S.K.; Murray, J.; Queen, K.; Tao, Y.; Paden, C.R.; Zhang, J. Severe Acute Respiratory Syndrome Coronavirus 2 from Patient with 2019 Novel Coronavirus Disease, United States. *Emerg. Infect. Dis.* **2020**, *26*. [CrossRef] [PubMed]
34. Sheahan, T.P.; Sims, A.C.; Zhou, S.; Graham, R.L.; Pruijssers, A.J.; Agostini, M.L.; Leist, S.R.; Schäfer, A.; Dinnon, K.H.; Stevens, L.J. An orally bioavailable broad-spectrum antiviral inhibits SARS-CoV-2 in human airway epithelial cell cultures and multiple coronaviruses in mice. *Sci. Transl. Med.* **2020**, *12*, eabb5883. [CrossRef] [PubMed]
35. Wyler, E.; Mösbauer, K.; Franke, V.; Diag, A.; Gottula, L.T.; Arsie, R.; Klironomos, F.; Koppstein, D.; Ayoub, S.; Buccitelli, C.; et al. Bulk and single-cell gene expression profiling of SARS-CoV-2 infected human cell lines identifies molecular targets for therapeutic intervention. *bioRxiv* **2020**. [CrossRef]
36. Müller, S.; Glaß, M.; Singh, A.K.; Haase, J.; Bley, N.; Fuchs, T.; Lederer, M.; Dahl, A.; Huang, H.; Chen, J. IGF2BP1 promotes SRF-dependent transcription in cancer in a m6A-and miRNA-dependent manner. *Nucleic Acids Res.* **2019**, *47*, 375–390. [CrossRef]
37. Love, M.I.; Huber, W.; Anders, S. Moderated estimation of fold change and dispersion for RNA-seq data with DESeq2. *Genome Biol.* **2014**, *15*, 550. [CrossRef]
38. Zhang, X.; Chu, H.; Wen, L.; Shuai, H.; Yang, D.; Wang, Y.; Hou, Y.; Zhu, Z.; Yuan, S.; Yin, F. Competing endogenous RNA network profiling reveals novel host dependency factors required for MERS-CoV propagation. *Emerg. Microbes Infect.* **2020**, *9*, 733–746. [CrossRef]
39. Wei, P.; Xie, Y.; Abel, P.W.; Huang, Y.; Ma, Q.; Li, L.; Hao, J.; Wolff, D.W.; Wei, T.; Tu, Y. Transforming growth factor (TGF)-β1-induced miR-133a inhibits myofibroblast differentiation and pulmonary fibrosis. *Cell Death Dis.* **2019**, *10*, 1–17. [CrossRef]
40. Fukuhara, T.; Kambara, H.; Shiokawa, M.; Ono, C.; Katoh, H.; Morita, E.; Okuzaki, D.; Maehara, Y.; Koike, K.; Matsuura, Y. Expression of microRNA miR-122 facilitates an efficient replication in nonhepatic cells upon infection with hepatitis C virus. *J. Virol.* **2012**, *86*, 7918–7933. [CrossRef]
41. Schult, P.; Roth, H.; Adams, R.L.; Mas, C.; Imbert, L.; Orlik, C.; Ruggieri, A.; Pyle, A.M.; Lohmann, V. microRNA-122 amplifies hepatitis C virus translation by shaping the structure of the internal ribosomal entry site. *Nat. Commun.* **2018**, *9*, 1–14. [CrossRef]
42. Wen, W.; He, Z.; Jing, Q.; Hu, Y.; Lin, C.; Zhou, R.; Wang, X.; Su, Y.; Yuan, J.; Chen, Z. Cellular microRNA-miR-548g-3p modulates the replication of dengue virus. *J. Infect.* **2015**, *70*, 631–640. [CrossRef] [PubMed]

43. Trobaugh, D.W.; Gardner, C.L.; Sun, C.; Haddow, A.D.; Wang, E.; Chapnik, E.; Mildner, A.; Weaver, S.C.; Ryman, K.D.; Klimstra, W.B. RNA viruses can hijack vertebrate microRNAs to suppress innate immunity. *Nature* **2014**, *506*, 245–248. [CrossRef] [PubMed]
44. Yang, D.; Leibowitz, J.L. The structure and functions of coronavirus genomic 3′ and 5′ ends. *Virus Res.* **2015**, *206*, 120–133. [CrossRef] [PubMed]
45. Walls, A.C.; Park, Y.-J.; Tortorici, M.A.; Wall, A.; McGuire, A.T.; Veesler, D. Structure, function, and antigenicity of the SARS-CoV-2 spike glycoprotein. *Cell* **2020**, *181*, 281–292.e6. [CrossRef] [PubMed]
46. Wu, C.; Liu, Y.; Yang, Y.; Zhang, P.; Zhong, W.; Wang, Y.; Wang, Q.; Xu, Y.; Li, M.; Li, X. Analysis of therapeutic targets for SARS-CoV-2 and discovery of potential drugs by computational methods. *Acta Pharm. Sin. B* **2020**, *10*, 766–788. [CrossRef] [PubMed]
47. Lambert, D.W.; Yarski, M.; Warner, F.J.; Thornhill, P.; Parkin, E.T.; Smith, A.I.; Hooper, N.M.; Turner, A.J. Tumor necrosis factor-α convertase (ADAM17) mediates regulated ectodomain shedding of the severe-acute respiratory syndrome-coronavirus (SARS-CoV) receptor, angiotensin-converting enzyme-2 (ACE2). *J. Biol. Chem.* **2005**, *280*, 30113–30119. [CrossRef]
48. Jia, H.P.; Look, D.C.; Shi, L.; Hickey, M.; Pewe, L.; Netland, J.; Farzan, M.; Wohlford-Lenane, C.; Perlman, S.; McCray, P.B. ACE2 receptor expression and severe acute respiratory syndrome coronavirus infection depend on differentiation of human airway epithelia. *J. Virol.* **2005**, *79*, 14614–14621. [CrossRef]
49. Tseng, C.-T.K.; Tseng, J.; Perrone, L.; Worthy, M.; Popov, V.; Peters, C.J. Apical entry and release of severe acute respiratory syndrome-associated coronavirus in polarized Calu-3 lung epithelial cells. *J. Virol.* **2005**, *79*, 9470–9479. [CrossRef]
50. Jopling, C.L.; Yi, M.; Lancaster, A.M.; Lemon, S.M.; Sarnow, P. Modulation of hepatitis C virus RNA abundance by a liver-specific MicroRNA. *Science* **2005**, *309*, 1577–1581. [CrossRef]
51. Liu, X.; Zhu, L.; Liao, S.; Xu, Z.; Zhou, Y. The porcine microRNA transcriptome response to transmissible gastroenteritis virus infection. *PLoS ONE* **2015**, *10*, e0120377. [CrossRef]
52. Janssen, H.L.; Reesink, H.W.; Lawitz, E.J.; Zeuzem, S.; Rodriguez-Torres, M.; Patel, K.; van der Meer, A.J.; Patick, A.K.; Chen, A.; Zhou, Y. Treatment of HCV infection by targeting microRNA. *N. Engl. J. Med.* **2013**, *368*, 1685–1694. [CrossRef] [PubMed]
53. Lanford, R.E.; Hildebrandt-Eriksen, E.S.; Petri, A.; Persson, R.; Lindow, M.; Munk, M.E.; Kauppinen, S.; Ørum, H. Therapeutic silencing of microRNA-122 in primates with chronic hepatitis C virus infection. *Science* **2010**, *327*, 198–201. [CrossRef] [PubMed]
54. Huang, J.; Wang, F.; Argyris, E.; Chen, K.; Liang, Z.; Tian, H.; Huang, W.; Squires, K.; Verlinghieri, G.; Zhang, H. Cellular microRNAs contribute to HIV-1 latency in resting primary CD4+ T lymphocytes. *Nat. Med.* **2007**, *13*, 1241–1247. [CrossRef] [PubMed]
55. Wen, B.-P.; Dai, H.-J.; Yang, Y.-H.; Zhuang, Y.; Sheng, R. MicroRNA-23b inhibits enterovirus 71 replication through downregulation of EV71 VPl protein. *Intervirology* **2013**, *56*, 195–200. [CrossRef]
56. Bai, X.T.; Nicot, C. miR-28-3p is a cellular restriction factor that inhibits human T cell leukemia virus, type 1 (HTLV-1) replication and virus infection. *J. Biol. Chem.* **2015**, *290*, 5381–5390. [CrossRef]
57. Zhang, G.-L.; Li, Y.-X.; Zheng, S.-Q.; Liu, M.; Li, X.; Tang, H. Suppression of hepatitis B virus replication by microRNA-199a-3p and microRNA-210. *Antivir. Res.* **2010**, *88*, 169–175. [CrossRef]
58. Zhang, Z.; Wu, Q.; Zhang, T. Pangolin homology associated with 2019-nCoV. *bioRxiv* **2020**. [CrossRef]
59. Nao, N.; Yamagishi, J.; Miyamoto, H.; Igarashi, M.; Manzoor, R.; Ohnuma, A.; Tsuda, Y.; Furuyama, W.; Shigeno, A.; Kajihara, M. Genetic predisposition to acquire a polybasic cleavage site for highly pathogenic avian influenza virus hemagglutinin. *MBio* **2017**, *8*, e02298-16. [CrossRef]
60. Hillen, H.S.; Kokic, G.; Farnung, L.; Dienemann, C.; Tegunov, D.; Cramer, P. Structure of replicating SARS-CoV-2 polymerase. *Nature.* **2020**, *584*, 154–156. [CrossRef]
61. Yin, W.; Mao, C.; Luan, X.; Shen, D.-D.; Shen, Q.; Su, H.; Wang, X.; Zhou, F.; Zhao, W.; Gao, M. Structural basis for inhibition of the RNA-dependent RNA polymerase from SARS-CoV-2 by remdesivir. *Science* **2020**, *368*, 1499–1504. [CrossRef]
62. Sun, J.-Z.; Wang, J.; Wang, S.; Yuan, D.; Li, Z.; Yi, B.; Hou, Q.; Mao, Y.; Liu, W. MicroRNA miR-320a and miR-140 inhibit mink enteritis virus infection by repression of its receptor, feline transferrin receptor. *Virol. J.* **2014**, *11*, 210. [CrossRef] [PubMed]

63. Duy, J.; Koehler, J.W.; Honko, A.N.; Schoepp, R.J.; Wauquier, N.; Gonzalez, J.-P.; Pitt, M.L.; Mucker, E.M.; Johnson, J.C.; O'Hearn, A. Circulating microRNA profiles of Ebola virus infection. *Sci. Rep.* **2016**, *6*, 1–13. [CrossRef] [PubMed]
64. Domingo, E.; Holland, J. RNA virus mutations and fitness for survival. *Annu. Rev. Microbiol.* **1997**, *51*, 151–178. [CrossRef] [PubMed]
65. Wang, X.-S.; Gong, J.-N.; Yu, J.; Wang, F.; Zhang, X.-H.; Yin, X.-L.; Tan, Z.-Q.; Luo, Z.-M.; Yang, G.-H.; Shen, C. MicroRNA-29a and microRNA-142-3p are regulators of myeloid differentiation and acute myeloid leukemia. *Blood* **2012**, *119*, 4992–5004. [CrossRef]
66. Lu, X.; Li, X.; He, Q.; Gao, J.; Gao, Y.; Liu, B.; Liu, F. miR-142-3p regulates the formation and differentiation of hematopoietic stem cells in vertebrates. *Cell Res.* **2013**, *23*, 1356–1368. [CrossRef]
67. Wang, F.; Wang, X.-S.; Yang, G.-H.; Zhai, P.-F.; Xiao, Z.; Xia, L.-Y.; Chen, L.-R.; Wang, Y.; Wang, X.-Z.; Bi, L.-X. miR-29a and miR-142-3p downregulation and diagnostic implication in human acute myeloid leukemia. *Mol. Biol. Rep.* **2012**, *39*, 2713–2722. [CrossRef]
68. Xu, L.; Xu, Y.; Jing, Z.; Wang, X.; Zha, X.; Zeng, C.; Chen, S.; Yang, L.; Luo, G.; Li, B. Altered expression pattern of miR-29a, miR-29b and the target genes in myeloid leukemia. *Exp. Hematol. Oncol.* **2014**, *3*, 17. [CrossRef]
69. Park, S.-Y.; Lee, J.H.; Ha, M.; Nam, J.-W.; Kim, V.N. miR-29 miRNAs activate p53 by targeting p85α and CDC42. *Nat. Struct. Mol. Biol.* **2009**, *16*, 23. [CrossRef]
70. Gong, J.; Yu, J.; Lin, H.; Zhang, X.; Yin, X.; Xiao, Z.; Wang, F.; Wang, X.; Su, R.; Shen, C. The role, mechanism and potentially therapeutic application of microRNA-29 family in acute myeloid leukemia. *Cell Death Differ.* **2014**, *21*, 100–112. [CrossRef]
71. Xiong, Y.; Fang, J.H.; Yun, J.P.; Yang, J.; Zhang, Y.; Jia, W.H.; Zhuang, S.M. Effects of MicroRNA-29 on apoptosis, tumorigenicity, and prognosis of hepatocellular carcinoma. *Hepatology* **2010**, *51*, 836–845. [CrossRef]
72. Pekarsky, Y.; Croce, C.M. Role of miR-15/16 in CLL. *Cell Death Differ.* **2015**, *22*, 6–11. [CrossRef]
73. Cimmino, A.; Calin, G.A.; Fabbri, M.; Iorio, M.V.; Ferracin, M.; Shimizu, M.; Wojcik, S.E.; Aqeilan, R.I.; Zupo, S.; Dono, M. miR-15 and miR-16 induce apoptosis by targeting BCL2. *Proc. Natl. Acad. Sci. USA* **2005**, *102*, 13944–13949. [CrossRef] [PubMed]
74. Ofir, M.; Hacohen, D.; Ginsberg, D. MiR-15 and miR-16 are direct transcriptional targets of E2F1 that limit E2F-induced proliferation by targeting cyclin E. *Mol. Cancer Res.* **2011**, *9*, 440–447. [CrossRef]
75. Han, H.; Zhang, Z.; Yang, X.; Yang, W.; Xue, C.; Cao, X. miR-23b suppresses lung carcinoma cell proliferation through CCNG1. *Oncol. Lett.* **2018**, *16*, 4317–4324. [CrossRef]
76. Luna, J.M.; Scheel, T.K.; Danino, T.; Shaw, K.S.; Mele, A.; Fak, J.J.; Nishiuchi, E.; Takacs, C.N.; Catanese, M.T.; de Jong, Y.P. Hepatitis C virus RNA functionally sequesters miR-122. *Cell* **2015**, *160*, 1099–1110. [CrossRef]
77. Fulzele, S.; Sahay, B.; Yusufu, I.; Lee, T.J.; Sharma, A.; Kolhe, R.; Isales, C.M. COVID-19 Virulence in Aged Patients Might Be Impacted by the Host Cellular MicroRNAs Abundance/Profile. *Aging Dis.* **2020**, *11*, 509. [CrossRef]
78. Chow, J.T.-S.; Salmena, L. Prediction and Analysis of SARS-CoV-2-Targeting MicroRNA in Human Lung Epithelium. *Genes* **2020**, *11*, 1002. [CrossRef]
79. Cullen, B.R. Viruses and microRNAs. *Nat. Genet.* **2006**, *38*, S25–S30. [CrossRef]
80. Demirci, M.D.S.; Adan, A. Computational analysis of microRNA-mediated interactions in SARS-CoV-2 infection. *PeerJ* **2020**, *8*, e9369. [CrossRef] [PubMed]
81. Morales, L.; Oliveros, J.C.; Fernandez-Delgado, R.; Robert tenOever, B.; Enjuanes, L.; Sola, I. SARS-CoV-encoded small RNAs contribute to infection-associated lung pathology. *Cell Host Microbe* **2017**, *21*, 344–355. [CrossRef]
82. Arisan, E.D.; Dart, A.; Grant, G.H.; Arisan, S.; Cuhadaroglu, S.; Lange, S.; Uysal-Onganer, P. The Prediction of miRNAs in SARS-CoV-2 Genomes: Hsa-miR Databases Identify 7 Key miRs Linked to Host Responses and Virus Pathogenicity-Related KEGG Pathways Significant for Comorbidities. *Viruses* **2020**, *12*, 614. [CrossRef] [PubMed]
83. Periwal, N.; Sarma, S.; Arora, P.; Sood, V. In-silico analysis of SARS-CoV-2 genomes: Insights from SARS encoded non-coding RNAs. *bioRxiv* **2020**. [CrossRef]
84. Choi, E.-J.; Kim, H.B.; Baek, Y.H.; Kim, E.-H.; Pascua, P.N.Q.; Park, S.-J.; Kwon, H.-i.; Lim, G.-J.; Kim, S.; Kim, Y.-I. Differential microRNA expression following infection with a mouse-adapted, highly virulent avian H5N2 virus. *BMC Microbiol.* **2014**, *14*, 252. [CrossRef]

85. Zhang, S.; Ouyang, X.; Jiang, X.; Gu, D.; Lin, Y.; Kong, S.; Xie, W. Dysregulated serum microRNA expression profile and potential biomarkers in hepatitis C virus-infected patients. *Int. J. Med Sci.* **2015**, *12*, 590. [CrossRef]
86. Ling, T.-Y.; Kuo, M.-D.; Li, C.-L.; Alice, L.Y.; Huang, Y.-H.; Wu, T.-J.; Lin, Y.-C.; Chen, S.-H.; Yu, J. Identification of pulmonary Oct-4+ stem/progenitor cells and demonstration of their susceptibility to SARS coronavirus (SARS-CoV) infection in vitro. *Proc. Natl. Acad. Sci. USA* **2006**, *103*, 9530–9535. [CrossRef]
87. Mallick, B.; Ghosh, Z.; Chakrabarti, J. MicroRNome analysis unravels the molecular basis of SARS infection in bronchoalveolar stem cells. *PLoS ONE* **2009**, *4*, e7837. [CrossRef]
88. Kuba, K.; Imai, Y.; Rao, S.; Gao, H.; Guo, F.; Guan, B.; Huan, Y.; Yang, P.; Zhang, Y.; Deng, W. A crucial role of angiotensin converting enzyme 2 (ACE2) in SARS coronavirus–induced lung injury. *Nat. Med.* **2005**, *11*, 875–879. [CrossRef]

6

First Evidence for the Disease-Stage, Cell-Type, and Virus Specificity of microRNAs during Human Immunodeficiency Virus Type-1 Infection

Lauren Fowler [1], Viviane Conceicao [1], Suneth S. Perera [1], Priyanka Gupta [1], Choo Beng Chew [2], Wayne B. Dyer [3] and Nitin K. Saksena [1,*]

1 Westmead Millennium Institute & Westmead Hospital, University of Sydney, Westmead, Sydney, NSW 2145, Australia; lauren_fowler@live.com.au (L.F.); vickanc@hotmail.com (V.C.); perera.suneth@gmail.com (S.S.P.); pgup7905@uni.sydney.edu.au (P.G.)

2 Research and Development, Australian Red Cross Blood Service, 17 O'Riordan Street, Alexandria, Sydney, NSW 2015, Australia; choobengchew@hotmail.com

3 ICPMR, Department of Virology, Westmead Hospital, Westmead, Sydney, NSW 2145, Australia; Wayne.dyer@sydney.edu.au

* Correspondence: nitin.saksena@sydney.edu.au or nitin.saksena@bigpond.com

Academic Editor: Kenth Gustafsson

Abstract: The potential involvement of host microRNAs (miRNAs) in HIV infection is well documented, and evidence suggests that HIV modulates and also dysregulates host miRNAs involved in maintaining the host innate immune system. Moreover, the dysregulation of host miRNAs by HIV also effectively interferes directly with the host gene expression. In this study, we have simultaneously evaluated the expression of host miRNAs in both CD4+ and CD8+ T-cells derived from HIV-positive (HIV+) individuals (viremic and aviremic individuals while receiving highly active antiretroviral therapy (HAART), therapy-naïve long-term non-progressors (LTNP), and HIV-negative (HIV–) healthy controls. miRNAs were run on Affymetrix V2 chips, and the differential expression between HIV+ and HIV− samples, along with intergroup comparisons, was derived using PARTEK software, using an FDR of 5% and an adjusted p-value < 0.05. The miR-199a-5p was found to be HIV-specific and expressed in all HIV+ groups as opposed to HIV– controls. Moreover, these are the first studies to reveal clearly the highly discriminatory miRNAs at the level of the disease state, cell type, and HIV-specific miRNAs.

Keywords: CD4+ T-cells; CD8+ T-cells; micro-RNA; human immunodeficiency virus; acquired immunodeficiency syndrome; non-progression; gene expression

---

1. Introduction

MicroRNAs (miRNAs) are small (21–22 nt), non-coding RNA fragments found in many organisms, from plants to humans, which function to negatively regulate gene expression. In humans, roles for miRNA have been identified in metabolism, growth, and development. Dicer, a protein critical for miRNA processing, has been associated with proper development of the limbs, lungs, and hair follicles and in T-cell differentiation [1,2]. Dysregulation of human miRNA has been implicated in the pathogenesis of many cancers [3]. As negative regulators of gene expression, miRNAs can function as tumor suppressors that, if dysregulated, can augment the chance of cancer development [4]. For example, a chromosomal deletion associated with the development of chronic lymphocytic leukaemia (CLL) causes a loss of intrinsic miRNA tumor suppressors [5]. Currently, over 800 miRNAs have been identified in the human genome with diverse functions. miRNAs function by binding to partially complementary sites in the messenger RNA (mRNA) of other genes, inhibiting the production of the proteins synthesized from these genes. A single miRNA can interact with hundreds and thousands of genes in regulating various physiological processes, thus adding to the complex regulation of the human genome.

Viruses are known to subvert and manipulate the human gene machinery, and miRNAs are no exception. Host miRNAs are dysregulated by viruses, and both DNA and RNA viruses have been found to encode and express miRNAs in the infected host, demonstrating the importance of the role of host- and virus-encoded miRNAs in host-virus interactions. The miRNA expression profile of cells is altered in both clinical PBMC samples and cultured cells. Cells transfected with an infectious HIV-1 strain have significantly altered miRNA profiles compared to control cells [6]. In 2008, Houzet *et al.* showed that the vast majority of host miRNAs are downregulated in HIV infection of peripheral blood mononuclear cells (PBMCs). When individuals were separated into groups based on their T-cell counts and viral loads (VLs), specific miRNA "profiles" could be seen for each of the classes. Furthermore, many miRNA changes in patient cells could not be accounted for by infection alone, indicating a complex role for miRNA in gene regulation [7].

In 2007, Huang *et al.* showed overexpression of host miRNAs in resting T-cells that target sequences in the 3′ end of HIV-1 RNA, silencing viral mRNA and enforcing latency [8]. Furthermore, Witwer *et al.* showed that PBMC miRNA profiles could distinguish elite suppressors (ES) and uninfected controls from viremic HIV-1 infected patients [9]. Their results demonstrated correlations between miRNA expression, CD4+ T-cell count and viral load. Some miRNAs found to differ in expression have previously been shown to correlate with HIV-1 latency, including miR-29s, miR-125b, and miR-150. Their analysis identified several miRNAs that have not been previously described in association with HIV infection, including miR-31, which distinguishes controls and ES and regulates a protein with implications for T-cell differentiation. Although this study has also shown that HIV-1-positive ES are characterized by a PBMC miRNA profile that in general resembles that of uninfected individuals, they also reiterate that the ES, on the basis of miRNA expression, are a heterogeneous group. This suggests that different mechanisms, shaped or marked by different miRNA expression patterns, underlie sustained and durable control in therapy-naïve HIV-infected individuals. In a recent International AIDS Society (IAS) meeting, Zhu *et al.* showed a set of 18 differentially expressed miRNAs, which could identify the outcome of HIV disease at the chronic stage more accurately. Six out of 18 miRNAs were significantly related to faster rate of CD4+ T-cell decline [10]. Studies of larger cohorts of individuals are needed to address miRNA specific to different stages of HIV disease and explain the underlying genomic basis of natural control of HIV in therapy-naive ES.

Since all the studies to date have been performed on whole PBMC or tissue, we endeavored to address disease- and cell-type-specific miRNAs and their role in HIV pathogenesis. We have adopted a novel approach for this study, which simultaneously analyzes miRNAs from the CD4+ and CD8+ T-cells from viremic, aviremic BDL patients, and elite controllers. This study is unique in showing the HIV disease-stage and cell-type specificity of miRNA during HIV infection and its natural control in elite controllers.

2. Results

2.1. *Patient Samples Used in Microarray Analysis*

Patients were classed into disease groups based on their HIV plasma viral load (VL) and also the antiretroviral drug treatment, as shown in Table 1. Prior to the microarray analysis, RNA quality and integrity was checked with an Agilent Bioanalyzer. All RNA samples with an RNA integrity number (RIN) above 8 were deemed appropriate for microarray analysis. The results are shown below in Table 1 for each sample and specific cell types analyzed.

Table 1. Clinical profiles of the study patients, and RIN.

Patient	Sex	Plasma Viral Load	Disease Stage	Cell Type	RIN
VIR1	M	42,200	Viremic	CD4+ CD8+	9.6 9.7
VIR2	M	58	Viremic	CD4+ CD8+	9.5 9.6
VIR3	M	226	Viremic	CD4+ CD8+	9.5 9.7
VIR4	M	5240	Viremic	CD4+ CD8+	9.5 9.7
VIR5	M	90,300	Viremic	CD4+ CD8+	9.1 9.3
VIR6	M	112,000	Viremic	CD4+ CD8+	9.7 9.7
AVIR1	M	BDL	Aviremic	CD4+ CD8+	9.6 9.4
AVIR2	M	BDL	Aviremic	CD4+ CD8+	9.8 9.9
AVIR3	M	BDL	Aviremic	CD4+ CD8+	10 10
AVIR4	F	BDL	Aviremic	CD4+ CD8+	9.6 9.9
AVIR5	M	BDL	Aviremic	CD4+ CD8+	9.7 9.6
LTNP1	M	BDL	LTNP	CD4+ CD8+	9.7 9.8
LTNP2	M	BDL	LTNP	CD4+ CD8+	8.8 9.8
LTNP3	M	BDL	LTNP	CD4+ CD8+	8.6 8.1
LTNP4	M	BDL	LTNP	CD4+ CD8+	10 8.9
HD1	M	-	Negative	CD4+ CD8+	9.7 9.8
HD2	M	-	Negative	CD4+ CD8+	10 9.7
HD3	F	-	Negative	CD4+ CD8+	9.6 9.2

RIN = RNA Integrity Number. Plasma HIV viral load is in per mL plasma. BDL = Below Detectable Levels of plasma HIV as seen in therapy-naive long-term non-progressors (LTNPs) and the therapy-experienced aviremic group (AVIR). VIR = HIV-infected patients with detectable viremia while on highly active antiretroviral therapy (HAART). HD = HIV-healthy donors.

2.2. Assessment of the Quality and Integrity of the Dataset by Principal Component Analysis (PCA)

Further, upon data retrieval from microarray analysis, the quality and integrity of the miRNA dataset were also assessed using the principal component analysis (PCA) available in the Partek Genomics Suite. Data in Figure 1A, shows the analysis of all HIV-infected groups against the HIV-negative (HIV–) healthy donors. This data reflects the high integrity of samples through a clear distinction between HIV-infected and uninfected healthy donors, deeming these samples fit for inclusion in downstream analysis. Following the broad HIV-positive (HIV+) *vs.* HIV– analysis (Figure 1), we examined the inter-group contrasts using the PCA for their integrity based on the cell types (CD4+ and CD8+ T-cells), as shown in Figure 1B,C. Again, excellent segregation was apparent for all four contrasts (long-term non-progressor (LTNP), aviremic, viremic and HIV– groups) in both CD4+ and CD8+ T-cells. From these data, it is clear that the miRNA profiles of the four disease states examined were distinct and separable. One interesting observation was that the segregation of groups based on cell phenotype was better resolved for all four groups examined in CD4+ T-cells (Figure 2a–d).

In contrast, the CD8+ T-cells, although indicating segregation of all four groups, showed considerable closeness between viremic, aviremic, and LTNP groups, which was expected, as these three groups were HIV+. Taken together, the data represented in Figure 1, suggest unambiguous data integrity between contrasts, which provided a solid platform to study differentially expressed (DE) miRNA between different HIV disease groups.

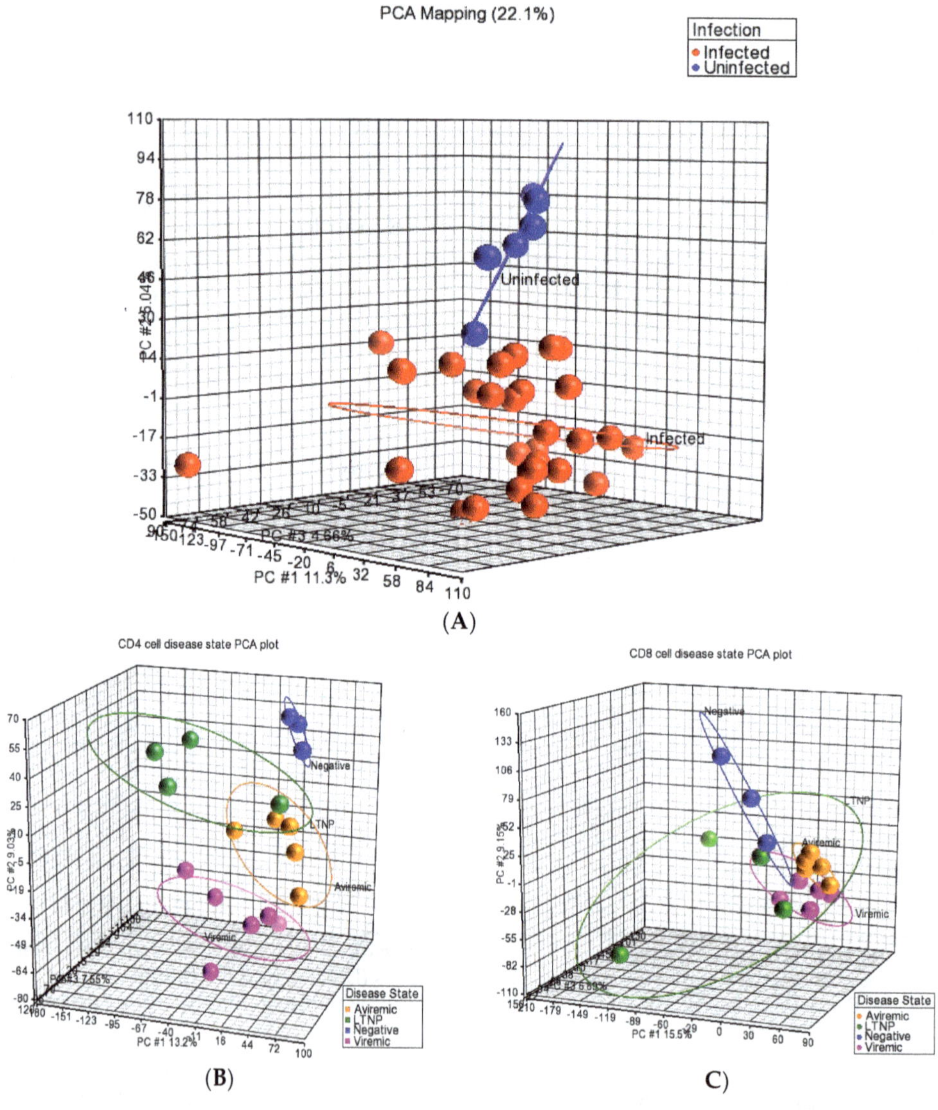

Figure 1. (**A**) Principal component analysis (PCA) of samples highlighting concordance and clustering of all infected cell samples, compared to uninfected samples. *X* Axis (Component 1) = 11.3%, *Y* Axis (Component 2) = 6.04%, *Z* Axis (Component 3) = 4.66%. **Red** = HIV-positive (HIV+), **Blue** = HIV-negative (HIV–), healthy donors; PCA of CD4+ (**B**) and CD8+ (**C**) T-cells respectively, highlighting disease group concordance. **Pink** = viremic, **Yellow** = aviremic, **Green** = LTNP, **Blue** = negative. Viremic $n = 6$, aviremic $n = 5$, LTNP $n = 4$, negative $n = 3$ for each cell type. (**B**) *X* axis (Component 1) = 13.2%, *Y* axis (Component 2) = 9.03%, *Z* axis (Component 3) = 7.55%. (**C**) *X* axis (Component 1) = 15.5%, *Y* axis (Component 2) = 9.15%, *Z* axis (Component 3) = 6.89%.

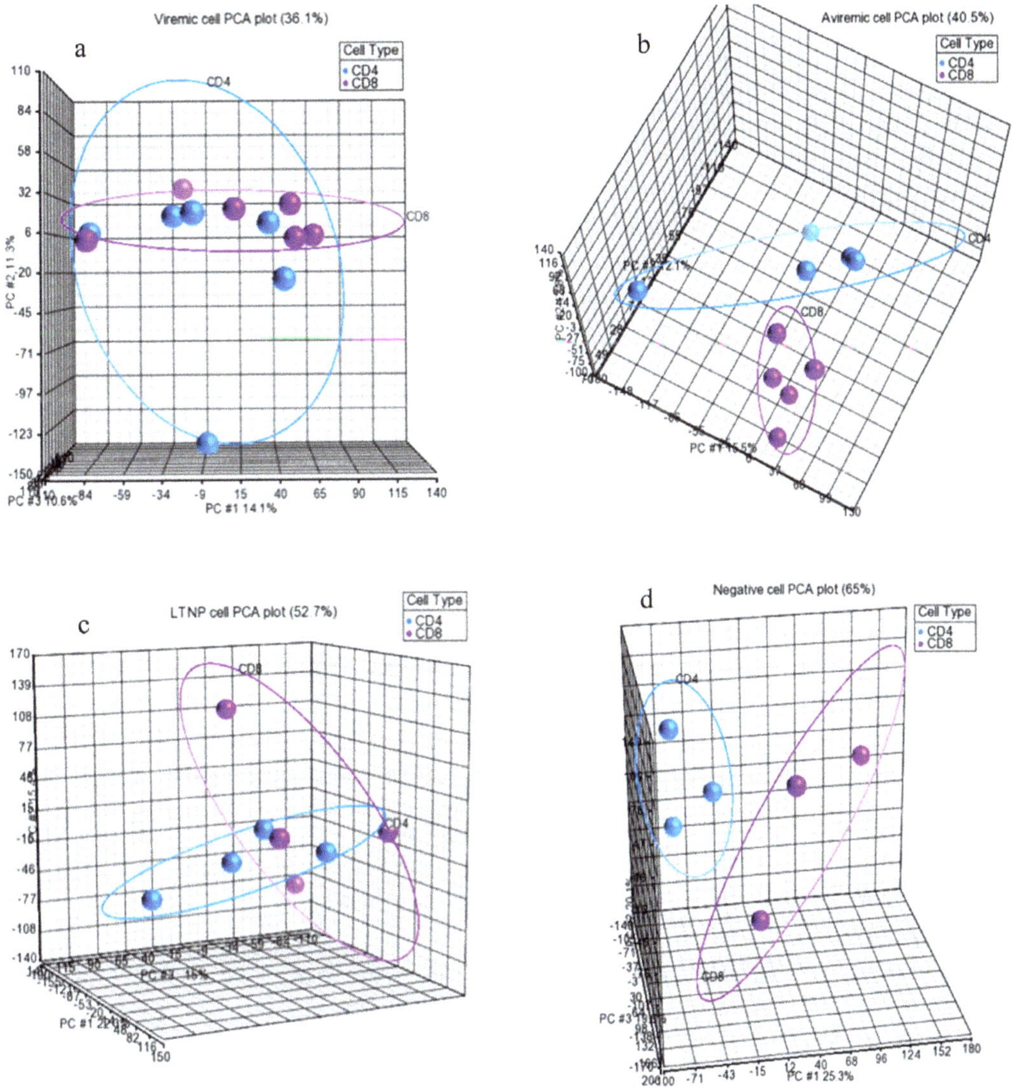

Figure 2. PCA for each disease state, comparing CD4+ and CD8+ T-cells. Analysis of CD4+ and CD8+ cells from patient groups viremic (**a**); aviremic (**b**); LTNP (**c**) and negative (**d**). Azure = CD4+ T-cells, **Purple** = CD8+ T-cells. (**a**) Viremic CD4+ ($n = 6$) *vs.* CD8+ ($n = 6$). *X* Axis (Component 1) = 14.1%, *Y* Axis (Component 2) = 11.3%, *Z* Axis (Component 3) = 10.6%. (**b**) Aviremic CD4+ ($n = 5$) *vs.* CD8+ ($n = 5$). *X* axis (Component 1) = 15.5%, *Y* axis (Component 2) = 12.8%, *Z* axis (Component 3) = 12.1%. (**c**) LTNP CD4+ ($n = 4$) *vs.* CD8+ ($n = 4$). *X* axis (Component 1) = 22.3%, *Y* axis (Component 2) = 15.4%, *Z* axis (Component 3) = 15%. (**d**) Negative CD4+ ($n = 3$) *vs.* CD8+ ($n = 3$). *X* axis (Component 1) = 25.3%, *Y* axis (Component 2) = 20.1%, *Z* axis (Component 3) = 19.6%.

The underlying reasons for this distinction between CD4+ and CD8+ T-cells in their ability to segregate HIV disease and non-disease groups remain unclear. PCA was also used to determine cell specificity in relation to disease group by directly comparing the CD4+ and CD8+ T-cells of each disease state. This analysis indicated a clear segregation of CD4+ and CD8+ T-cell type based on disease state, implying cell-specific functional segregation during HIV disease stages (Figure 2a–c) and in a healthy state, as can be seen in Figure 2d. Although good segregation of CD4+ and CD8+ T-cells was seen based on disease group for viremic (a), aviremic (b), and negative (d) comparisons, the data shown for the LTNPs in Figure 2c show a close relationship and little segregation between CD4+ and CD8+ T-cells, suggesting a possible functional significance of CD4+ and CD8+ T-cells and cooperation between them in mounting sustained response during natural control of HIV in the LTNPs.

2.3. Frequency Histograms

Frequency histograms representative of the CD4+ and CD8+ T-cell datasets (Figure 3a,b) show that normalization applied by the Partek Genomics Suite was effective and that there was negligible patient-to-patient variability. High concordance between all 18 samples based on CD4+ (Figure 3a) and CD8+ T-cells was evident (Figure 3b), which supports the validity of the processed and normalized data for further analysis.

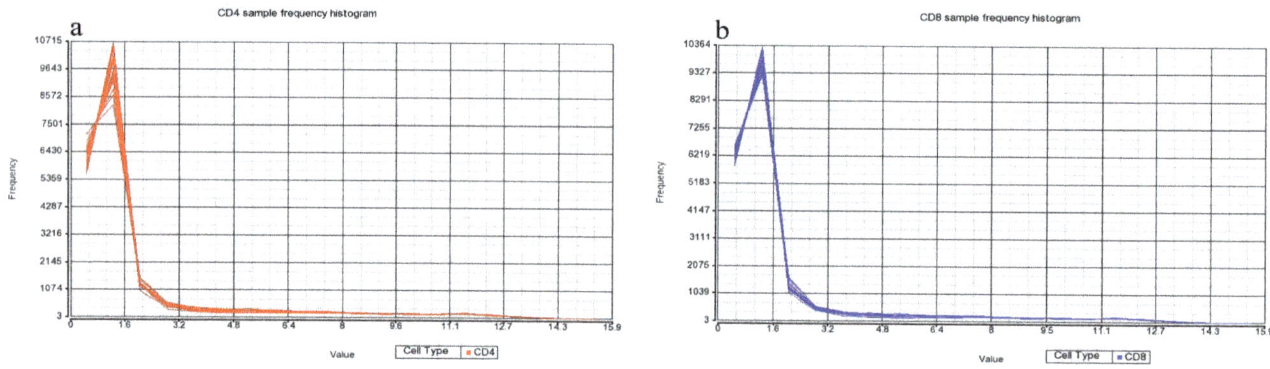

Figure 3. Frequency histograms of CD4+ (**a**) and CD8+ (**b**) cell samples, showing the concordance of data from each T-cell type. CD4+/CD8+ $n = 18$.

2.4. Differential Expression of miRNAs between HIV Positive Samples and Negative Controls

Upon a quality and integrity check using various algorithms discussed above, we analyzed differential miRNA expressions between disease groups and disease groups based on cell types. Heat maps, coupled with hierarchical clustering, were used throughout to visualize the validity and integrity of differential expressions between each contrast.

In the first contrast, all of the HIV+ groups (viremic, aviremic, LTNP) were taken together and compared against the HIV– healthy donors using the false discovery rate (FDR). Two DE miRNAs were found specific to CD4+ T-cells (miR-199a-5p and miR-720) (Figures 4a and 5), and two specific to the CD8+ T-cells (miR-941-3s and miR-1323) (Figures 4b and 5, Table 2), with significant fold changes (FCs) supported by strong p-values (Table 2).

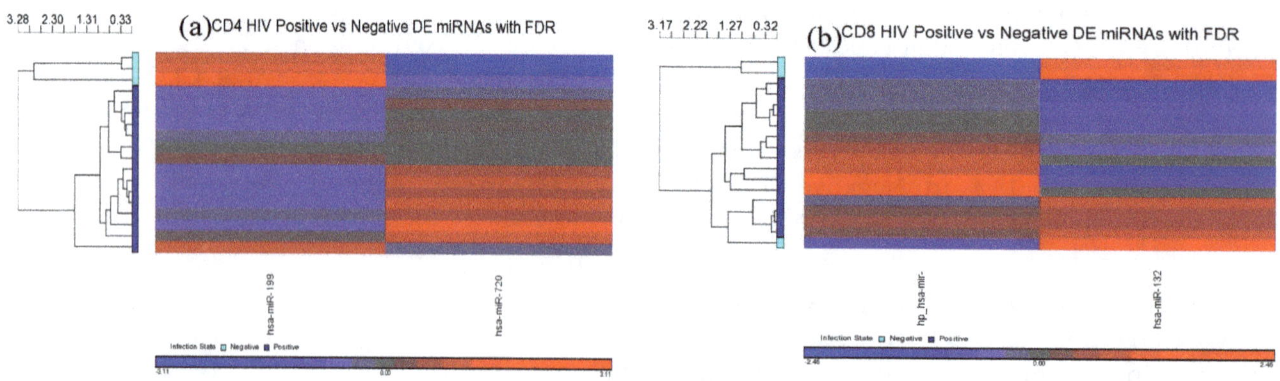

Figure 4. (**a**,**b**) Heat maps indicating differentially expressed (DE) microRNAs (miRNAs) in HIV+ CD4+ (**a**) and CD8+ (**b**) T-cells compared to negative controls. In each case, uninfected $n = 3$ and infected $n = 15$.

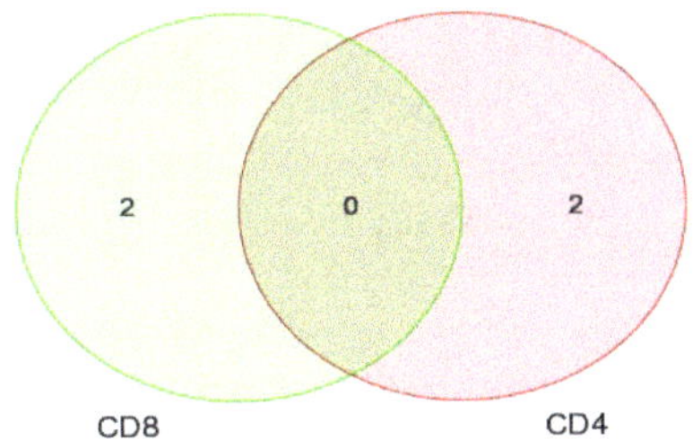

Figure 5. Venn diagram showing miRNAs specific to CD4+ and CD8+ T-cells with the absence of any overlapping miRNAs between these two cell types. Uninfected $n = 3$, and HIV-infected $n = 15$.

Table 2. DE miRNAs in HIV infected CD4+ and CD8+ T-cells using the false discovery rate (FDR) method.

Cell Type	miRNA	Adjusted p-Value	Fold Change
CD4+	miR-199a-5p	1.22×10^{-6}	−5.6
	miR-720	6.61×10^{-5}	3.69
CD8+	miR-941-3-s	5.83×10^{-5}	2.77
	miR-1323	9.14×10^{-5}	−3.72

Negative sign (−) signifies downregulation, whereas no sign signifies upregulation.

When HIV-infected cells were compared to uninfected cells (Figure 5) using the FDR method, miR-199a-5p was found to be downregulated in CD4+ T-cells compared to negative cells over 5-fold, whereas miR-720 was upregulated nearly 4-fold. In CD8+ T-cells, the miR-941-3-s was upregulated nearly 3-fold, where the miR-1323 was downregulated nearly 4-fold (Table 2 and Figure 5). No overlapping miRNAs between HIV+ and HIV− cells were seen, and the differential expression was unique to CD4+ and CD8+ T-cell subsets, as shown in the Venn diagram (Figure 6), suggesting that the DE miRNA are cell-specific signatures that are unique to a T-cell subset within the infected individual. This is the first evidence demonstrating cell-specific miRNA during HIV disease stages.

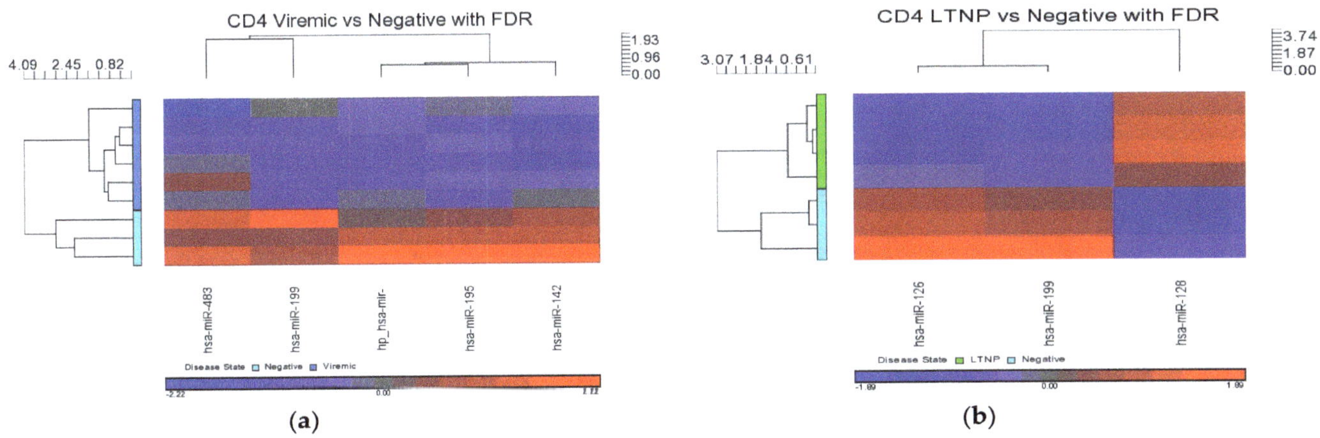

Figure 6. Heat maps indicating DE miRNAs in CD4+ T-cell disease group comparisons of viremic (**a**) and LTNP (**b**) compared to negative controls. FDR method was used. Viremic $n = 6$, LTNP $n = 4$, negative $n = 3$.

Apart from using the FDR method with $p < 0.05$, we also visualized the dataset using no FDR with $p < 0.01$. It is important to state that, when no FDR was applied to the analysis of differentially expressed miRNAs in infected *vs.* uninfected CD4+ and CD8+ T-cells, more miRNAs were observed, as expected. However, in this analysis, the top-ranked highly significant miRNAs that were observed in the FDR method were also fully covered in the non-FDR analysis. This dual-analysis served to ensure that statistical stringency did not occlude biologically relevant miRNAs (Table A1).

2.5. *A Closer Examination of Differentially Expressed miRNAs in CD4+ T-Cells and Their Association with Disease Groups*

2.5.1. CD4+ Disease Groups *vs.* the HIV Negative Group

Independent inter-group comparisons were carried out for the DE miRNAs for each cell type. First, the CD4+ T-cells from unique disease groups (viremic, aviremic, and LTNPs) were individually compared against CD4+ T-cells from the HIV– controls using the FDR $p < 0.05$. Five DE miRNAs were identified in the viremic CD4+ T-cells (miR-199a-5p, miR-142-5p, miR-520f, miR-195, and miR-483-5p), and three DE miRNAs were identified in CD4+ LTNPs (miR-1995p, miR-1280, and miR-126) (Figure 6). All of these showed strong *p*-values and fold changes (Table 3). Interestingly, miR-199a-5p was found to be overlapping between these two contrasts (Figure 7).

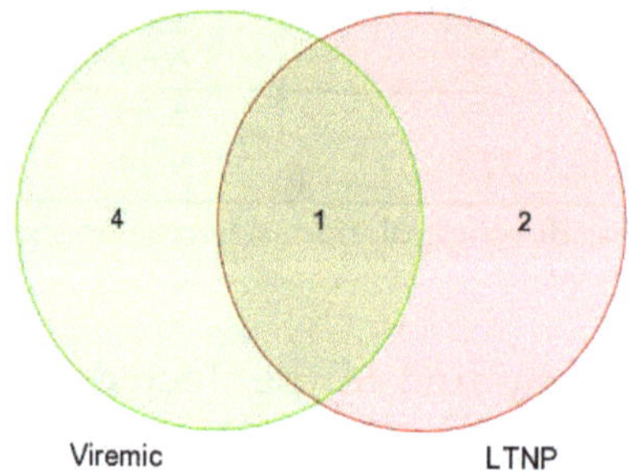

Figure 7. Venn diagram of DE miRNA in CD4+ T-cells when disease groups are compared to HIV– controls. FDR method was used. This diagram shows the lack of results for aviremic samples and overlapping expression of one miRNA; miR-199a-5p.

Table 3. DE miRNAs in CD4+ T-cell disease groups compared to HIV– controls.

Disease State Compared	miRNA	Adjusted *p*-Value	Fold Change
Viremic	miR-199a-5p	4.61×10^{-6}	−6.31
	miR-142-5p	1.49×10^{-4}	−3.7
	miR-520f	1.02×10^{-4}	−2.96
	miR-195	8.72×10^{-5}	−6.44
	miR-483-5p	7.50×10^{-5}	−4.28
Aviremic		no results	
LTNP	miR-199a-5p	7.66×10^{-6}	−6.85
	miR-1280	9.42×10^{-6}	3.71
	miR-126	1.78×10^{-6}	−51.73

Viremic $n = 6$, aviremic $n = 6$ LTNP $n = 4$, negative $n = 3$. FDR method was used. Negative fold change indicates downregulation in disease state compared to negative. Positive fold change indicates upregulation in disease state compared to negative.

No DE miRNAs were found in the aviremic CD4+ T-cells when compared with the HIV– controls using the FDR method. However, when the non-FDR method was used, DE miRNAs were observed in the aviremic CD4+ T-cells (Table A2), including the miRNA miR-199a-5p, which was lacking for this comparison due to statistical stringency introduced by the FDR method (Table 3). It is important to reiterate that the miRNA-199-5p is one of the most notable miRNAs, as it was unique to HIV-infected CD4+ T-cells. Viremic and LTNP comparisons demonstrated this miRNA as significant using the FDR data, but significance in the aviremic comparison appeared only to relax the statistical stringency to non-FDR with $p < 0.01$. This suggests that, although miR-199a-5p was differentially expressed in HIV infection in all three disease states, it was not expressed significantly enough to be picked up for CD4+ aviremic *vs.* HIV− comparison due to statistical stringency. Thus, taken together, the FDR (Table 3, Figure 7) and the non-FDR data (Table A2) concur in that miR-199a-5p was a signature unique to HIV+ CD4+ T-cells and may have a considerable biological significance. Interestingly, the fold changes for each contrast (CD4 HIV+ *vs.* HIV− = −5.6; viremic *vs.* negative = −6.31; LTNP *vs.* negative = −6.85; and aviremic *vs.* negative = −4.13 (non-FDR)) indicate that miR-199a-5p was downregulated in the viremic state and that its upregulation appears to be characteristic of an HIV− state in healthy individuals (Table 2).

2.5.2. CD4+ T Cells Inter-Disease Group Comparisons

Within the CD4+ T-cell subset, differential expression of miRNAs was intercompared between disease states. Three comparisons were examined: viremic *vs.* aviremic, viremic *vs.* LTNP, and LTNP *vs.* aviremic groups, respectively. No differential expression was observed in the viremic *vs.* aviremic or LTNP *vs.* aviremic comparisons when the FDR method was used. In the viremic *vs.* LTNP comparison, seven DE miRNAs were detected (miR-509-3p, miR-345, miR-1972, miR-1975, miR-3163, miR-768-3p, and miR-1201)—six downregulated and one upregulated. This upregulated miRNA, miR-345, demonstrated a 11.6-fold change with a p-value of 3.79×10^{-5} (Figure 8 and Table 4), but the downregulation of miRNAs during viremia was the main feature in this comparison.

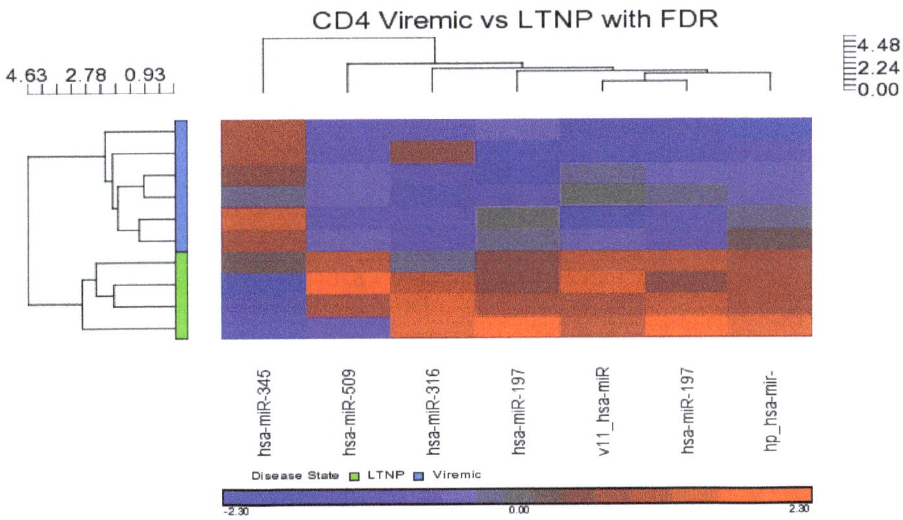

Figure 8. Heat map showing DE miRNAs in CD4+ viremic *vs.* LTNP disease group comparison. Viremic n = 6, LTNP n = 4.

Table 4. DE miRNAs in CD4+ T-cells derived through inter-disease group comparisons.

Comparison	miRNA	p-Value	Fold Change
Viremic *vs.* Aviremic		No DE miRNA	
Viremic *vs.* LTNP	miR-509-3p	7.77×10^{-5}	−7.57
	miR-345	3.79×10^{-5}	11.66
	miR-1972	5.31×10^{-5}	−7.9
	miR-1975	2.03×10^{-5}	−2.62
	miR-3163	1.10×10^{-4}	−3.8
	miR-768-3p	9.30×10^{-6}	−2.06
	mir-1201_s	8.31×10^{-5}	−2.14
LTNP *vs.* Aviremic		No DE miRNA	

FDR method, $p < 0.05$. No sign indicates upregulation in the former disease state compared to the latter, whereas negative sign (−) indicates downregulation in the former disease state compared to the latter.

When the non-FDR method with p-value < 0.01 was used, differential expression of a large number of miRNAs was detected for each comparison listed in Table A3.

2.6. *A Closer Examination of Differentially Expressed miRNAs in CD8+ T-Cells and Their Association with Disease Groups*

2.6.1. CD8+ Disease Groups *vs.* the HIV Negative Group

Similar to the CD4+ T-cells, CD8+ T-cell samples were also examined according to their disease groups against the CD8+ T-cells from HIV– controls.

Using the FDR method and $p < 0.05$, three miRNAs were found to be DE in viremic CD8+ T-cells, (miR-1323, miR-541 and miR-570), and all three showed downregulation (Table 5 and Figure 9). One miRNA (miR-1298) was specific to the aviremic group, showing 2-fold downregulation (Table 5). This is the first miRNA that is specific to CD8+ T-cells of HIV patients that control HIV through highly active antiretroviral therapy (HAART) therapy, and may have biological significance.

Table 5. DE miRNAs in CD8+ T-cell disease groups compared to HIV– controls.

Disease State Compared	miRNA	Adjusted p-Value	Fold Change
Viremic	miR-1323	0.0002	−4.23
	miR-541	0.0001	−3.95
	miR-570	8.79×10^{-6}	−2.18
Aviremic	miR-1298	2.42×10^{-5}	−2.02
LTNP	miR-572	7.06×10^{-6}	−14.86
	miR-126	5.46×10^{-6}	−31.25
	miR-151-5p	1.41×10^{-5}	2.35
	miR-486-5p	7.96×10^{-5}	6.77
	miR-885-3p	1.89×10^{-5}	−29.22

FDR method with $p < 0.05$ was used. Viremic $n = 6$, aviremic $n = 5$, LTNP $n = 4$, negative $n = 3$. No sign indicates upregulation and negative sign (−) indicates downregulation in the disease state compared to negative controls.

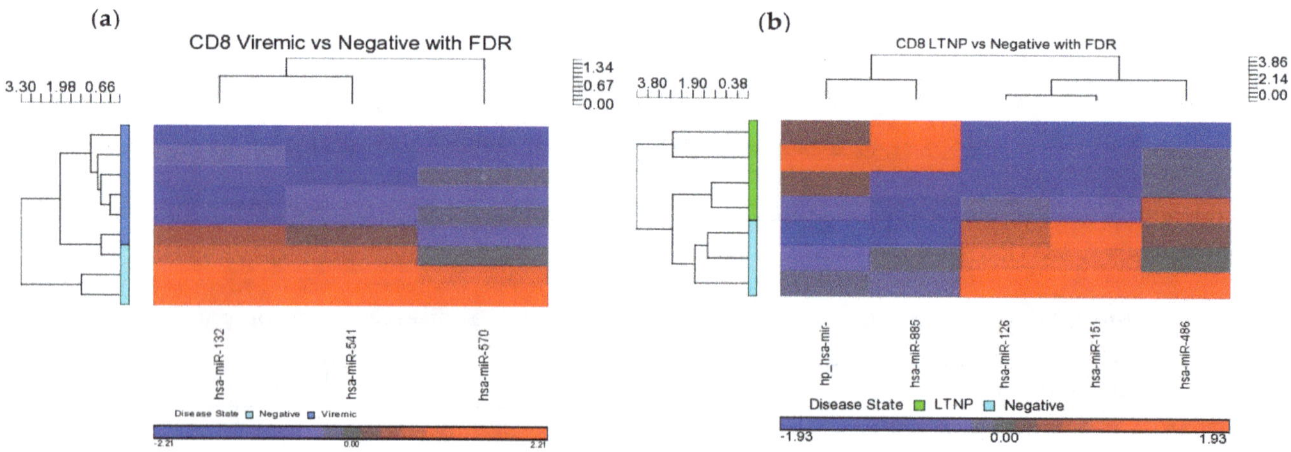

Figure 9. Heat maps indicating DE miRNAs in CD8+ disease group comparisons of viremic (**a**) and LTNP (**b**) compared to negative controls. Viremic $n = 6$, LTNP $n = 4$, negative $n = 3$.

Further, five DE miRNAs were identified in the LTNP and HIV– controls, with miR-572, miR-126, and miR-885-3p being downregulated, and miR-151-5p and miR-486-5p being upregulated, and supported by strong p-values (Table 5 and Figure 9). Interestingly, several downregulated miRNAs (572, 126, and 885-3p) showed noticeably high fold changes (−15, −31, and −29 fold, respectively), and all of them were specific to the CD8+ T-cells from the LTNP group (Table 5).

Most notable was that, when all three CD8+ disease groups were intercompared using a Venn diagram, no overlapping miRNAs were found (Figure 10), suggesting the identification of all disease-state-specific miRNAs in each case were supported by strong p-values (Table 5). This is the first evidence of HIV disease-state-specific miRNAs based on the CD8+ T-cells and is consistent with the role of CD8+ T-cells in containing viremia. Therefore, these miRNAs may have considerable biological significance in gene regulation in CD8+ T-cells.

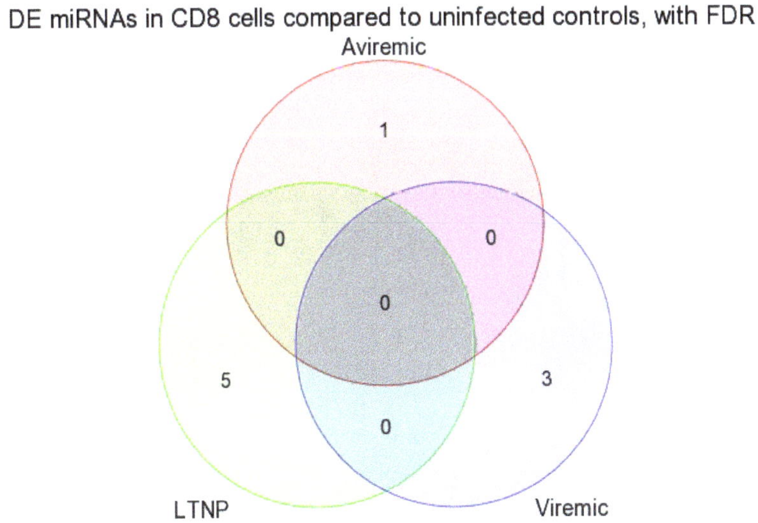

Figure 10. Venn diagram of DE miRNA in CD8+ T-cells when disease groups were compared to HIV– controls. FDR method with $p < 0.05$ was used. Viremic $n = 6$, aviremic $n = 5$, LTNP $n = 4$, negative $n = 3$.

As expected, the non-FDR method resulted in differential expression of more miRNAs than the FDR method between disease groups based on this cell type (Table A4). Again, there was a lack of evidence for overlapping miRNA between disease states using the non-FDR method.

2.6.2. CD8+ Inter-Disease Group Comparisons

Within the CD8+ subset, differential expression of miRNAs was intercompared between disease states. The three comparisons examined were viremic *vs.* aviremic, viremic *vs.* LTNP, and LTNP *vs.* aviremic. Differential expression was seen in both the viremic *vs.* LTNP and LTNP *vs.* aviremic comparisons, but not for the viremic *vs.* aviremic comparison.

Based on the lack of differential expression seen between viremic and aviremic cells in both CD4+ and CD8+ T-cell comparisons, it is apparent that there was little separation between the viremic and aviremic states in the presence or absence of viremia while on HAART, suggesting that the same miRNAs may regulate viremic and aviremic phases of HIV infection during therapy. Differential expression in the CD8+ T-cells was observed only in viremic *vs.* LTNP and LTNP *vs.* aviremic comparisons. In the viremic *vs.* LTNP contrast, 10 DE miRNAs (5 upregulated and 5 down regulated) were identified (Table 6; Figure 11), whereas in the LTNP *vs.* aviremic contrast, 61 DE miRNAs were identified (34 were down regulated, and the remaining 27 miRNAs were upregulated), all of which were supported by strong p-values (Table 6 and Figure 11).

Table 6. DE miRNAs in CD8+ T-cells when disease groups are intercompared.

Comparison	miRNA	*p*-Value	Fold Change	Disease State	miRNA	*p*-Value	Fold Change
Viremic *vs.* Aviremic		No Results		**LTNP *vs.* Aviremic (cont'd)**	miR-17	7.47×10^{-5}	−2.49
Viremic *vs.* LTNP	miR-151-5p	2.22×10^{-5}	8.22		miR-20a	4.10×10^{-5}	−5.1
	miR-486-5p	4.53×10^{-6}	19.01		Let-7d *	6.75×10^{-4}	−2.23
	miR-885-3p	2.17×10^{-5}	−5.95		miR-1268	4.17×10^{-5}	4.28
	miR-3141	4.88×10^{-5}	−2.26		miR-548u	1.23×10^{-3}	2.05
	miR-20b	1.60×10^{-4}	7.5		miR-199b-3p	4.34×10^{-4}	−5.46
	miR-106a	4.98×10^{-5}	2.44		miR-3126-3p	2.70×10^{-4}	2.14
	miR-17	4.70×10^{-5}	2.48		miR-502-3p	6.25×10^{-4}	−3.71
	miR-548u	1.79×10^{-4}	−2.3		miR-326	8.02×10^{-4}	2.09
	miR-3126-3p	1.29×10^{-4}	−2.18		miR-1289-2	4.14×10^{-5}	2.48
	mir-1289-2	7.24×10^{-5}	−2.31		miR-1207-5p	9.97×10^{-4}	2.41
LTNP *vs.* Aviremic	miR-1909	1.40×10^{-3}	5.57		miR-126	9.90×10^{-4}	−8.34
	miR-151-5p	2.08×10^{-5}	−9.03		miR-145	1.30×10^{-3}	−16.72
	miR-1469	6.34×10^{-4}	3.35		miR-494	5.62×10^{-4}	4.33
	miR-663	8.00×10^{-4}	4.18		miR-30d	4.93×10^{-4}	−4.66
	miR-3185	7.88×10^{-4}	6.92		miR-27a	5.93×10^{-4}	−5.92
	miR-1908	7.06×10^{-4}	4.31		miR-31	2.45×10^{-4}	−11.48
	miR-151-3p	6.66×10^{-4}	−6.34		miR-26a	7.81×10^{-4}	−2.21
	miR-762	9.57×10^{-5}	5.5		miR-30b	5.32×10^{-4}	−12.55
	miR-3196	2.49×10^{-4}	4.25		miR-19b	3.54×10^{-4}	−7.8
	miR-1228 *	9.00×10^{-5}	6.2		miR-4270	6.97×10^{-4}	2.68
	miR-4281	1.11×10^{-4}	4.14		miR-1975	6.73×10^{-6}	2.95
	miR-92b *	1.20×10^{-3}	4.37		miR-769-5p	8.46×10^{-4}	−6.87
	miR-638	4.71×10^{-5}	4.38		miR-28-3p	1.72×10^{-3}	−2.25
	miR-1915	1.25×10^{-4}	4.71		miR-629	1.15×10^{-3}	−12.53
	miR-486-5p	1.90×10^{-5}	−16.05		miR-106b	3.94×10^{-4}	−6.18
	miR-885-3p	5.09×10^{-5}	5.7		miR-30e *	1.51×10^{-3}	−3.78
	miR-2861	8.69×10^{-5}	5.53		miR-1979	1.32×10^{-4}	7.32
	miR-3141	6.97×10^{-6}	2.65		miR-30c	4.30×10^{-4}	−2.56
	miR-146b-5p	4.17×10^{-4}	−7.07		miR-30e	9.59×10^{-4}	−9.65
	miR-149 *	4.05×10^{-5}	7.49		miR-194	1.50×10^{-3}	−7.99
	miR-532-5p	1.05×10^{-3}	−3.86		miR-28-5p	8.44×10^{-4}	−4.92
	miR-572	1.72×10^{-3}	4.58		miR-23b	3.02×10^{-4}	−2.5
	miR-1910	1.50×10^{-3}	3.53		miR-25	6.61×10^{-4}	−2.31
	miR-20b	1.34×10^{-4}	−8.38		let-7g	2.73×10^{-4}	−2.22
	miR-106a	1.16×10^{-4}	−2.38		miR-182	1.62×10^{-3}	−2.35

* FDR method. Viremic $n = 6$, aviremic $n = 5$, LTNP $n = 4$.

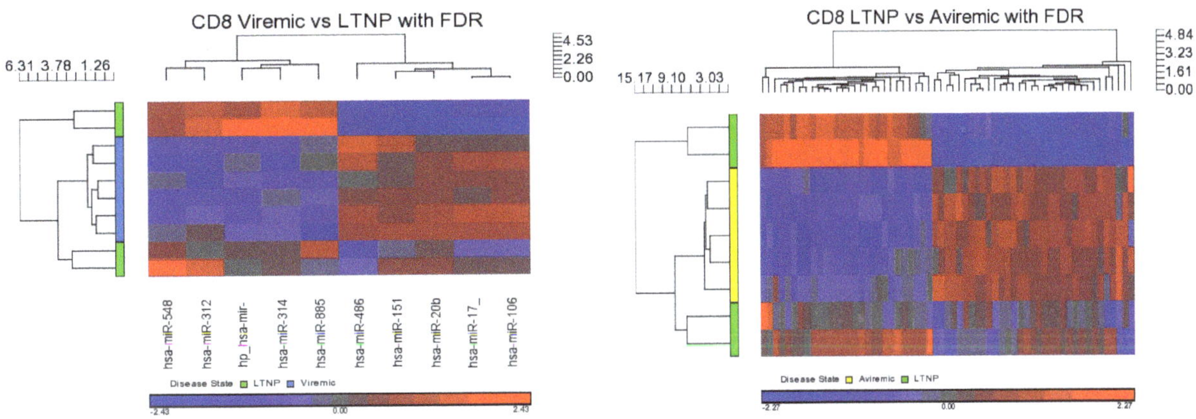

(a) (b)

Figure 11. Heat maps indicating DE miRNAs in CD8+ viremic *vs.* LTNP (**a**) and LTNP *vs.* aviremic (**b**) disease group comparisons. FDR method with $p < 0.05$ was used. Viremic $n = 6$, aviremic $n = 5$, LTNP $n = 4$.

Non-FDR data was also visualized for the CD8+ T-cells, and, as expected, a large number of miRNAs were detected for each comparison (see Table A5).

2.7. Comparison of Differential Expression of miRNA between CD4+ and CD8+ T-Cells

The main reason for these comparisons was to derive specificity of CD4+ or CD8+ T-cells in predicting a disease association. To do this, simultaneous comparison of the relationship between miRNA expression CD4+ and CD8+ T-cells in different disease groups for both cell types was carried out. Table 7 shows the side-by-side comparisons, which concur with the data already stated in Tables 3 and 5. Since few results were obtained for the comparisons involving aviremic cells, Venn diagrams were generated comparing differential expression of miRNAs in viremic (Figure 12a) and LTNP (Figure 12b) cells only.

Table 7. DE miRNAs in viremic, aviremic, and LTNP cells, both CD4+ and CD8+T-cells.

CD4			CD8		
miRNA	***p*-Value**	**Fold Change**	**miRNA**	***p*-Value**	**Fold Change**
		Viremic *vs.* Negative			
miR-199a-5p	4.61×10^{-6}	−6.31	miR-1323	1.72×10^{-4}	−4.23
miR-142-5p	1.49×10^{-4}	−3.7	miR-570	1.27×10^{-4}	−3.95
hp_mir-520f	1.02×10^{-4}	−2.96	miR-541	8.79×10^{-6}	−2.18
miR-195	8.72×10^{-5}	−6.44	-	-	-
miR-483-5p	7.50×10^{-5}	−4.28	-	-	-
		Aviremic *vs.* Negative			
	N/A		hp_mir-1298	2.42×10^{-5}	−2.02
		LTNP *vs.* Negative			
miR-199a-5p	7.66×10^{-6}	−6.85	miR-151-5p	7.06×10^{-6}	−14.86
miR-1280	9.42×10^{-6}	3.71	miR-486-5p	5.46×10^{-6}	−31.25
miR-126	1.78×10^{-6}	−51.73	mir-572	1.41×10^{-5}	2.35
-	-	-	miR-885-3p	7.96×10^{-5}	6.77
-	-	-	miR-126	1.89×10^{-5}	−29.22

FDR method was used.

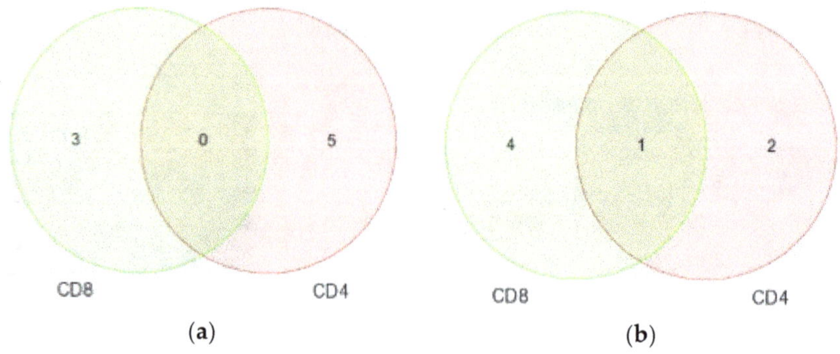

Figure 12. Venn diagrams showing DE miRNAs in the CD4+ and CD8+ cells of viremic (**a**) and LTNP (**b**) disease groups. FDR method with $p < 0.05$ was used.

For the first contrast, we found three DE miRNA for the CD4+ T-cells and five for the CD8+ T-cells during viremia. No overlapping miRNAs were seen. In contrast, for the LTNP CD4+ and CD8+ T-cells, four DE miRNAs in CD4+ T-cells and five in CD8+ T-cells during the non-progressive phase of the HIV disease were identified. The miRNA-199a-5p was found to be specific to the HIV+ CD4+ T-cells. Interestingly, miRNA-126 was the only miRNA found overlapping between CD4+ and CD8+ T-cells during the non-progressive HIV disease, which, according to the current literature, is the first miRNA possibly related to the non-progression of HIV disease (Table 7 and Figure 12b). No differential expression was seen for the aviremic comparison. This suggests a degree of similarity in the miRNA profiles of the CD4+ and CD8+ T-cells of aviremic patients undergoing HAART (Figure 12a,b).

2.8. Direct Comparison of DE miRNAs of the Same Disease State between CD4+ and CD8+ T-Cells

A direct comparison of CD4+ with CD8+ T-cells for the same disease group was carried out to obtain disease-stage-specific miRNAs common to both cell types. We performed comparisons between viremic CD4+ *vs.* CD8+ T-cells, aviremic CD4+ *vs.* CD8+ T-cells, LTNP CD4+ *vs.* CD8+ T-cells, and negative CD4+ *vs.* CD8+ T-cells using the FDR method to identify DE miRNAs for each disease group. Using the FDR method and an adjusted $p < 0.05$, only one miRNA was differentially expressed. This miRNA was miR-151-5p, and was differentially expressed between the CD4+ and CD8+ T-cells of the LTNP group with a fold change of 10.26 and a p-value of 1.95×10^{-5}, suggesting that it is a possible biomarker of non-progressive HIV disease based on both cell types.

Since defining a disease-stage-specific miRNA based on two different T-cell types is an important biological attribute in clarifying disease and cell type association, we also opted for using the non-FDR method with $p < 0.01$. By relaxing this statistical stringency, differential expression was observed for all contrasts (Figure 13a–d and Table 8). A number of disease-stage-specific miRNAs were identified for viremic, aviremic, LTNP, and HIV– stages based on both CD4+ and CD8+ T-cells (Table 8), including the previously mentioned miRNA miR-151-5p in the LTNP comparison. Notably, even HIV– individuals showed cell-specific miRNAs, suggesting distinct miRNAs guiding cellular immunity functions of CD4+ and CD8+ T-cells during the disease-free state. This is the first evidence showing this phenomenon.

Table 8. DE miRNAs found (using the non-FDR method) by comparing CD4+ with CD8+ T-cells for each disease group.

Disease State	miRNA	*p*-Value	Fold Change	Disease State	miRNA	*p*-Value	Fold Change
Aviremic	miR-151-3p	7.60×10^{-3}	3.7	**Viremic**	miR-3185	5.68×10^{-3}	−3.76
	miR-125b	3.93×10^{-3}	4.97		miR-1909	1.90×10^{-3}	−4.17
	miR-99a	1.39×10^{-3}	7		miR-125a-5p	5.26×10^{-3}	3.34
	miR-196a	7.99×10^{-4}	−5.52		miR-2277	7.56×10^{-4}	−2.63
	miR-152	6.15×10^{-3}	−5.79		let-7e	1.69×10^{-3}	3.05
LTNP	miR-151-5p	1.96×10^{-5}	10.27		miR-125a-3p	4.31×10^{-3}	3.45
	miR-1469	5.31×10^{-3}	−2.73		miR-572	4.01×10^{-3}	−3.27
	miR-486-5p	1.45×10^{-4}	12.22		miR-99b	3.82×10^{-3}	4.02
	miR-151-3p	7.78×10^{-3}	4.3		miR-3195	1.30×10^{-3}	−6.46
	miR-125b	9.64×10^{-5}	13.35		miR-885-5p	2.92×10^{-3}	−2.99
	miR-663	4.66×10^{-3}	−3.43		miR-4286	8.22×10^{-3}	−3.25
	miR-1909	5.81×10^{-4}	−7.25	**Negative**	miR-1469	4.61×10^{-3}	−3.25
	miR-1228 *	7.88×10^{-3}	−3.34		miR-663	5.08×10^{-3}	−4.09
	miR-885-3p	9.17×10^{-4}	−4.15		miR-99a	9.83×10^{-3}	7.11
	miR-1908	8.85×10^{-3}	−3.12		miR-199a-5p	4.15×10^{-3}	3.23
	miR-125a-5p	5.16×10^{-5}	10.29		miR-125a-3p	3.35×10^{-3}	6.1
	let-7e	1.19×10^{-3}	4.12		miR-570	7.42×10^{-3}	−2.8
	miR-125a-3p	9.93×10^{-4}	6.03		mir-520f	3.35×10^{-3}	2.43
	miR-92b *	1.76×10^{-3}	−4.44		miR-921	5.02×10^{-4}	2.05
	miR-1231	2.26×10^{-3}	−5.56		miR-142-5p	1.40×10^{-3}	3.39
	miR-99b	7.07×10^{-4}	7.8		miR-1911 *	1.52×10^{-3}	−2.25
	miR-146b-5p	5.95×10^{-3}	4.64		miR-1256	7.49×10^{-3}	−2.08
	miR-509-3p	1.65×10^{-3}	5.31		mir-941-3_s	1.12×10^{-3}	2.96
	miR-3195	4.66×10^{-3}	−7.14		miR-127-3p	4.70×10^{-3}	3.11
	miR-181a-2 *	6.21×10^{-3}	−7.8		miR-628-5p	3.58×10^{-3}	−3.9
	miR-3187	2.22×10^{-3}	−2.98		mir-620	4.41×10^{-3}	2.24
	miR-675	7.50×10^{-3}	−2.63		miR-601	2.40×10^{-3}	3.96
	miR-637	1.47×10^{-3}	−2.5		miR-513a-5p	4.45×10^{-3}	2.07
	miR-486-3p	8.41×10^{-3}	4.41				
	miR-532-3p	3.21×10^{-3}	3.64				
	miR-320e	6.89×10^{-3}	2.67				

* Less functionally defined miRNAs. Non-FDR method. For each cell type, viremic $n = 6$, aviremic $n = 5$, LTNP $n = 4$ and negative $n = 3$. Negative fold change implicates downregulation in CD4+ T-cells (Upregulation in CD8+) and positive fold change implies upregulation in CD4+ T-cells (downregulation in CD8+).

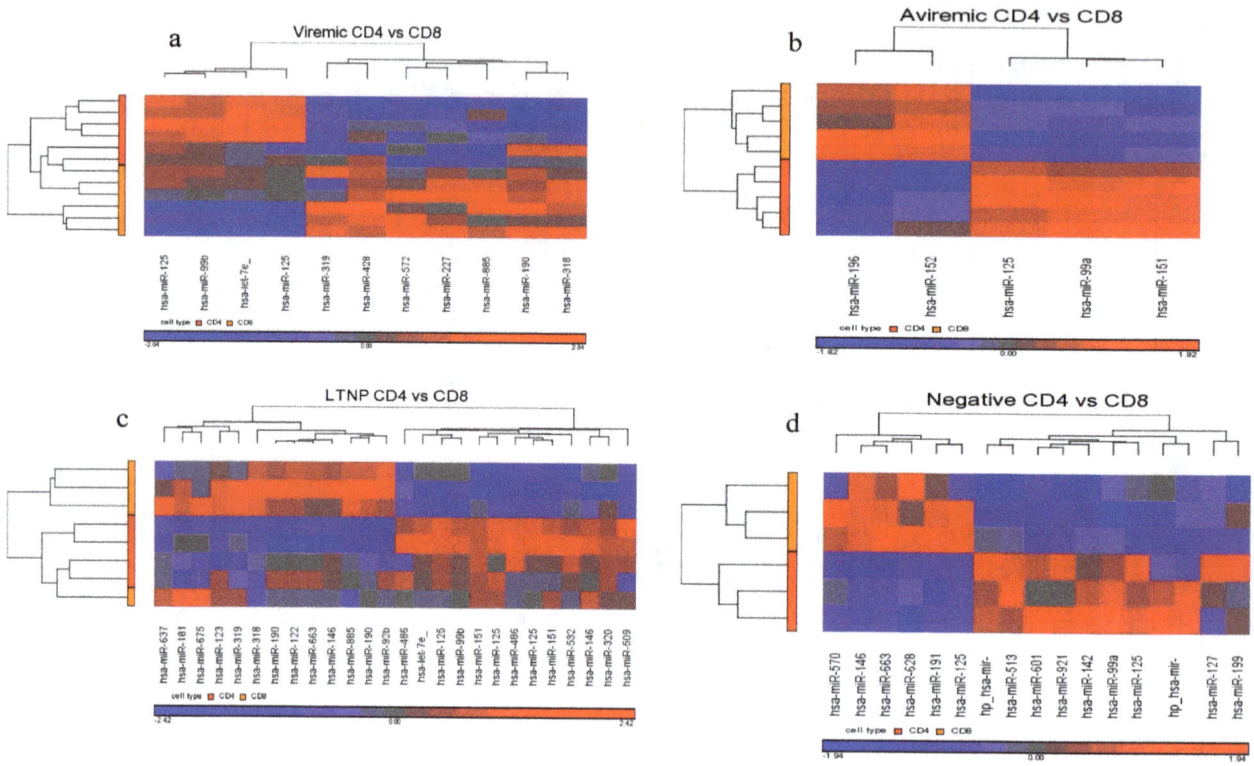

Figure 13. Heat map representation of miRNAs DE between CD4+ and CD8+ T-cells from disease groups viremic (**a**); aviremic (**b**); LTNP (**c**); and negative (**d**). Non-FDR method with $p < 0.01$ was used. For each cell type, viremic $n = 6$, aviremic $n = 5$, LTNP $n = 4$, and negative $n = 3$.

2.9. qRT-PCR Validation

Statistical significance of individual miRNA was used to pick two highly significant and important miRNAs to analyze by qRT-PCR. The fold changes determined for each comparison by qRT-PCR after an averaging of duplicates and calculation are shown in Table 9.

Table 9. qRT-PCR results for miRNAs of high significance chosen for validation along with associated microarray results.

miRNA	Analysis Where Significant	Adjusted *p*-Value	Fold Change from Microarray	Hypothesis/ Importance	Additional Significances from Non-FDR Analysis	Fold Change from qRT-PCR
126	CD4 LTNP *vs.* negative	1.76×10^{-6}	−51.73	LTNP infection marker based on both CD4+ and CD8+	CD4 HIV+ *vs.* HIV− (CD4 viremic and LTNP *vs.* negative, LTNP *vs.* aviremic)	−103.76
	CD8 LTNP *vs.* negative	5.46×10^{-6}	−31.25			−14.26
199a-5p	CD4 HIV+ *vs.* HIV−	1.22×10^{-6}	−5.6	HIV Infection marker for CD4+ T-cells	CD4 HIV+ *vs.* HIV− (CD4 viremic, aviremic and LTNP *vs.* negative)	−209.48
	CD4 viremic *vs.* negative	4.61×10^{-6}	−6.31			−2.73
	CD4 LTNP *vs.* negative	7.66×10^{-6}	−6.85			−50.34

The results of qRT-PCR shown in Table 9 show consistent trends between the results of microarray analysis and qRT-PCR analysis for two miRNAs (126 and 199a-5p). Raw amplification values are presented in Appendix, Table A6. For miR-126, the significant downregulation in LTNP individuals seen in the microarray (FC-50 for CD4+ T-cells and -30 for CD8+ T-cells) was supported by qRT-PCR analysis. Large negative fold changes were identified, particularly in CD4+ T-cells, where the fold change determined was greater than −100 fold.

miRNA-199a-5p was significant and found to be specific to HIV+ CD4+ T-cells in viremic and LTNP groups using the FDR method ($p < 0.05$), but was seen also in the aviremic groups only upon relaxing the FDR ($p < 0.01$).

The results of qRT-PCR for miR-199a-5p showed a consistent trend with the results of the microarray. In the CD4+ viremic *vs.* negative analysis, downregulation (FC-2.7) observed in qRT-PCR was consistent with the trend (FC-6.31). In the CD4+ T-cells—HIV+ *vs.* HIV− and LTNP *vs.* HIV negative comparisons—the qRT-PCR results also showed consistent trends of expression with the microarray. In the HIV+ *vs.* HIV− comparison, the microarray result of −5 fold downregulation was consistent with the qPCR result of −209. Again, in the LTNP *vs.* negative CD4+ T-cells, the downregulation detected in the microarray analysis (−7 fold) was also consistent in trend seen in qRT-PCR, where the detected fold change was −50 fold.

qRT-PCR analysis was also performed for the comparison CD4+ aviremic *vs.* negative for miR-199a-5p. This comparison was significant for the microarray only in the non-FDR analysis. The fold change determined by qRT-PCR was −2.15 fold (data not shown), which is again in agreement with the trend of downregulation seen for this miRNA in CD4+ HIV-infected T-cells in the microarray.

3. Discussion

Along with their association with pathological and physiological pathways, host miRNAs play a significant role in many viral infections, including HIV. In this work, we describe the first direct examination of virus-mediated cellular miRNA expression manipulation in the CD4+ and CD8+ T-cell fractions taken from HIV+ patients at differing stages of HIV disease. This disease-stage stratification was primarily based on the detectable virus in the plasma of diseased individuals and the length of infection. The main rationale behind conducting a parallel analysis of CD4+ and CD8+ T-cells for miRNAome was based on the roles of these cells in HIV infection. CD4+ T-cells are the primary targets of HIV [11], whereas the CD8+ T-cells mediate antiviral control replication through cytotoxic and non-cytotoxic mechanisms. The effectiveness of this response directly correlated with prognosis [12,13]. It is evident that the HIV infection of host cells leads to an alteration of the global RNA interference machinery [14]. Although it is possible that HIV-induced changes in cellular miRNA expression results from combinatorial molecular interactions among proteins, transcripts, and genomes, the mechanisms behind these changes during host-virus interactions remain to be understood. Here, we demonstrate the first evidence for cell-, disease-stage- and HIV-specific host miRNA during HIV disease stages.

Previous studies have shown unique and variable global modifications in microRNA expressions exhibited by different host cell types and cell lines in response to HIV-1 infection [6,7,15–17]. These findings showed a consistent and prominent trend of downregulation of cellular miRNAs in HIV-infected individuals. For instance, in 2008, Houzet *et al.* showed a prominent trend towards downregulation of miRNA in HIV+ individuals, with 59 of 62 DE miRNAs downregulated in HIV+ PBMCs [7]. This finding is in agreement with other studies performed on cell lines [6,15], which found that when HELA cell lines are transfected with HIV-1 in a pNL4-3 vector, >43% of cellular miRNAs were downregulated, with the majority of the remainder unchanged. Our study concurs with these findings and further demonstrates that the downregulation of cellular miRNAs was systematic across both CD4+ and CD8+ T-cells and diverse disease groups. This trend of cellular miRNA downregulation was confirmed in our qRT-PCR expression trends, suggesting the functional significance of downregulation of cellular miRNAs during HIV infection and guiding the disease course.

A large body of work on miRNAs has been done on cell lines rather than primary patient cells. It is important to iterate that miRNA signatures of PBMCs from HIV+ patient blood are more than 50% discordant with those of *in vitro*-infected PBMCs [7]. This implies that a significant portion of the miRNA signature changes seen in HIV+ patients cannot be accounted for by infection alone. These changes could be attributed to *in vivo* interaction between cell types and other combinatorial interactions, which together play a functional role in determining the dysregulation of host miRNAs during HIV infection. Thus, the analysis of primary CD4+ and CD8+ T-cells we have performed from diverse HIV disease groups carry considerable functional relevance in the context of HIV infection *in vivo*.

To obtain differentially and statistically significant miRNA candidates, we employed the Benjamini–Hochberg FDR corrective testing with a p-value cut-off of <0.05 throughout. The Benjamini–Hochberg algorithm is designed to sort through a list of p-values and determine the likelihood of a false rejection of the null hypothesis (a false positive) and so "adjust" the p-value to control the likelihood of the false positive. Employing this algorithm provided data of greater integrity, and this is the data we have made use of in interpreting our DE dataset. Due to the stringency of this method of analysis, we failed to detect differential expression of miRNAs in some comparisons, while the DE miRNAs detected for other groups were statistically strong. To clarify any possible occlusion of biologically relevant miRNAs, we also analyzed the dataset without FDR corrective testing. To minimize false positives in the absence of FDR correction, the p-value cut-off was then set at <0.01. This allowed the detection of other DE miRNAs for the groups we failed to detect DE miRNAs using FDR corrective testing. Our hypothesis for using this rationale was that the statistical stringency possibly occludes biologically relevant miRNAs, as several of these miRNAs identified at $p < 0.01$ are not only of functional relevance, but are also known in the literature.

3.1. *Signature miRNA in HIV Disease Class Prediction*

Through the use of PCA, we demonstrated that the miRNA profiles of HIV+ CD4+ and CD8+ T-cells were distinct from T-cells of HIV– individuals as all HIV-infected groups (viremic, aviremic, and LTNPs) were more closely related as a group and segregated from the uninfected group. PCA also highlighted that the individual samples from the same disease stages cluster together, and the sample classes with the greater similarities were more closely associated. Despite the close relationship of all HIV+ samples, PCA could also distinguish between HIV disease sub-groups, with the LTNP, aviremic, and HIV groups showing a common miRNA signature specific to them. This is plausible, as these three groups have undetectable levels of plasma viremia. Our results can be likened to the results shown by Witwer *et al.* on PBMCs, where the miRNA signatures of LTNP HIV+ individuals with low viral load and good CD4+ T-cell counts showed greater similarities to uninfected controls than HAART-treated HIV+ patients.

In support of these findings, several previous papers have suggested that a broad statistical analysis of the miRNAs that are differentially expressed in HIV+ individuals distinguish patients according to their CD4+ T-cell counts, viremia, or HIV exposure level [7,9,15]. In 2008, Houzet *et al.* showed that HIV+ patients could be categorized into disease groups based on DE miRNAs present in the PBMC fraction of their blood. Further, the miRNA signature of CD4+ T-cells from multiply exposed uninfected (MEU) individuals can distinguish them from therapy-naïve infected CD4+ T-cells [15]. Bignami *et al.* also showed that miRNA signatures of LTNPs, MEUs, and therapy-naïve HIV+ patient CD4+ T-cells cluster based on their exposure level and disease state [15]. As mentioned, Witwer *et al.* showed that, through examination of DE miRNAs, LTNPs with low CD4+ T-cell counts and high plasma viremia correlate more closely with HIV+ viremic individuals, and those with high CD4+ T-cell counts correlate with HIV− controls [9]. Thus, the characteristics of an miRNA signature allowed accurate assignment of samples into disease states using a variety of predictive programs [9], which is what our analysis has shown at both cell-type and disease-stage levels.

3.2. *DE miRNAs in CD8+ T-Cell Predict HIV Disease Stage Better than CD4+ T-Cells*

When DE miRNAs were analyzed in the CD8+ T-cells of HIV+ individuals from each of the disease groups (viremic, aviremic, and LTNP), it was shown that some miRNAs were highly specific for these disease groups when compared with negative controls. Namely, miR-1323, -541, and -570 were associated with the CD8+ viremic group, miR-1298 was associated with the CD8+ aviremic group, and the miRNA-572, -126, 151-5p, -486-5p, and -885-3p with the CD8+ LTNP group. This is the first direct evidence of HIV disease-stage-specific miRNAs in CD8+ T-cells, which is consistent with the role of CD8+ T-cells in containing viremia. A clear segregation of DE miRNA in each comparison demonstrates that the CD8+ T-cells are better suited to predicting HIV disease stage and cell-specific miRNAs than their CD4+ T-cell counterparts.

3.3. Unique miRNAs in Non-Progressive HIV Disease

Our analysis performed in CD4+ and CD8+ T-cell types suggests that miRNAs-199a-5p, -1280, and -126 were differentially expressed in the CD4+ T-cells of LTNPs when compared to negative controls, and the miRNAs-151-5p, -486-5p, -572, -885-3p, and -126 were differentially expressed in the CD8+ T-cells of LTNPs when compared to negative controls. MiR-126 was differentially expressed in both CD4+ and CD8+ T-cells in the LTNP disease state. Consistent expression trends between microarray and qRT-PCR for miR-126 suggests that this miRNA is significantly downregulated in both the CD4+ and CD8+ T-cells of LTNPs, suggesting that it is a possible biomarker for non-progressive HIV disease. This, to our knowledge, is the first report of a miRNA that correlates with the LTNP disease state and may be involved in the regulation of cellular immunity and possibly the natural control of HIV in LTNPs.

MiR-199a-5p was differentially expressed in HIV+ CD4+ T-cells belonging to all HIV+ groups analyzed (viremic, aviremic, and LTNPs). Stronger downregulation was observed in LTNP *vs.* negative disease group than in either the viremic or aviremic CD4+ T-cell comparisons against HIV− individuals. Thus, it is likely that miR-199a-5p plays a role in predicting HIV+ status based on CD4+ T-cells. We hypothesize that, although it is expressed and downregulated in the CD4+ T-cells of all HIV+ groups, the degree of downregulation of miR199a-5p in different groups imparts susceptibility or protection, as evident from the CD4+ T-cells from the LTNPs that showed the highest degree of downmodulation of miRNA199-5p, which may play a protective role against the virus. Supporting this conclusion, it has previously been shown that miR-199a (along with miRNAs-143, -303-3p, and -335) is differentially expressed in the PBMCs of class 1 (HIV+, high CD4+ T-cell count, low viral load) individuals, which agrees with our work with miRNA 199a showing the strongest downregulation in the LTNPs group. Our analysis using three phenotypically different HIV+ groups clarifies and unambiguously demonstrates a clear association of miR-199a with HIV positivity based uniquely on CD4+ T-cells, a trend which was further confirmed by qRT-PCR analysis, implicating miR-199a to be an HIV-specific miRNA.

About other differentially expressed miRNAs, miRNA-155 in HIV+ CD4+ T-cells was recently observed to distinguish LTNPs from therapy-naive HIV+ patients [15]. They suggest that high miR-155 in CD4+ T-cells may correlate with HIV-1 pathogenesis. Our analysis showed similar trends to those of studies by Bignami *et al.* and Witwer *et al.*

3.4. The miR-125 Family Is Important in the Maintenance of Latency and Aviremia

The 3′ ends of HIV-1 mRNAs (Nef) are targeted by cellular miR-125b (along with miR-28, miR-150, miR-223, and miR-382), all of which are enriched in resting CD4+ T-cells compared to active CD4+ T-cells [8]. These miRNAs have inhibitory effects and enforce viral latency in resting HIV+ CD4+ lymphocytes by reducing protein translation and viral production [8]. Moreover, miR-125b has been identified as one of the constitutive miRNA signatures of naïve CD4+ T-cells [18]. MiR-125a-3p, along with miR-23 and miR-155, is shown to be upregulated in HIV+ therapy-naïve CD4+ T-cells, normal in LTNP CD4+ T-cells, and downregulated in the CD4+ T-cells of MEU samples [15]. A recent study involving PBMC miRNA profiling [9] did not find miR-125b correlating with replication control and latency; however, this study involved only the whole PBMCs and not the individual cell subsets [18,19], which may harbor unique miRNA characteristics shown in our study.

Interestingly, our results show that the family of miR-125 was consistently upregulated in the comparison of diseased CD4+ T- and CD8+ T-cells using a non-FDR statistical analysis. This upregulation was markedly higher in the CD4+ T-cells of LTNPs than those of viremic or aviremic individuals, followed by its upregulation in aviremic individuals, suggesting a possible role for this family of miRNAs in aviremia and natural control and possibly viral latency. Our results agree in part with the findings by Huang *et al.*, and we concur that miR-125 family members are potential candidates associated with latency in CD4+ T-cells [8]. More work is needed to prove the biological relevance of this miRNA. It is important to iterate that this differential expression was only achieved when less stringent corrective testing was applied. Nonetheless, the fold changes and *p*-values of miRNA candidates with the disease stage added reliability and confidence to these conclusions.

3.5. Possible Role of miR-29 Family in Viral Replication and Management of Viremia during HAART Treatment

miR-29 targets the viral Nef 3′UTR and downregulates Nef protein expression and interferes with HIV-1 replication, thereby inhibiting viral production and infectivity in HIV-infected T lymphocytes [7]. Low levels of miR-29b are associated with disease progression and viremia [9,10], whereas high expression of miR-29b positively correlates with CD4+ T-cell count. Furthermore, miR-29a specifically targets the HIV-1 3′UTR [10,20,21]. This implies a role for miR-29 members in prohibiting viremia and disease progression [9]. Recently, it has been discovered that miR-29 suppresses HIV-1 replication and binds directly to HIV-1 Nef mRNAs [22]. An indirect role for miR-29 has also been suggested in the modulation of cyclins in HIV infection [23].

Although our results did not immediately suggest a role for miR-29 in managing viremia and disease progression, it became clear upon examination that miR-29 was indeed upregulated in aviremic CD4+ T-cells, when compared to both viremic and LTNP CD4+ T-cells. Although it is peculiar that the miR-29s did not show association with the LTNP disease state, this is not unexpected. The biggest difference between HIV+ aviremic and LTNP individuals is the preservation of immune system strength and natural control of HIV in LTNP individuals, where aviremic individuals require HAART to control HIV replication. It is likely that miRNAs may have some role in maintaining below detectable levels of plasma viremia while on HAART therapy. On another note, it is important to iterate that we also found miR-1298 specific to the CD8+ aviremic group (with respect to negative control), showing 2-fold downregulation. This was obtained using the FDR corrective testing, and we believe it is the first miRNA specific to CD8+ T-cells of HIV patients that may be involved in control HIV during HAART therapy and may offer some biological significance. Thus, these two miRNAs may have some functional relevance during HAART therapy, and whether they come into play when therapy is effective in controlling plasma viremia remains unknown. Nonetheless, the two miRNAs in some ways predict the efficacy of HAART and should be explored in the context.

Similarly, in aviremic CD8+ T-cells, miR-29b/c were upregulated when compared to negative controls, whereas the miR-29a/b/c were upregulated in aviremic individuals when compared to LTNPs. Taken together, these data suggest that the expression of the cellular miR-29 family is triggered by viral replication and implies a potential role for miR-29 family members in both CD4+ and CD8+ T-cell subsets in HIV infection, although perhaps not in the management of viremia in the LTNP group. Determining whether miR-1298 and the miR-29 family have some functional synergy between them requires future functional validation.

3.6. miRNA-17/20 Are Downregulated in CD8+ T Cells in LTNPs: A Tentative Link to Their Role in Viral Replication

miR-17/20 target 3′UTR of PCAF acetyltransferase and are known to indirectly inhibit HIV infection [7,8]. It has been shown that the transfection of miR-17-5p and -20 into the PBMCs of HIV-1 infected donors leads to reduced HIV-1 production, and that transfection of antisense inhibitor enhances PCAF expression and HIV-1 production [24]. PCAF1 is an important cofactor for tat in HIV-1 gene expression and has potential sites for these miRNAs in its 3′UTR. Recently, miR-17a has been associated with enhanced viral production and predicted to bind cellular factors that indirectly influence HIV-1 replication [22].

We found that miR-17 and miR-20a were both downregulated in the CD8+ T-cells of LTNPs when compared directly with the CD8+ T-cells from aviremic patients. MiR-17 was downregulated in CD8+ T-cells from the LTNPs when compared against the viremic patients. This upregulation of miRNAs could be related to the inhibition of HIV-1 in a more severe disease state (aviremic on HAART compared to LTNPs), rather than the downregulation of these miRNAs by HIV.

As can be noted in several sections of the discussion, the correlation of DE miRNAs observed in our study could only be compared with other studies when non-FDR corrective testing was used. Thus, it is important to reiterate that, since we performed a bimodal analysis of our dataset using FDR ($p < 0.05$) and non-FDR ($p < 0.01$) statistical analysis, we were able to observe both the strongly DE miRNAs responsible for clear segregation of disease groups using FDR analysis, as well as the gentler trends of miRNA expression between disease classes and cell types in the non-FDR analysis. Our observations regarding the two datasets raise two possibilities: that either most studies in the miRNA field are not as stringently controlled as our primary FDR analysis, or that the stringency we have employed in deriving significantly DE miRNA is too strict. If this is the case, high stringency analysis may occlude biologically significant miRNA candidates, as was apparent when functionally important miRNA deemed important in the literature were confirmed and supported by our non-FDR dataset.

In summary, the findings presented herein provide evidence that HIV infection plays a significant role in altering cellular miRNA expression and, conversely, that cell-specific miRNA expression may have implications for the control of HIV infection. We report the findings of a variety of miRNAs that are differentially expressed in a cell-, disease-stage- and HIV-infection-specific manner, and offer support to the findings of several other studies. This is the first evidence showing tight miRNA regulation during HIV disease in different cell types and disease stages, suggesting an essential role for miRNA-guided cell phenotype in HIV pathogenesis and the development of AIDS. A unique aspect of these findings is the evidence for miRNAs that are involved in HIV resistance, as seen in non-progression. Further study of the roles of these miRNAs, their cognate mRNAs, and the reason for their dysregulation in HIV infection may shed light on the mechanisms controlling the progression or non-progression of HIV disease, and open up a pathway to a new line of therapeutics for the effective treatment and control of HIV.

4. Materials and Methods

4.1. Fresh PBMC Isolation from Whole Blood

Up to 10 mL of freshly collected whole blood in EDTA was diluted to 35 mL with DPBS (Dulbecco's phosphate buffered saline, Lonza BioWhittaker, Walkersville, MD, USA) and layered carefully onto 15 mL of Ficoll (Ficoll-Paque PLUS, 1.077 ± 0.001 g/mL, GE Healthcare, Uppsala, Sweden) in 50-mL tubes, without disrupting the Ficoll border. Separation of the PBMC buffy coat was achieved via centrifugation at 1800 rpm for 20 min at room temperature. The buffy coat containing peripheral blood mononuclear cells (PBMCs) was extracted to another 50-mL tube via transfer pipette. The separated PBMCs were washed twice with DPBS by centrifugation at 1500 rpm for 7 min at 4 °C, then once in 10 mL of cold (4 °C) magnet activated cell sorting method (MACS) buffer (DPBS pH 7.2, 2 mM EDTA (Amresco, Sydney, Australia), 0.5% FBS (Fetal Bovine Serum, SAFC Biosciences, Lenexa, KS, USA) via centrifugation at 1200 rpm for 10 min at 4 °C. The supernatant was aspirated retaining the PBMC pellet.

4.2. Processing of Cryopreserved PBMCs

PBMCs from the LTNPs used in this study were previously cryopreserved. These cells frozen in liquid nitrogen were raised to −80 °C for one day before revival. Cells were slowly thawed on ice and were transferred using a transfer pipette into 15-mL tubes containing 5 mL of RPMI 1640 (Lonza

BioWhittaker) 10% FBS in 15-mL tubes. The cell suspension was centrifuged for 10 min at 1000 rpm at room temperature to pellet the revived cells. The supernatant was aspirated and decanted retaining the clean cell pellet in preparation for the CD4+ and CD8+ cell separation.

4.3. *Macs Column Separation of CD4+ and CD8+ T Cells from Whole PBMC*

The PBMCs were centrifuged at 1000g to remove cell debris. The cell pellet was further washed twice with phosphate buffered saline (PBS) to remove excess antigen. After aspirating all of the PBS, the cell pellet was further washed with 10ml of MACS buffer (50 mL PBS + 250 μL of Fetal Bovine Serum (FBS) + 200 μL of 2 mM EDTA) by centrifuging at 300 *g* for 10 min at 4 °C. Cells were washed with PBS, and MACS® Column Technology was used for separating PBMCs into different cell subsets. Using MACS® MicroBeads, $CD4^+$ helper T-cells and $CD8^+$ cytotoxic T-cells were extracted according to the manufacturer's specifications using the positive selection MACS, Miltenyi Biotech (Marburg, Germany). Flow cytometric verification of cell purity of CD4+ and CD8+ T-cells was evaluated using flow cytometry, as previously described by us (Potter *et al.*, 2006).

4.4. *RNA Extraction and miRNA Enrichment*

Both total RNA and miRNA were extracted from the brain cortex using the miRNeasy Mini Kit (Qiagen Pty Ltd., Clifton Hill, Victoria, Australia) with an integrated step of on-column DNase digestion as per the manufacturer's protocol for the kit.

4.5. *RNA Quality Check and Quantification*

RNA quality was checked by RNA micro-electrophoresis on a 2100 Bioanalyzer apparatus (Agilent Technologies, Oxford, UK) and by using the RNA 6000 Nano Labchip Technology Kit. All reagents were equilibrated to room temperature for 30 min before use, and the quantity and quality of RNA were assessed based on the RNA integration numbers (RIN) for each sample. All RNA samples used in q-PCR and miRNA analysis had an RIN of >5, and anything <5 was deemed inappropriate for the miRNA chip, but were considered for the q-PCR.

4.6. *FlashTag Biotin HSR RNA Labeling and Affymetrix GeneChip miRNA Arrays*

MicroRNA was prepared for Affymetrix GeneChip arrays using the FlashTag® Biotin HSR RNA Labeling Kit for Affymetrix® GeneChip® miRNA Arrays (Genisphere, Hatfield, PA, USA). miRNA array preparation was conducted according to instructions provided by Genisphere for the Affymetrix GeneChip miRNA array procedure (FlashTag Biotin HSR RNA Labeling Kit for Affymetrix GeneChip miRNA Arrays), including the use of reagents from this kit (vials numbered). Hybridization of prepared samples, incubation, washing, staining, and scanning was completed by the sample submission service at the Ramaciotti Center for Gene Function Analysis, University of New South Wales, Sydney.

4.7. *Analysis*

4.7.1. Sample Classification and Criteria

Patients were classed into disease groups based on their HIV plasma VL determined by M3 Prep/Cobas TaqMan HIV-Version 2 kit (ROCHE, Inc., Pleasanton, CA, USA) to test HIV plasma VL, at the Westmead Hospital ICPMR (Institute of Clinical Pathology and Medical Research). Diverse HIV disease groups were analyzed, including HAART-experienced viremic (VL above 40 copies/mL plasma) and aviremic (below detectable levels of plasma viremia) patients. An LTNP group, comprising of HIV+ therapy-naïve individuals who showed high CD4+ and CD8+ T-cell counts and below detectable plasma viremia in the absence of HAART, was also used. All comparisons between disease groups were made against HIV– healthy individuals. Thus, the main criterion for analysis was the stratification of patients into viremic, aviremic, LTNP, and HIV– healthy groups, with the aim of defining disease-stage-specific and cell-specific (CD4+ and CD8+ T-cells) miRNA.

4.7.2. Normalization, QC, and PCA Analysis

The Analysis was performed in the Partek Genomics Suite using inbuilt data normalization for Affymetrix CEL files (consisting of background correction, quantile normalization, Log_2 transformation, and median polish summarization of probesets). The quality of the miRNA dataset and its integrity was assessed using PCA. PCA is a mathematical operation involving applied linear algebra, which transforms datasets into a new coordinate system such that the most variable components of the dataset are presented on axes where each component is independent of the last. This allows a complex dataset to be simplified and visualized by its most important gradients. In turn, the analysis also allows the determination of which factors are the most important in distinguishing one cluster from another. PCA plots were generated, to first test the integrity of infected samples in comparison to uninfected samples, and then PCA was used to inter-compare the disease states for each cell type and the cell types for each disease state (Figures 1–3). Frequency histograms were also generated for both the CD4+ and CD8+ T-cell subsets, showing the effectiveness of normalization applied by the Partek Genomics Suite and concordance of the datasets (Figure 3).

4.7.3. Statistical Analysis

Samples were analyzed using the Partek Genomics Suite software, applying two forms of corrective and statistical testing. Primarily, the analysis of differential expression was completed using *t*-tests followed by Benjmini–Hochberg FDR corrective testing. The Benjamini–Hochberg algorithm sorts a list of *p*-values in ascending order and determines how likely value is to be a false rejection of the null hypothesis (a false positive) at each point in the list. In this way, it develops a list of corrected or adjusted *p*-values for the detected differentially expressed miRNAs. A fold change of 2 (up- or downregulation) was also applied to deem any miRNA to be biologically and statistically significant. Although not a part of the result section, data were also processed without FDR corrective testing to see the difference between two datasets. This was to determine whether the statistical manipulation of the data using the FDR method occluded biologically or functionally significant miRNAs. For non-FDR data, a stricter $p < 0.01$ significance threshold was applied for significant changes, along with the fold change of 2, to ensure biological significance. This method will herein be referred to as the non-FDR method and is used as a comparative set of data for the FDR results.

Notably, even though non-FDR testing showed a higher number of DE miRNAs than FDR-controlled corrective testing, all microRNAs deemed significant using strict FDR corrective testing were also picked up as highly significant miRNAs in the less stringent, non-FDR method, confirming the reliability of FDR corrective testing.

Comparisons of miRNA expression were made in an organized and structured manner. First, differential expression of miRNAs between HIV+ and HIV– individuals was considered. Then, comparisons were made between CD4+ and CD8+ T-cells before examining the differential expression between disease states within CD4+ and CD8+ T-cells. Lastly, results observed in independent CD4+ and CD8+ T-cell analyses were inter-compared. Analyses included clustering, the generation of heat maps, and intergroup comparisons through the generation of Venn diagrams.

4.7.4. Quantitative Real-Time PCR Validation of Selected and Significant DE miRNAs

Two miRNAs of significance and importance were selectively corroborated using quantitative real-time PCR (qRT-PCR) based on their adjusted *p*-values, fold changes, and differential expression. cDNA was generated from RNA samples using the Qiagen miScript Reverse Transcription Kit (Qiagen, Sydney, Australia). A master mix containing 160 μL of buffer RT and 40 μL of miScript reverse transcriptase mix was produced, and 5 μL was added to each sample of 250-ng RNA. The samples were made with RNase free water (up to 20 μL) and incubated at 37 °C for 60 min, and then 95 °C for 5 min to inactivate the RT enzyme using the Hybaid Thermal Cycler for PCR. Expression levels of two differentially expressed miRNAs (hsa-miR-199a and miR-126) were validated by qRT-PCR using the Qiagen miScript PCR system (Qiagen, Valencia, CA, USA). Master mixes containing SYBR green PCR master-mix (10 µL/assay), miScript universal primer (2 µL/assay), and RNase-free water (5 µL/assay) was produced for each specific primer. Unique miScript primers specific for the miRNAs of interest were added (2 µL/assay). Data was analyzed using Microsoft Excel. RNU6B control was used to normalize data. All the experiments were performed in duplicate, and relative expression levels of these miRNAs were determined by the $\Delta\Delta{-}C_T$ method (see Table 9 for qRTPCR results).

5. Conclusions

MicroRNAs are the master regulators of gene expression; therefore, this analysis of differentially expressed miRNA in CD4+ and CD8+ T-cells at different stages of HIV infection provides highly valuable data on how these two T-cell subsets interact during HIV infection. Our analysis provides data on how miRNAs may guide disease progression: leading to gradual disruption and decline of the immune system in viremic individuals or providing possible resistance against HIV infection, as seen in LTNPs.

Determining the role of differentially expressed miRNAs in HIV infection is essential. We believe that an understanding of the dysregulated miRNAs in HIV infection will prove crucial for the development of miRNA-based therapies or the targeting of cellular miRNA as treatment for HIV infection. Before this is possible, a clear understanding of miRNA gene targets is needed. This is a challenge, as one miRNA can target thousands of genes, and HIV infection of cells can induce large changes in gene expression and overall cell state, meaning that the relationships between miRNAs and mRNAs in HIV infection may be very complex. As such, more work is needed to map precisely the mRNA and miRNA interactions that guide HIV pathogenesis. Such analysis will not only allow the development of new strategies for HIV control but may also lead to the identification of miRNA candidates that can be used as a new generation of therapeutic targets to control HIV. For example, a miR-122 agonist, miravirsen, is the first miRNA-targeted drug to be developed for the treatment of viral accumulation in hepatocytes during HCV infection [25].

With an increase of research in the area of miRNA expression profiling and functional analysis, several confounding factors have emerged. Currently, a variety of papers in this area all used different cell types, different definitions of HIV disease states, different methods of analysis and statistical evaluation, and diverse microarray platforms [7–9,15,21,23]. With all these variances in methods used by different studies, it is no surprise that the majority of research in this area is not currently in agreement. A consensus on miRNAs that is relevant in HIV disease will provide a durable framework for future work involving functional assays and miRNA/mRNA interaction analysis. Further, these clarifications will provide a clear view to developing new generation therapeutics to treat HIV disease. What is significant is that these analyses show miRNA that is specific to the natural control of HIV disease in LTNPs, which may be of vital importance in designing a new generation of prognostics and diagnostics for HIV disease.

Acknowledgments: No funding was received for this project, as it was done as a pilot project for the Hons thesis of Lauren Fowler submitted to the University of Sydney, Australia.

Author Contributions: L.F. performed the experiments, analyzed the dataset, wrote the manuscript, and presented it for her Hons degree defense; V.C. and S.P. assisted with the cell processing and separation, and the cell sorting on the FACS and PC3 handling and processing of patient samples; P.G. guided the bioinformatics part of the work and functional analysis using RT PCR; W.D. and C.B.C. provided all the clinical samples from their repository along with the clinical profiles; N.K.S. conceptualized the study, planned it, guided it, and was actively involved in the analysis of dataset and the writing of the manuscript.

Abbreviations

AIDS	acquired immunodeficiency syndrome
LTNP	long-term non-progressors
qRT-PCR	quantitative reverse transcriptase PCR
PCR	polymerse chain reaction
PBMC	peripheral blood mononuclear cells
miRNA	micro-ribonucleic acid
FDR	false discovery rate
PCA	principal component analysis
DE	differential expression
RIN	RNA integration number
BDL	below detectable levels

Appendix

Supplementary results are presented below, for comparison with FDR method derived results presented in results section. These results are for assessing differential expression of miRNAs between cell types and different HIV disease groups.

Table A1. Differentially expressed miRNAs in HIV+ CD4+ and CD8+ cells compared to negative controls. Non-FDR method. Infected n = 15, Uninfected n = 3 for each cell type.

Cell type	miRNA	p-Value	Fold Change
CD4	miR-199a-5p	1.22×10^{-6}	−5.60
	miR-720	6.61×10^{-5}	3.69
	miR-1469	7.90×10^{-3}	2.91
	miR-127-3p	9.48×10^{-5}	−3.34
	miR-130a	2.10×10^{-3}	−3.77
	miR-487b	5.00×10^{-4}	−2.64
	miR-129-3p	3.00×10^{-4}	−2.40
	miR-483-5p	7.00×10^{-4}	−3.28
	miR-520f	3.00×10^{-4}	−2.58
	miR-601	2.40×10^{-3}	−2.77
	miR-126	5.40×10^{-3}	−10.61
	miR-584	7.80×10^{-3}	−2.78
	miR-432	5.80×10^{-3}	−2.73
	miR-142-5p	1.20×10^{-3}	−2.89
	miR-1280	2.00×10^{-3}	2.33
	miR-139-5p	3.00×10^{-3}	−2.78
	miR-3137	4.40×10^{-3}	−2.01
	miR-606	3.50×10^{-3}	−2.57
	miR-548i-4	3.50×10^{-3}	−2.40
	miR-195	3.40×10^{-3}	−4.21
CD8	let-7e	4.00×10^{-3}	−3.02
	miR-941-3_s	5.83×10^{-5}	2.77
	miR-1323	9.14×10^{-5}	−3.72
	miR-4321	6.00×10^{-3}	2.42
	miR-572	9.20×10^{-3}	3.78
	miR-570	9.00×10^{-4}	−3.17
	miR-1256	1.10×10^{-3}	−2.12
	miR-4284	1.90×10^{-3}	4.95
	miR-890	2.90×10^{-3}	−2.58
	miR-589 *	9.10×10^{-3}	2.57
	miR-603	6.70×10^{-3}	−2.36

* Functionally undefined miRNA.

Table A2. Differentially expressed miRNAs of CD4+ T-cells in each disease state when compared to negative controls.

Disease State	miRNA	*p*-Value	Fold Change
LTNP	miR-126	1.78×10^{-6}	−51.73
	miR-1469	3.85×10^{-4}	4.24
	miR-151-3p	1.09×10^{-3}	−7.4
	miR-3141	1.28×10^{-3}	2.06
	miR-663	1.35×10^{-3}	4.68
	miR-99a	6.41×10^{-3}	−7.04
	miR-3185	4.20×10^{-3}	6.19
	miR-762	1.40×10^{-3}	4.54
	miR-1228 *	2.05×10^{-3}	4.69
	miR-3196	2.80×10^{-3}	3.62
	miR-149 *	9.18×10^{-3}	5.74
	miR-4281	4.99×10^{-4}	4.12
	miR-638	1.51×10^{-3}	3.42
	miR-1908	5.29×10^{-3}	3.75
	miR-199a-5p	7.66×10^{-6}	−6.85
	miR-1915	1.24×10^{-3}	4.16
	miR-1268	3.76×10^{-3}	2.94
	miR-2861	9.38×10^{-4}	4.82
	miR-1280	9.42×10^{-6}	3.71
	miR-199b-3p	3.89×10^{-3}	−4.6
	miR-195	6.18×10^{-4}	−5.44
	miR-1910	4.18×10^{-3}	3.57
	miR-3175	6.93×10^{-4}	8.47
	miR-3163	7.54×10^{-3}	2.75
LTNP (cont'd)	miR-1972	3.33×10^{-3}	5.2
	miR-363	2.38×10^{-3}	−11.54
	miR-520f	7.11×10^{-4}	−2.68
	miR-142-5p	2.40×10^{-4}	−3.88
	miR-130a	9.21×10^{-3}	−3.61
	miR-1207-5p	5.93×10^{-3}	2.25
	miR-720	2.56×10^{-4}	4.47
	miR-302d	1.10×10^{-4}	−2.19
	miR-199a-3p	4.81×10^{-3}	−5.31
	miR-182	1.39×10^{-3}	−2.69
	miR-197	2.79×10^{-3}	2.17
	miR-155 *	2.09×10^{-3}	2.7
	miR-663b	4.23×10^{-4}	4.61
	mir-151_x	1.97×10^{-3}	−2
	miR-1260b	3.25×10^{-3}	2.73
	miR-127-3p	6.51×10^{-4}	−3.77
	miR-620	2.38×10^{-3}	−2.25
	miR-548i-4	9.59×10^{-3}	−2.41
	miR-100	3.02×10^{-3}	−6.28
	miR-663b_x	3.31×10^{-3}	2
	miR-487b	2.88×10^{-3}	−2.84
	miR-584	3.31×10^{-3}	−4.14
	miR-519e	4.35×10^{-3}	−2.18
	miR-134	9.11×10^{-3}	−2.15
Viremic	miR-126	4.42×10^{-4}	−11.24
	miR-1469	9.01×10^{-4}	3.43
	miR-663	3.15×10^{-3}	3.66
	miR-99a	6.66×10^{-4}	−10.45
	miR-1228 *	4.92×10^{-3}	3.61
	miR-149 *	5.54×10^{-3}	3.72
	miR-199a-5p	4.61×10^{-6}	−6.31
	miR-1280	4.41×10^{-4}	2.45
	miR-195	8.72×10^{-5}	−6.44
	miR-3175	1.40×10^{-3}	6.29
	miR-483-5p	7.50×10^{-5}	−4.28
	miR-548q	1.04×10^{-3}	−2.82
	miR-520f	1.02×10^{-4}	−2.96
	miR-142-5p	1.49×10^{-4}	−3.7
	miR-130a	2.84×10^{-4}	−5.83
	miR-378	2.36×10^{-3}	2.28
	miR-720	4.92×10^{-4}	3.69
	miR-574-3p	2.43×10^{-3}	3.19
	miR-23a *	8.67×10^{-3}	3.48
	miR-510	9.44×10^{-3}	−2.59
	miR-127-3p	1.02×10^{-3}	−3.24
	miR-3172	9.57×10^{-3}	−2.9
	miR-548i-4	4.96×10^{-3}	−2.44
	miR-129-3p	7.83×10^{-4}	−2.58
Viremic Cont'd	miR-3137	8.47×10^{-4}	−2.56
	miR-487b	6.03×10^{-3}	−2.41
	miR-574-5p	9.77×10^{-3}	2.64
	miR-1308	7.95×10^{-3}	2.71
	miR-601	1.98×10^{-3}	−3.38
	miR-23b *	8.30×10^{-3}	−2.03
	miR-765	2.19×10^{-3}	−2.35
	miR-1225-3p	5.74×10^{-3}	2.11
	miR-214	9.80×10^{-3}	−2.08
Aviremic	miR-199a-5p	2.34×10^{-4}	−4.13
	miR-483-5p	1.20×10^{-3}	−3.21
	mir-520f	4.79×10^{-3}	−2.13
	miR-720	2.26×10^{-3}	3.16
	miR-127-3p	1.76×10^{-3}	−3.14
	miR-548i-4	8.26×10^{-3}	−2.36
	miR-129-3p	1.70×10^{-3}	−2.47
	miR-487b	2.33×10^{-3}	−2.78
	miR-606	3.81×10^{-3}	−3.05
	miR-517a_x	7.91×10^{-3}	−2.25
	miR-134	9.60×10^{-3}	−2.06

* Non-FDR method. Viremic $n = 6$, aviremic $n = 5$, LTNP $n = 4$, negative $n = 3$.

Table A3. Differentially expressed miRNAs when disease classes of CD4 cells are intercompared.

CD4	miRNA	p-Value	Fold Change	Disease State	miRNA	p-Value	Fold Change	Disease State	miRNA	p-Value	Fold Change
Viremic *vs* Aviremic	miR-99a	1.68×10^{-4}	−9.76	Viremic *vs* LTNP	miR-143	8.46×10^{-3}	6.73	LTNP *vs* Aviremic	miR-223	3.66×10^{-3}	−5.08
	miR-1307	3.11×10^{-3}	2.15		miR-3163	1.10×10^{-4}	−3.8		miR-197	7.76×10^{-4}	2.19
	miR-744	8.69×10^{-5}	2.48		miR-768-3p	9.30×10^{-6}	−2.06		miR-1268	2.69×10^{-3}	2.68
	miR-142-5p	5.24×10^{-3}	−2.17		miR-1979	8.61×10^{-3}	−3.39		miR-199b-3p	3.86×10^{-4}	−5.56
	miR-574-3p	2.46×10^{-3}	2.69		miR-1301	8.42×10^{-3}	2.1		miR-142-5p	7.12×10^{-3}	−2.27
	miR-25 *	1.73×10^{-3}	2.05		miR-4284	7.78×10^{-3}	−3.51		miR-199a-3p	3.15×10^{-4}	−4.7
	miR-29c *	1.08×10^{-4}	−2.39		miR-1973	3.77×10^{-4}	−4.94		miR-363	7.80×10^{-4}	−11.28
	miR-3175	2.37×10^{-3}	4.41		miR-1201_s	8.31×10^{-5}	−2.14		miR-155 *	6.37×10^{-4}	2.69
	miR-195	2.82×10^{-3}	−3.13		miR-548e	4.75×10^{-3}	−2.07		miR-345	2.02×10^{-3}	−5.92
	miR-29b	7.44×10^{-3}	−5.34		miR-514b-5p	3.95×10^{-4}	−3.45		miR-1280	2.29×10^{-4}	2.46
	miR-30e *	1.47×10^{-3}	−3.33		miR-663b	1.96×10^{-3}	−3.02		miR-126	2.32×10^{-5}	−18.55
	miR-19a	2.30×10^{-3}	−2.11		miR-18a	8.66×10^{-3}	4.77		miR-145	2.96×10^{-4}	−25.89
	miR-23a *	7.59×10^{-4}	4.18	LTNP *vs* Aviremic	miR-1469	9.73×10^{-3}	2.39		miR-1972	1.81×10^{-4}	6.99
Viremic *vs* LTNP	miR-2277	6.17×10^{-3}	−2.34		miR-3185	4.43×10^{-3}	4.91		miR-26a	1.72×10^{-3}	−2.08
	miR-509-3p	7.77×10^{-5}	−7.57		miR-99a	3.11×10^{-3}	−6.57		miR-19b	1.18×10^{-3}	-6.2
	miR-223	2.36×10^{-3}	5.2		miR-151-3p	1.50×10^{-3}	−5.46		miR-1975	1.94×10^{-4}	2.31
	miR-510	7.00×10^{-4}	−3.28		miR-762	4.66×10^{-3}	3.17		miR-143	5.40×10^{-4}	−15.36
	miR-155 *	7.69×10^{-4}	−2.55		miR-3196	3.50×10^{-3}	3		miR-3163	2.74×10^{-3}	2.76
	miR-345	3.79×10^{-5}	11.66		miR-1228 *	7.26×10^{-3}	3.18		miR-3175	1.15×10^{-3}	5.95
	miR-500	3.74×10^{-3}	3.25		miR-4281	9.55×10^{-3}	2.41		miR-29b	7.37×10^{-3}	−6.41
	miR-145	1.16×10^{-3}	15.52		miR-638	5.02×10^{-3}	2.55		miR-30e *	4.53×10^{-3}	−3.21
	miR-503	8.85×10^{-3}	2.58		miR-1915	7.36×10^{-3}	2.74		miR-30e	2.02×10^{-3}	−8.1
	miR-424 *	4.85×10^{-3}	4.44		miR-2861	8.03×10^{-3}	2.9		miR-514b-5p	1.71×10^{-3}	3.04
	miR-3152	2.93×10^{-3}	−3.18		miR-149 *	8.41×10^{-3}	3.24		miR-663b	2.30×10^{-3}	3.09
	miR-1972	5.31×10^{-5}	−7.9		miR-1910	1.72×10^{-3}	3.47		miR-18a	4.15×10^{-3}	−6.03
	miR-1975	2.03×10^{-5}	−2.62		miR-509-3p	2.35×10^{-4}	6.81		let-7f	9.09×10^{-3}	−2.36

* Non-FDR method. Viremic $n = 6$, Aviremic $n = 5$, LTNP $n = 4$, Negative $n = 3$.

Table A4. Differentially expressed miRNAs of CD8+ T-cells in each disease state when compared to negative controls.

CD8	miRNA	*p*-Value	Fold Change	Disease State	miRNA	*p*-Value	Fold Change	Disease State	miRNA	*p*-Value	Fold Change	Disease State	miRNA	*p*-Value	Fold Change
	miR-126	5.37×10^{-3}	−6.26		miR-345	4.24×10^{-3}	5.88		miR-151-5p	7.06×10^{-6}	−14.86		miR-92b *	4.74×10^{-4}	6.31
	miR-345	7.94×10^{-4}	7.94		miR-181b	5.59×10^{-4}	−2.54		miR-126	1.89×10^{-5}	−29.22		miR-1979	3.18×10^{-4}	8.16
	miR-541	8.79×10^{-6}	−2.18		miR-223	9.75×10^{-3}	4.7		miR-1469	1.42×10^{-3}	3.55		miR-1231	3.43×10^{-3}	5.82
	let-7e	5.30×10^{-3}	−3.28		miR-3121	7.04×10^{-3}	−2.37		miR-486-5p	5.46×10^{-6}	−31.25		miR-572	2.25×10^{-4}	8.3
	miR-3163	3.55×10^{-3}	−2.82		miR-30e *	9.76×10^{-3}	3.13		miR-1975	1.89×10^{-3}	2.15		miR-99b	8.93×10^{-3}	−5.15
	miR-548u	1.37×10^{-3}	−2.12		miR-143	7.01×10^{-3}	9.16		miR-151-3p	2.12×10^{-4}	−10.35		miR-146b-5p	2.23×10^{-3}	−6.51
	miR-572	4.74×10^{-3}	4.14		miR-570	4.30×10^{-4}	−3.58		miR-125b	5.61×10^{-4}	−10.99		miR-4321	6.84×10^{-4}	3.72
	miR-514b-5p	3.25×10^{-3}	−2.96		miR-30e	5.68×10^{-3}	7.42		miR-663	3.81×10^{-3}	3.93		mir-1224	2.28×10^{-3}	2.16
	miR-3121	7.81×10^{-4}	−2.96		miR-29b	4.10×10^{-3}	8.92		miR-3185	7.77×10^{-3}	5.35		miR-1307	9.31×10^{-3}	2.3
	miR-570	1.27×10^{-3}	−3.95		mir-1298	2.42×10^{-5}	−2.02		miR-762	2.57×10^{-3}	4.11		miR-31	2.36×10^{-3}	−9.14
	miR-3124	7.18×10^{-3}	−4		miR-21	2.14×10^{-3}	13.59		miR-1909	6.48×10^{-3}	5.07		miR-4284	5.10×10^{-4}	7.66
	mir-16-2	4.34×10^{-4}	−2.19		miR-4284	5.78×10^{-3}	4.4		miR-1228 *	1.44×10^{-3}	4.99		miR-3188	4.37×10^{-3}	4.7
	miR-3152	1.05×10^{-3}	−4.14		miR-29c	7.63×10^{-3}	10.28		miR-3196	4.44×10^{-3}	3.37		miR-3187	8.03×10^{-3}	2.72
	miR-4284	6.63×10^{-3}	4.09		miR-1323	3.04×10^{-4}	−4.13		miR-885-3p	7.96×10^{-5}	6.77	LTNP (cont'd)	miR-23b	2.01×10^{-3}	−2.36
Viremic	miR-23a *	9.03×10^{-3}	3.45	Aviremic	miR-4317	9.33×10^{-3}	2.16	LTNP	miR-149 *	3.55×10^{-4}	6.8		miR-1225-5p	6.70×10^{-4}	12.96
	miR-1323	1.72×10^{-4}	−4.23		miR-26b	5.84×10^{-3}	6.98		miR-4281	3.19×10^{-3}	3.19		miR-3180-1_s	4.34×10^{-3}	2.33
	miR-1256	2.62×10^{-4}	−2.51		miR-548i-4_x	1.74×10^{-3}	−2.1		miR-572	1.41×10^{-5}	2.35		miR-885-5p	6.62×10^{-3}	3.69
	miR-885-5p	7.07×10^{-3}	3.31		miR-941-3_s	1.36×10^{-3}	2.58		miR-638	6.71×10^{-3}	2.79		miR-885	6.75×10^{-3}	2.32
	miR-941-3_s	4.57×10^{-4}	2.79		miR-27a *	5.75×10^{-3}	6.21		miR-1908	5.91×10^{-3}	3.68		miR-941-3_s	5.07×10^{-4}	2.99
	miR-27a *	3.14×10^{-3}	6.76		miR-27a	7.48×10^{-3}	4.23		miR-1915	7.12×10^{-3}	3.16		miR-637	4.18×10^{-3}	2.4
	miR-603	4.66×10^{-4}	−3.31		miR-488	8.19×10^{-3}	−2.01		miR-1268	2.48×10^{-4}	4.19		miR-939	2.55×10^{-3}	3.91
	miR-488	1.99×10^{-3}	−2.25		miR-140-5p	6.69×10^{-3}	8.78		miR-2861	3.15×10^{-3}	3.95		miR-30d	4.75×10^{-3}	−3.91
	miR-628-5p	1.24×10^{-3}	−3.79		miR-1263	5.08×10^{-3}	−4.86		miR-125a-5p	1.87×10^{-3}	−6.12		miR-346	4.63×10^{-4}	6.9
	miR-16-2 *	8.92×10^{-3}	−2.88		miR-98	7.70×10^{-3}	2.46		miR-2277	2.02×10^{-3}	3.17		miR-4322	2.12×10^{-3}	3.1
	miR-1181	6.11×10^{-3}	2.12		miR-374b	3.28×10^{-3}	2.56		miR-17	2.03×10^{-3}	−2.14		miR-3172	7.73×10^{-3}	3.28
	miR-890	1.75×10^{-3}	−3.03		miR-890	3.56×10^{-3}	−2.87		miR-106a	2.71×10^{-3}	−2.07		miR-665	7.41×10^{-4}	6.02
	miR-200b *	6.99×10^{-3}	−2.18		miR-331-5p	7.47×10^{-3}	4.31		let-7e	1.49×10^{-3}	−4.46		miR-92a-1 *	1.71×10^{-3}	−4.54
	miR-335	4.81×10^{-3}	−7.7						miR-20b	4.23×10^{-4}	−8.91		miR-3162	3.53×10^{-3}	4.33
	miR-1272	7.98×10^{-3}	−3.12						miR-1910	1.35×10^{-3}	4.28		miR-1300	2.24×10^{-3}	6.4
									miR-3175	5.60×10^{-3}	5.38		miR-1244	8.05×10^{-3}	3.26

* Non-FDR method. Viremic $n = 6$, aviremic $n = 5$, LTNP $n = 4$, negative $n = 3$.

Table A5. Differentially expressed miRNAs when disease classes of CD8+ cells are intercompared.

CD8	miRNA	p-Value	Fold Change	Disease State	miRNA	p-Value	Fold Change
Viremic *vs* Aviremic	miR-181b	3.63×10^{-5}	2.64	Viremic *vs* LTNP (cont'd)	miR-3126-3p	1.29×10^{-4}	−2.18
	miR-1915	8.81×10^{-3}	2.42		miR-501-3p	2.70×10^{-3}	3.77
	miR-548q	6.20×10^{-3}	−2.05		miR-502-3p	1.23×10^{-3}	3.25
	miR-603	9.70×10^{-3}	−2.05		miR-345	3.45×10^{-4}	7.75
Viremic *vs* LTNP	miR-151-5p	2.22×10^{-5}	8.22		miR-1289-2	7.24×10^{-5}	−2.31
	miR-151-3p	1.11×10^{-3}	5.41		miR-202 *	7.47×10^{-3}	−2.53
	miR-762	9.96×10^{-3}	−2.71		miR-126	9.69×10^{-3}	4.67
	miR-1228 *	7.96×10^{-3}	−3		miR-145	1.62×10^{-3}	14.08
	miR-92b *	3.74×10^{-3}	−3.47		miR-378c	7.77×10^{-3}	2.49
	miR-486-5p	4.53×10^{-6}	19.01		miR-494	2.21×10^{-3}	−3.39
	miR-532-3p	7.63×10^{-3}	2.86		miR-132	1.90×10^{-3}	3.3
	miR-885-3p	2.17×10^{-5}	−5.95		miR-635	2.08×10^{-3}	−2.98
	miR-3124	2.42×10^{-3}	−4.29		miR-30d	4.95×10^{-3}	3.15
	miR-637	1.88×10^{-3}	−2.26		miR-378	3.94×10^{-3}	2.03
	miR-3141	4.88×10^{-5}	−2.26		miR-92a-1 *	7.78×10^{-4}	4.01
	miR-146b-5p	2.71×10^{-3}	4.7		miR-450b-5p	7.85×10^{-3}	−2.36
	miR-149 *	7.94×10^{-3}	−3.13		miR-130b	4.37×10^{-3}	2.03
	miR-532-5p	7.31×10^{-4}	3.85		miR-1975	6.75×10^{-4}	−2.05
	miR-652	1.09×10^{-3}	2.19		miR-629	9.89×10^{-3}	6.42
	miR-500 *	1.47×10^{-3}	2.34		miR-665	4.03×10^{-3}	−3.51
	miR-20b	1.60×10^{-4}	7.5		mir-548q	1.55×10^{-3}	−3.4
	miR-16-2 *	9.86×10^{-3}	−2.59		miR-3163	8.73×10^{-3}	−2.31
	miR-223	7.10×10^{-3}	4.19		miR-106b	2.95×10^{-3}	4.13
	miR-106a	4.98×10^{-5}	2.44		miR-509-3-5p	1.22×10^{-3}	−2.37
	miR-17	4.70×10^{-5}	2.48		miR-1979	1.77×10^{-3}	−4.46
	miR-20a	5.96×10^{-4}	3.48		miR-548x	5.20×10^{-3}	−3.13
	miR-501-5p	3.44×10^{-3}	2.35		miR-3172	5.90×10^{-3}	−2.83
	miR-3121	9.34×10^{-4}	−2.64		miR-28-5p	8.09×10^{-3}	3.22
	miR-1226 *	2.02×10^{-3}	−3.54		miR-27a *	3.74×10^{-3}	5.51
	miR-1268	7.37×10^{-4}	−2.98		miR-1973	3.88×10^{-3}	−3.47
	miR-4253	5.05×10^{-3}	−2.1		miR-3128_x	6.06×10^{-3}	−2.45
	miR-548u	1.79×10^{-4}	−2.3				

Table A5. *Cont.*

CD8	miRNA	p-Value	Fold Change	Disease State	miRNA	p-Value	Fold Change
	miR-1909	1.40×10^{-3}	5.57		miR-17	7.47×10^{-5}	−2.49
	miR-151-5p	2.08×10^{-5}	−9.03		miR-20a	4.10×10^{-5}	−5.1
	miR-125b	1.99×10^{-3}	−6.33		miR-501-5p	8.41×10^{-3}	−2.2
	miR-1469	6.34×10^{-4}	3.35		let-7d *	6.75×10^{-4}	−2.23
	miR-663	8.00×10^{-4}	4.18		miR-4322	9.04×10^{-3}	2.28
	miR-3185	7.88×10^{-4}	6.92		miR-664 *	8.44×10^{-3}	−2.38
	miR-1908	7.06×10^{-4}	4.31		miR-21	2.60×10^{-3}	−10.41
	miR-151-3p	6.66×10^{-4}	−6.34		miR-1268	4.17×10^{-5}	4.28
	miR-762	9.57×10^{-5}	5.5		miR-939	5.12×10^{-3}	3
	miR-3196	2.49×10^{-4}	4.25		miR-548u	1.23×10^{-3}	2.05
	miR-1228 *	9.00×10^{-5}	6.2		miR-199b-3p	4.34×10^{-4}	−5.46
	miR-4281	1.11×10^{-4}	4.14		miR-199a-3p	5.41×10^{-3}	−4.24
	miR-92b *	1.20×10^{-3}	4.37		miR-146a	6.01×10^{-3}	−3.17
	miR-638	4.71×10^{-5}	4.38		miR-363	3.14×10^{-3}	−7.99
	miR-1915	1.25×10^{-4}	4.71		miR-1225-5p	6.03×10^{-3}	5.74
	miR-486-5p	1.90×10^{-5}	−16.05		miR-3126-3p	2.70×10^{-4}	2.14
LTNP *vs* Aviremic	miR-885-3p	5.09×10^{-5}	5.7	LTNP *vs* Aviremic	miR-501-3p	2.58×10^{-3}	−4
	miR-2861	8.69×10^{-5}	5.53		miR-502-3p	6.25×10^{-4}	−3.71
	miR-637	4.30×10^{-3}	2.15		mir-326	8.02×10^{-4}	2.09
	miR-3141	6.97×10^{-6}	2.65		miR-345	2.36×10^{-3}	−5.74
	miR-146b-5p	4.17×10^{-4}	−7.07		miR-320e	8.38×10^{-3}	−2.47
	miR-149 *	4.05×10^{-5}	7.49		miR-548q	3.22×10^{-3}	−2.38
	mir-3185	8.85×10^{-3}	2.23		mir-1289-2	4.14×10^{-5}	2.48
	miR-532-5p	1.05×10^{-3}	−3.86		miR-1207-5p	9.97×10^{-4}	2.41
	miR-3188	2.72×10^{-3}	4.23		miR-126	9.90×10^{-4}	−8.34
	miR-572	1.72×10^{-3}	4.58		miR-500	4.10×10^{-3}	−3.35
	miR-1910	1.50×10^{-3}	3.53		mir-885	2.94×10^{-3}	2.28
	miR-500 *	3.87×10^{-3}	−2.2		miR-145	1.30×10^{-3}	−16.72
	miR-20b	1.34×10^{-4}	−8.38		miR-660	8.64×10^{-3}	−2.33
	miR-223	5.85×10^{-3}	−4.62		miR-494	5.62×10^{-4}	4.33
	miR-106a	1.16×10^{-4}	−2.38		miR-132	5.84×10^{-3}	−2.95
	miR-30d	4.93×10^{-4}	−4.66		miR-30e *	1.51×10^{-3}	−3.78
	miR-27a	5.93×10^{-4}	−5.92		miR-1979	1.32×10^{-4}	7.32

Table A5. *Cont.*

CD8	miRNA	*p*-Value	Fold Change	Disease State	miRNA	*p*-Value	Fold Change
	miR-140-3p	2.24×10^{-3}	−2.06		miR-30c	4.30×10^{-4}	−2.56
	miR-339-3p	2.06×10^{-3}	−3.79		miR-3172	8.10×10^{-3}	2.82
	miR-31	2.45×10^{-4}	−11.48		miR-346	3.84×10^{-3}	3.85
	miR-26a	7.81×10^{-4}	−2.21		miR-26b	3.91×10^{-3}	−6.56
	miR-1263	9.55×10^{-3}	3.77		miR-30e	9.59×10^{-4}	−9.65
	miR-30b	5.32×10^{-4}	−12.55		miR-128	7.31×10^{-3}	−5.32
	miR-29c	7.64×10^{-3}	−8.5		miR-194	1.50×10^{-3}	−7.99
	miR-19b	3.54×10^{-4}	−7.8		miR-19a	2.13×10^{-3}	−2.31
	miR-17 *	8.67×10^{-3}	−5.42		miR-28-5p	8.44×10^{-4}	−4.92
	miR-4270	6.97×10^{-4}	2.68		miR-210	7.28×10^{-3}	−4.66
	miR-1975	6.73×10^{-6}	2.95		miR-27a *	7.32×10^{-3}	−5.07
	miR-143	7.91×10^{-3}	−7.39		miR-23b	3.02×10^{-4}	−2.5
	miR-769-5p	8.46×10^{-4}	−6.87		miR-25	6.61×10^{-4}	−2.31
	miR-28-3p	1.72×10^{-3}	−2.25		miR-625	4.34×10^{-3}	−2.26
	miR-629	1.15×10^{-3}	−12.53		miR-4317	2.44×10^{-3}	−2.33
	miR-29a	4.11×10^{-3}	−4.22		miR-923	2.67×10^{-3}	3.15
	miR-106b	3.94×10^{-4}	−6.18		let-7f	2.37×10^{-3}	−2.79
	miR-22	7.87×10^{-3}	−4.23		let-7g	2.73×10^{-4}	−2.22
	miR-195	6.29×10^{-3}	−3.12		miR-182	1.62×10^{-3}	−2.35
	miR-29b	3.99×10^{-3}	−7.52				

* Non-FDR method. Viremic $n = 6$, Aviremic $n = 5$, LTNP $n = 4$, Negative $n = 3$.

Table A6. qRT-PCR raw amplification values.

Sample	miR-199a	miR-126	U6B	Sample	miR-199a	miR-126	U6B	Sample	miR-199a	miR-126	U6B
V1 CD4 (1)	31.45	27.85	18.16	AV1 CD4 (1)	29.68	26.13	17.12	LTNP2 CD4 (1)	36.16	32.71	17.19
V1 CD4 (2)	30.69	28.52	17.33	AV1 CD4 (2)	29.42	26.07	17.01	LTNP2 CD4 (2)	33.88	31.94	15.51
V1 CD8 (1)	32.34	29.16	16.45	AV1 CD8 (1)	30.38	27	16.62	LTNP2 CD8 (1)	33.19	29.53	15.7
V1 CD8 (2)	31.5	28.84	15.59	AV1 CD8 (2)	29.75	26.99	16.04	LTNP2 CD8 (2)	33.26	29.41	15.66
V2 CD4 (1)	31.36	27	16.79	AV2 CD4 (1)	30.66	27.18	16.56	LTNP3 CD4 (1)	35.12	30.62	16.34
V2 CD4 (2)	30.78	28.65	17.92	AV2 CD4 (2)	30.45	27.53	16.27	LTNP3 CD4 (2)	34.48	30.96	16.59
V2 CD8 (1)	33.16	29.12	15.93	AV2 CD8 (1)	31.59	28.65	16.37	LTNP3 CD8 (1)	35.09	33.4	17.11
V2 CD8 (2)	32.2	29.62	16.17	AV2 CD8 (2)	30.74	27.62	16.06	LTNP3 CD8 (2)	34.67	32.64	16.6
V3 CD4 (1)	30.72	25.73	16.55	AV3 CD4 (1)	29.87	No Ct	16.15	LTNP4 CD4 (1)	35.17	33.15	15.36
V3 CD4 (2)	30.34	25.68	16.75	AV3 CD4 (2)	30.3	26.24	15.74	LTNP4 CD4 (2)	34.89	32.93	15.85
V3 CD8 (1)	33.17	28	15.98	AV3 CD8 (1)	33.04	28.85	16.43	LTNP4 CD8 (1)	31.87	28.82	15.12
V3 CD8 (2)	33.57	28.3	16.6	AV3 CD8 (2)	33.55	29.27	17.06	LTNP4 CD8 (2)	31.2	29.22	15.01
V4 CD4 (1)	31.65	26.99	17.63	AV4 CD4 (1)	31.22	27.37	16.95	Neg1 CD4 (1)	29.86	26.12	16.42
V4 CD4 (2)	30.4	26.65	16.68	AV4 CD4 (2)	30.29	26.92	16.62	Neg1 CD4 (2)	29.68	25.67	15.76
V4 CD8 (1)	30.83	27.16	16.15	AV4 CD8 (1)	31.37	27.73	14.53	Neg1 CD8 (1)	32.43	28.76	16.46
V4 CD8 (2)	30.12	26.68	15.59	AV4 CD8 (2)	30.6	27.6	15.39	Neg1 CD8 (2)	32.97	28.65	16.79
V5 CD4 (1)	33.92	29.44	17.48	AV5 CD4 (1)	32.53	27.87	15.42	Neg2 CD4 (1)	29.86	26.09	15.62
V5 CD4 (2)	33.41	29.08	17.23	AV5 CD4 (2)	31.65	27.69	15.8	Neg2 CD4 (2)	29.75	26.25	15.74
V5 CD8 (1)	33.4	28.12	16.26	AV5 CD8 (1)	31.62	27.29	16.11	Neg2 CD8 (1)	31.8	27.93	15.63
V5 CD8 (2)	32.48	27.83	15.43	AV5 CD8 (2)	30.83	26.91	16.81	Neg2 CD8 (2)	31.62	27.89	15.63
V6 CD4 (1)	33.34	29.47	16.91	LTNP1 CD4 (1)	35.31	33.16	16.5	Neg3 CD4 (1)	25.99	22.73	14.86
V6 CD4 (2)	32.81	28.92	16.43	LTNP1 CD4 (2)	34.5	32.92	15.86	Neg3 CD4 (2)	26.35	22.53	14.28
V6 CD8 (1)	33.08	28.69	16.64	LTNP1 CD8 (1)	32.84	29.91	13.92	Neg3 CD8 (1)	30.74	26.2	18.15
V6 CD8 (2)	33.48	29.66	17.12	LTNP1 CD8 (2)	32.3	30.01	14.88	Neg3 CD8 (2)	30.82	26.91	17.48

References

1. Fowler, L.; Saksena, N.K. Micro-RNA: New players in HIV-pathogenesis, diagnosis, prognosis and antiviral therapy. *AIDS Rev.* **2013**, *15*, 3–14. [PubMed]
2. Houzet, L.; Jeang, K.T. MicroRNAs and human retroviruses. *Biochim. Biophys. Acta* **2011**, *1809*, 686–693. [CrossRef] [PubMed]
3. Calin, G.A.; Sevignani, C.; Dumitru, C.D.; Hyslop, T.; Noch, E.; Yendamuri, S.; Shimizu, M.; Rattan, S.; Bullrich, F.; Negrini, M.; *et al.* Human microRNA genes are frequently located at fragile sites and genomic regions involved in cancers. *Proc. Natl. Acad. Sci. USA* **2004**, *101*, 2999–3004. [CrossRef] [PubMed]
4. Croce, C.M. Causes and consequences of microRNA dysregulation in cancer. *Nat. Rev. Genet.* **2009**, *10*, 704–714. [CrossRef] [PubMed]
5. Calin, G.A.; Dumitru, C.D.; Shimizu, M.; Bichi, R.; Zupo, S.; Noch, E.; Aldler, H.; Rattan, S.; Keating, M.; Rai, K.; *et al.* Frequent deletions and downregulation of micro-RNA genes miR15 and miR16 at 13q14 in chronic lymphocytic leukemia. *Proc. Natl. Acad. Sci. USA* **2002**, *99*, 15524–15529. [CrossRef] [PubMed]
6. Yeung, M.L.; Bennasser, Y.; Myers, T.G.; Jiang, G.; Benkirane, M.; Jeang, K.T. Changes in microRNA expression profiles in HIV-1-transfected human cells. *Retrovirology* **2005**, *2*, 1–8. [CrossRef] [PubMed]
7. Houzet, L.; Yeung, M.L.; De Lame, V.; Desai, D.; Smith, S.M.; Jeang, K.T. MicroRNA profile changes in human immunodeficiency virus type 1 (HIV-1) seropositive individuals. *Retrovirology* **2008**, *5*, 118–127. [CrossRef] [PubMed]
8. Huang, J.; Wang, F.; Argyris, E.; Chen, K.; Liang, Z.; Tian, H.; Huang, W.; Squires, K.; Verlinghieri, G.; Zhang, H. Cellular microRNAs contribute to HIV-1 latency in resting primary CD4+ T lymphocytes. *Nat. Med.* **2007**, *13*, 1241–1247. [CrossRef] [PubMed]
9. Witwer, K.W.; Watson, A.K.; Blankson, J.N.; Clements, J.E. Relationships of PBMC microRNA expression, plasma viral load, and CD4+ T-cell count in HIV-1-infected elite suppressors and viremic patients. *Retrovirology* **2012**, *9*, 1–15. [CrossRef] [PubMed]
10. Zhu, L.; Qui, C.; Wang, J.; Yuan, S.; Zhang, X.; Xu, J. Poster MOPE026—MicroRNA profile identifies different outcomes of disease progression at later stage of HIV infection. In Proceedings of the XIX IAS Conference, Washington, DC, USA, 22 July 2012.
11. Cohen, M.; Shaw, G.; McMichael, A.; Haynes, B. Acute HIV-1 Infection. *N. Engl. J. Med.* **2011**, *364*, 1943–1954. [CrossRef] [PubMed]
12. Koup, R.A.; Safrit, J.T.; Cao, Y.; Andrews, C.A.; McLeod, G.; Borowsky, W.; Farthing, C.; Ho, D.D. Temporal association of cellular immune responses with the initial control of viremia in primary human immunodeficiency virus type 1 syndrome. *J. Virol.* **1994**, *68*, 4650–4655. [PubMed]
13. Borrow, P.; Lewicki, H.; Hahn, B.H.; Shaw, G.M.; Oldstone, M.B.A. Virus-specific CD8+ cytotoxic T-lymphocyte activity associated with control of viremia in primary human immunodeficiency virus type 1 infection. *J. Virol.* **1994**, *68*, 6103–6110. [PubMed]
14. Sanghvi, V.R.; Steel, L.F. The cellular TAR RNA binding protein, TRBP, promotes HIV-1 replication primarily by inhibiting the activation of double-stranded RNA-dependent kinase PKR. *J. Virol.* **2011**, *85*, 12614–12621. [CrossRef] [PubMed]
15. Bignami, F.; Pilotti, E.; Bertoncelli, L.; Ronzi, P.; Guilli, M.; Marmirolli, N.; Magnani, G.; Pinti, M.; Lopalco, L.; Mussini, C.; *et al.* Stable changes in CD4+ T lymphocyte miRNA expression after exposure to HIV-1. *Blood* **2012**, *119*, 6259–6267. [CrossRef] [PubMed]
16. Noorbakhsh, F.; Ramachandran, R.; Barsby, N.; Ellestad, K.K.; LeBlanc, A.; Dickie, P.; Baker, G.; Hollenberg, M.D.; Cohen, E.A.; Power, C. MicroRNA profiling reveals new aspects of HIV neurodegeneration: Caspase-6 regulates astrocyte survival. *FASEB J.* **2010**, *24*, 1799–1812. [CrossRef] [PubMed]
17. Bennasser, Y.; Jeang, K.T. HIV-1 Tat interaction with Dicer: Requirement for RNA. *Retrovirology* **2006**, *3*, 1–6. [CrossRef] [PubMed]
18. Rossi, R.L.; Rossetti, G.; Wenandy, L.; Curti, S.; Ripamonti, A.; Bonnal, R.J.; Birolo, R.S.; Moro, M.; Crosti, M.S.; Gruarin, P.; *et al.* Distinct microRNA signatures in human lymphocyte subsets and enforcement of the naive state in CD4+ T-cells by the microRNA miR-125b. *Nat. Immunol.* **2011**, *12*, 796–803. [CrossRef] [PubMed]
19. Allantaz, F.; Cheng, D.T.; Bergauer, T.; Ravindran, P.; Rossier, M.F.; Ebeling, M.; Badi, L.; Reis, B.; Bitter, H.; D'Asaro, M.; *et al.* Expression profiling of human immune cell subsets identifies miRNA-mRNA regulatory relationships correlated with cell type specific expression. *PLoS ONE* **2012**, *7*, e29979. [CrossRef] [PubMed]

20. Nathans, R.; Chu, C.Y.; Serquina, A.K.; Lu, C.C.; Cao, H.; Rana, T.M. Cellular microRNA and P bodies modulate host-HIV-1 interactions. *Mol. Cell* **2009**, *34*, 696–709. [CrossRef] [PubMed]
21. Sun, G.; Li, H.; Wu, X.; Covarrubias, M.; Scherer, L.; Meinking, K.; Luk, B.; Chomchan, P.; Alluin, J.; Gombart, A.F.; *et al.* Interplay between HIV-1 infection and host microRNAs. *Nucleic Acids Res.* **2012**, *40*, 2181–2196. [CrossRef] [PubMed]
22. Tan Gana, N.H.; Onuki, T.; Victoriano, A.F.; Okamoto, T. MicroRNAs in HIV-1 infection: An integration of viral and cellular interaction at the genomic level. *Front. Microbiol.* **2012**, *3*. [CrossRef] [PubMed]
23. Chiang, K.; Sung, T.L.; Rice, A.P. Regulation of cyclin T1 and HIV-1 Replication by microRNAs in resting CD4+ T lymphocytes. *J. Virol.* **2012**, *86*, 3244–3252. [CrossRef] [PubMed]
24. Triboulet, R.; Mari, B.; Lin, Y.L.; Chable-Bessia, C.; Bennasser, Y.; Lebrigand, K.; Cardinaud, B.; Maurin, T.; Barbry, P.; Baillat, V.; *et al.* Suppression of microRNA-silencing pathway by HIV-1 during virus replication. *Science* **2007**, *315*, 1579–1582.
25. Jopling, C.; Yi, M.; Lancaster, A.; Lemon, S.; Sarnow, P. Modulation of hepatitis C virus RNA abundance by a liver-specific microRNA. *Science* **2005**, *309*, 1577–1581. [CrossRef] [PubMed]

7

Complemented Palindromic Small RNAs First Discovered from SARS Coronavirus

Chang Liu [1,†], Ze Chen [2,3,†], Yue Hu [1], Haishuo Ji [4,5], Deshui Yu [4], Wenyuan Shen [1], Siyu Li [4], Jishou Ruan [6], Wenjun Bu [4,*] and Shan Gao [4,5,*]

1 Laboratory of Medical Molecular Virology, School of Medicine, Nankai University, Tianjin 300071, China; changliu@nankai.edu.cn (C.L.); huyue2016@mail.nankai.edu.cn (Y.H.); shenwy@mail.nankai.edu.cn (W.S.)
2 State Key Laboratory of Veterinary Etiological Biology and Key Laboratory of Veterinary Parasitology of Gansu Province, Lanzhou Veterinary Research Institute, Chinese Academy of Agricultural Science, Lanzhou 730046, China; chenze@caas.cn
3 Co-Innovation Center for Prevention and Control of Important Animal Infectious Diseases and Zoonoses, Yangzhou 225009, China
4 College of Life Sciences, Nankai University, Tianjin 300071, China; haishuo_ji@mail.nankai.edu.cn (H.J.); deshui_yu@mail.nankai.edu.cn (D.Y.); siyu_li@mail.nankai.edu.cn (S.L.)
5 Institute of Statistics, Nankai University, Tianjin 300071, China
6 School of Mathematical Sciences, Nankai University, Tianjin 300071, China; jsruan@nankai.edu.cn
* Correspondence: wenjunbu@nankai.edu.cn (W.B.); gao_shan@mail.nankai.edu.cn (S.G.)
† These authors contributed equally to this work.

Abstract: In this study, we report for the first time the existence of complemented palindromic small RNAs (cpsRNAs) and propose that cpsRNAs and palindromic small RNAs (psRNAs) constitute a novel class of small RNAs. The first discovered 19-nt cpsRNA UUAACAAGCUUGUUAAAGA, named SARS-CoV-cpsR-19, was detected from a 22-bp DNA complemented palindrome TCTTTAACAAGCTTGTTAAAGA in the severe acute respiratory syndrome coronavirus (SARS-CoV) genome. The phylogenetic analysis supported that this DNA complemented palindrome originated from bat betacoronavirus. The results of RNA interference (RNAi) experiments showed that one 19-nt segment corresponding to SARS-CoV-cpsR-19 significantly induced cell apoptosis. Using this joint analysis of the molecular function and phylogeny, our results suggested that SARS-CoV-cpsR-19 could play a role in SARS-CoV infection or pathogenesis. The discovery of cpsRNAs has paved a way to find novel markers for pathogen detection and to reveal the mechanisms underlying infection or pathogenesis from a different point of view. Researchers can use cpsRNAs to study the infection or pathogenesis of pathogenic viruses when these viruses are not available. The discovery of psRNAs and cpsRNAs, as a novel class of small RNAs, also inspire researchers to investigate DNA palindromes and DNA complemented palindromes with lengths of psRNAs and cpsRNAs in viral genomes.

Keywords: palindromic small RNA; complemented palindromic small RNA; small RNA; DNA complemented palindrome; severe acute respiratory syndrome coronavirus

---

1. Introduction

Small RNA sequencing (small RNA-seq or sRNA-seq) is used to obtain thousands of short RNA sequences with lengths that are usually less than 50 bp. With sRNA-seq, many novel non-coding RNAs (ncRNAs) have been discovered. For example, two featured series of ribosomal RNA (rRNA)-derived RNA fragments (rRFs) constitute a novel class of small RNAs [1]. Small RNA-seq has also been used for virus detection in plants [2–4] and invertebrates [5]. In 2016, Wang et al. first used sRNA-seq data from the National Center for Biology Information Sequence Read Archive database (NCBI SRA) to show that sRNA-seq can be used to detect and identify human viruses [6], however the detection results were not as robust as those of plant or invertebrate viruses. To improve virus detection in mammals, our strategy was to detect and compare featured RNA fragments in plants, invertebrates, and mammals using sRNA-seq data. In a previous study [7], we detected siRNA duplexes that were induced by plant viruses and analyzed these small interfering RNA (siRNA) duplexes as an important class of featured RNA fragments. In this study, we aimed to investigate siRNA duplexes that were induced by invertebrate and mammalian viruses and unexpectedly discovered another important class of featured RNA fragments: complemented palindromic small RNAs (cpsRNAs). Among all of the detected cpsRNAs, the first discovered cpsRNA, named SARS-CoV-cpsR-19, from the severe acute respiratory syndrome coronavirus (SARS-CoV) strain MA15 merited further study, because mice infected with SARS-CoV MA15 died from an overwhelming viral infection with viral-mediated destruction of pneumocytes and ciliated epithelial cells [8]. SARS-CoV-cpsR-19 was detected from a DNA complemented palindrome TCTTTAACAAGCTTGTTAAAGA in the SARS-CoV genome. In our previous study of mitochondrial genomes, we reported palindromic small RNAs (psRNAs) for the first time [9]. Both of psRNAs and cpsRNAs could comprise a novel class of small RNAs due to their special sequence structures.

In this study, we compared the features of the siRNA duplexes that were induced by mammalian viruses with those that were induced by plant and invertebrate viruses. We found that the detected siRNA duplexes that were induced by mammalian viruses had significantly lower percentages of total sequenced reads than those that were induced by plant and invertebrate viruses, and it seemed that they were produced only from a few sites in the viral genomes. One possible reason could be that a large proportion of the sRNA-seq data is from other small RNA fragments, owing to the presence of a number of double-stranded RNA (dsRNA)-triggered nonspecific responses, such as type I interferon (IFN) synthesis [10]. Another possible reason could be that the missing siRNA duplexes or siRNA fragments function in cells by interacting with host RNAs or proteins. Based on this idea, we hypothesized that SARS-CoV-cpsR-19 or other cpsRNAs from SARS-CoV MA15 could play a role in SARS-CoV infection or pathogenesis.

To test our hypothesis, we conducted the joint analysis of the molecular function and phylogeny. First, we investigated the origins of SARS-CoV-cpsR-19 by the phylogenetic analysis of coronavirus genome sequences that were associated with bats, palm civets, rats, mice, monkeys, dogs, bovines, hedgehogs, giraffes, waterbucks, and equines. Subsequently, we performed RNA interference (RNAi) experiments to test the possible cellular effects that were induced by SARS-CoV-cpsR-19. The phylogenetic analysis supported that the DNA complemented palindrome TCTTTAACAAGCTTGTTAAAGA could originate from bat betacoronaviruses. The results of the RNAi experiments showed that one 19-nt segment corresponding to SARS-CoV-cpsR-19 significantly induced cell apoptosis. This study aimed to provide a different point of view for pathogen detection and pathogenesis studies.

2. Materials and Methods

2.1. Datasets and Data Analysis

All sRNA-seq data were downloaded from the NCBI SRA database. In our previous study, 6 mammalian viruses (HPV-18, HBV, HCV, HIV-1, SMRV, and EBV) were detected from 36 runs of sRNA-seq data [6]. In this study, 11 invertebrate viruses were detected from 51 runs of sRNA-seq data (Supplementary file 1) and 2 mammalian viruses (H1N1 and SARS-CoV) were detected from 20 runs of sRNA-seq data (NCBI SRA: SRP012018) [11]. In total, 11 invertebrate viruses and 8 mammalian viruses were detected from 107 runs of sRNA-seq data using VirusDetect [4], and their genome sequences were downloaded from the NCBI GenBank database. Among 107 runs of sRNA-seq data, four runs (NCBI SRA: SRR452404, SRR452406, SRR452408, and SRR452410) had been sequenced from lung tissue in mice that were infected with SARS-CoV MA15 [11] and they were used to detect SARS-CoV. The cleaning and quality control of the sRNA-seq data were performed using pipeline Fastq_clean [12], which was optimized to clean the raw reads from the Illumina platforms. Using the software Bowtie v0.12.7 [13] with one mismatch, we aligned all of the cleaned sRNA-seq reads to viral genome sequences and obtained alignment results in sequence alignment map (SAM) format for the detection of siRNA duplexes using the program duplexfinder [7]. Statistical computation and plotting were performed using the software R v2.15.3 (R Core Team, Vienna, Austria) with the Bioconductor packages [14]. The *ORF3b* gene from human betacoronavirus (GenBank: DQ497008.1), 20 homologous sequences from the bat betacoronaviruses, and nine homologous sequences from the civet betacoronaviruses (Supplementary file 1) were aligned using ClustalW2 [15] with curation. After the removal of the identical sequences, the *ORF3b* gene from human betacoronavirus, eight homologous sequences from bat betacoronaviruses, and two homologous sequences from civet betacoronaviruses were used for phylogenetic analysis. Since these homologous sequences had high identities (from 85.16% to 99.78%) to the *ORF3b* gene from DQ497008, the Neighbor Joining (NJ) method was used for phylogenetic analysis.

2.2. RNAi and Cellular Experiments

Based on the short hairpin RNA (shRNA) design protocol [1], the 16-nt, 18-nt, 19-nt, 20-nt, and 22-nt segments from the DNA complemented palindrome TCTTTAACAAGCTTGTTAAAGA and their control "CGTACGCGGAATACTTCGA" were selected as the target sequences for pSIREN-RetroQ vector construction (Clontech, Mountain View, CA, USA). PC-9 cells were provided by Dr. Qingsong Wang from Tianjin Medical University and they were divided into six groups—namely, 16, 18, 19, 20, 22, and control—for transfection using plasmids containing the 16-nt, 18-nt, 19-nt, 20-nt, and 22-nt segments and the control sequences. Each group had three replicate samples for plasmid transfection and cell apoptosis measurement. Each sample was processed following the procedure described below. The PC-9 cells were washed with phosphate buffer saline (PBS) and were trypsinized 12 h prior to transfection. Gbico RPMI-1640 medium (Thermo Fisher Scientific, Wattham, MA, USA) was added to the cells, which were then centrifuged at 1000 rpm for 10 min at 4 °C to remove the supernatant. Gbico RPMI-1640 medium containing 10% fetal bovine serum was added to adjust the solvent to reach a volume of 2 μL and contain 2×10^5 cells. These cells were seeded into one well of a 6-well plate for plasmid transfection. Transfection of 2 μg of plasmid was performed using 5 μL Lipofectamine 2000 (Life technology, Carlsbad, CA, USA), following the manufacturer's instructions. Cell apoptosis was measured with the FITC Annexin V Apoptosis Detection Kit I (BD Biosciences, Franklin Lakes, NJ, USA), following the procedure described below. The cells were washed with PBS, were trypsinized, and were collected using a 5-mL culture tube 48 or 72 h after transfection. The culture tube was then centrifuged at 1000 rpm for 10 min at 4 °C to remove the supernatant. The cells were washed twice with cold PBS and were resuspended in 1X Binding Buffer at a concentration of 1×10^6 cells/mL. Then, 100 μL of the solution (1×10^5 cells) was transferred to a new culture tube with 5 μL of FITC-Annexin V and 5 μL PI. The cells were gently vortexed and were incubated for 15 min at room temperature in the dark. 1X Binding Buffer (400 μL) was added to the tube. Finally, the sample was analyzed using a FACSCalibur flow cytometer (BD Biosciences, Franklin Lakes, NJ, USA) within 1 h. Apoptotic cells were quantified by summing the count of the early apoptotic cells (FITC-Annexin V+/PI−) and the late apoptotic cells (FITC-Annexin V+/PI+). To confirm the results using Annexin V/PI staining and detection, the expression levels of three cell-apoptosis marker genes (*BAX*, *BCL2*, and *CASP3*) were also measured by quantitatice PCR (qPCR) assays using 16-nt, 18-nt, 19-nt, 20-nt, and 22-nt segments to produce siRNA duplexes by pSIREN-RetroQ plasmid transfection. The cells were collected 24 h after transfection. For each gene, three RNAi samples and three control samples were tested for relative quantification. After transfection, RNA extraction, complementary DNA (cDNA) synthesis, and cDNA amplification were performed following the same procedure described below. For each sample, total RNA was isolated using RNAiso Plus Reagent (TaKaRa Bio Inc., Kusatsu, Shiga, Japan) and the cDNA was synthesized by

M-MuLV (New England Biolabs, Hitchin, UK). The cDNA product was amplified by qPCR (Eppendorf, Hamburg, Germany) using *GAPDH* as internal control and the qPCR reaction mixture using Real Time PCR Easy (Foregene, Chengdu, China) was incubated at 95 °C for 10 min, followed by 35 PCR cycles (10 s at 95 °C, 20 s at 55 °C, and 30 s at 72 °C for each cycle). The primers of five genes are listed in Table 1.

Table 1. Primers for quantitative PCR (qPCR) assays.

Gene Symbol	Forward Primer	Reverse Primer
GAPDH	ACATCGCTCAGACACCATG	TGTAGTTGAGGTCAATGAAGGG
BAX	AGTAACATGGAGCTGCAGAG	AGTAGAAAAGGGCGACAACC
BCL2	GTGGATGACTGAGTACCTGAAC	GCCAGGAGAAATCAAACAGAGG
CASP3	ACTGGACTGTGGCATTGAG	GAGCCATCCTTTGAATTTCGC
MDL1 *	CCCAATCCACATCAAAACCC	GGACGAGAAGGGATTTGACTG

* The primers of *MDL1* were used to amplify both *MDL1* and *MDL1AS* as they are sense–antisense transcripts from the same loci [9] and were predicted to be markers which can indicate the activities of mitochondria and even the whole cells.

3. Results

3.1. Comparison of siRNA Duplexes Induced by Plant, Invertebrate, and Mammalian Viruses

In total, 11 invertebrate and 8 mammalian viruses (HPV-18, HBV, HCV, HIV-1, SMRV, and EBV that were detected in the previous study, and H1N1 and SARS-CoV that were detected in this study) were used to detect siRNA duplexes (see Materials and Methods). Next, we compared the features of the siRNA duplexes that were induced by invertebrate viruses (Figure 1A) with those that were induced by plant viruses (Figure 1B). The results showed that duplex length was the principal factor that determined the read count in both plants and invertebrates. The 21-nt siRNA duplexes were the most abundant duplexes in both plants and invertebrates, followed by the 22-nt siRNA duplexes in plants, whereas it was the 20-nt siRNA duplexes in invertebrates. The 21-nt siRNA duplexes with 2-nt overhangs were the most abundant 21-nt duplexes in plants, whereas the 21-nt siRNA duplexes with 1-nt overhangs were the most abundant 21-nt duplexes in invertebrates, however they had a very similar read count to that of the 21-nt siRNA duplexes with 2-nt overhangs. The 18-nt, 19-nt, 20-nt, and 22-nt siRNA duplexes in invertebrates had much higher percentages of total sequenced reads than those in plants. In addition, 18-nt and 19-nt siRNA duplexes had very similar read counts, and the 20-nt and 22-nt siRNA duplexes had very similar read counts in invertebrates. As the siRNA duplexes that were induced by mammalian viruses had significantly lower percentages of total sequenced reads, comparison of the siRNA-duplex features between mammals and invertebrates or plants could not provide meaningful results using our data.

However, as an unexpected result of the comparison, we discovered cpsRNAs from both invertebrate and mammalian viruses. As this study was to focus on the study of cpsRNAs from SARS-CoV, we did not include cpsRNAs from other animal viruses. One 19-nt cpsRNA UUAACAAGCUUGUUAAAGA from a DNA complemented palindrome TCTTTAACAAGCTTGTTAAAGA (DQ497008: 25962-25983), located in the *ORF3b* gene of the SARS-CoV MA15 genome (GenBank: DQ497008.1), was detected in four runs of sRNA-seq data (see Materials and Methods). This 19-nt cpsRNA was named SARS-CoV-cpsR-19 and the DNA complemented palindrome contained 22 nucleotides which perfectly matched its reverse complement sequence. From this DNA complemented palindrome, we also detected one 18-nt and one 21-nt cpsRNA named SARS-CoV-cpsR-18 and SARS-CoV-cpsR-21 (Figure 2A), respectively. however we did not detect cpsRNAs of other lengths (e.g., 16-nt, 20-nt, or 22-nt) using our data. We speculated that 18-nt, 19-nt, and one 21-nt cpsRNA could be derived from siRNA duplexes (Figure 2B), as the presence of these overhanging nucleotides is a hallmark of small RNAs that are produced by silencing-related ribonucleases (e.g., Drosha or Dicer). However, we did not have any evidence to prove the existence of siRNA duplexes.

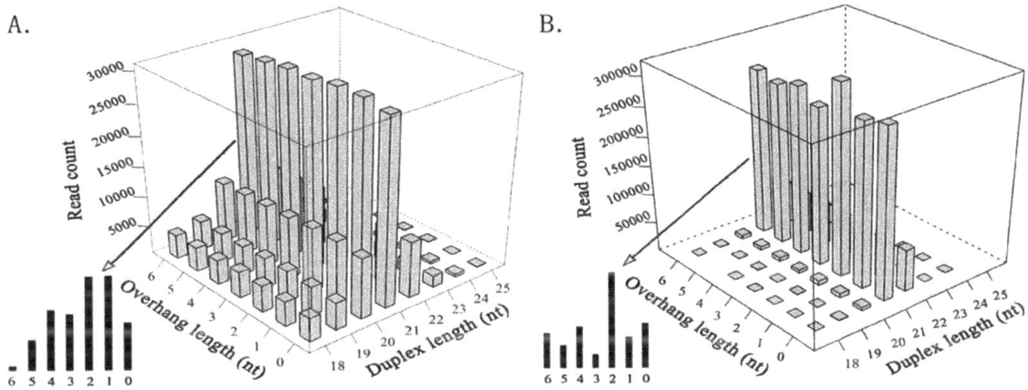

Figure 1. Comparison of small interfering RNA (siRNA) duplexes that were induced by plant and invertebrate viruses. All of the cleaned small RNA sequencing (sRNA-seq) reads were aligned to viral genome sequences using the software Bowtie v0.12.7 [13] with one mismatch. The detection of siRNA duplexes was performed using the program duplexfinder [7]. (**A**). The read count of siRNA duplexes varies with the duplex length and the overhang length, using data from 11 invertebrate viral genomes. (**B**). The read count of siRNA duplexes varies with the duplex length and the overhang length, using data from seven plant viral genomes [7].

3.2. *Discovery of psRNAs and cpsRNAs*

Palindromes have been discovered in the published genomes of most species and play important roles in biological processes. The well-known samples of DNA palindromes include restriction enzyme sites, methylation sites, and palindromic motifs in T cell receptors [16]. In this study, we classified DNA palindromes into DNA palindromes and DNA complemented palindromes, from which psRNAs and cpsRNAs could be produced, respectively. A DNA palindrome is classically defined as a nucleic acid sequence that is reverse complementary to itself, while small RNAs that are reverse complementary to themselves are defined as cpsRNAs in this study. Accordingly, a typical psRNA should have a sequence that is 100% identical to its reverse sequence, however most psRNAs are semipalindromic or heteropalindromic, such as hsa-tiR-MDL1-16 AAAGACACCCCCCACA from a DNA palindrome AAAGACACCCCCCACAGTTT (NC_012920: 561-580) [9]. As a heteropalindromic psRNA, hsa-tiR-MDL1-16 is derived by cleavage after it starts transcription of the long heavy strand (H-strand) primary transcript at the position 561 of the human mitochondrial genome (Figure 2A). All cpsRNAs that are discovered from SARS-CoV MA15 are semipalindromic or heteropalindromic, such as SARS-CoV-cpsR-19 and SARS-CoV-cpsR-21. Although SARS-CoV-cpsR-21 contains almost 100% of the total nucleotides which contribute to the matches (Figure 2A), most cpsRNAs have mismatches or insertions/deletions (InDels). One example is a new Epstein-Barr virus (EBV) microRNA precursor (pre-miRNA) with a length of 89 nt, as was reported in our previous study [6]. This pre-miRNA sequence contains only 87.64% (78/89) of the total nucleotides which contribute to the matches.

In this study, we also found that DNA palindromes with sizes ranging from 14 to 31 nt and DNA complemented palindromes with sizes ranging from 14 to 53 nt existed ubiquitously in the animal virus genomes, however only a few of them were detected as transcribed or processed into psRNAs or cpsRNAs. For example, we only detected psRNAs from two (CTACTGACCAGTCATC and AAGGTCTCCCTCTGGAA) of 14 DNA palindromes and cpsRNA from two (GCAAATTGCACAATTTGC and TCTTTAACAAGCTTGTTAAAGA) of 29 DNA complemented palindromes (Table 2) in the SARS-CoV genome (GenBank: DQ497008.1) using four runs of sRNA-seq data (see Materials and Methods). One possible reason could be that only a few of cpsRNAs had been ligated to adapters during the sRNA-seq library preparation process as they could form hairpins

(Figure 2A) at room temperature (~20 °C). From Table 2, it can be seen that the T_m (melting temperature) of cpsRNA hairpins in the SARS-CoV genome is distributed ranging from 14 °C to 26 °C. This finding provided a clue to explain why the detected siRNA duplexes that were induced by mammalian viruses had significantly lower percentages of total sequenced reads than those that were induced by plant and invertebrate viruses, and could help to improve the virus detection of mammalian viruses.

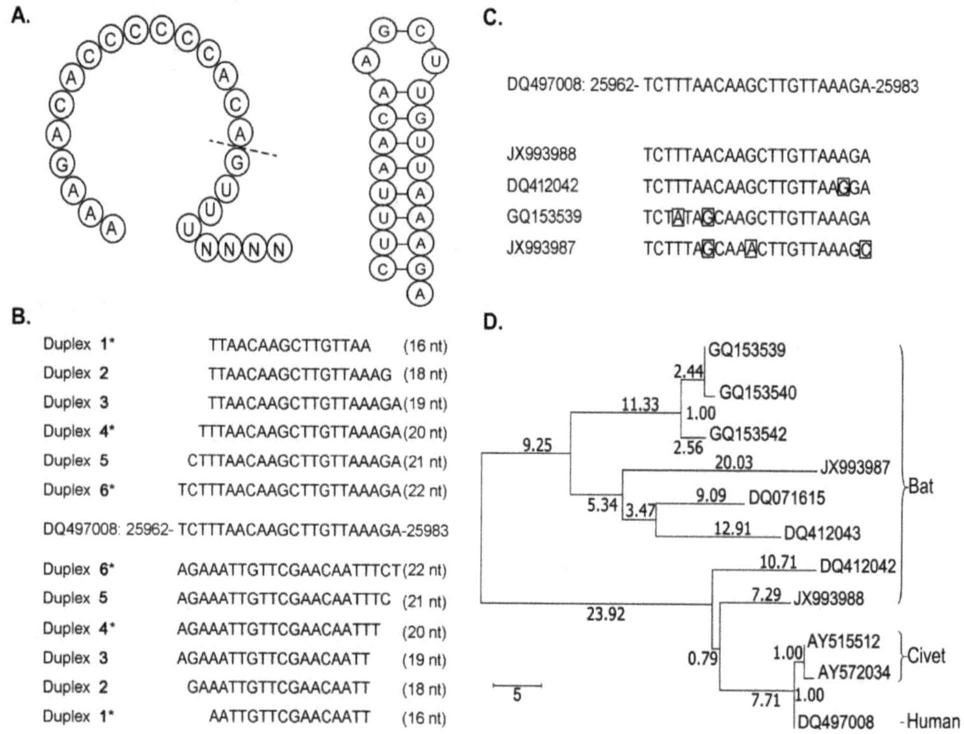

Figure 2. Clues to the origins of SARS-CoV-cpsR-19. All of the genome sequences are represented by their GenBank accession numbers (e.g., DQ497008). (**A**). hsa-tiR-MDL1-16 is derived by cleavage after it starts transcription of the long H-strand primary transcript at the position 561 of human mitochondrial genome (left). SARS-CoV-cpsR-21 can form a hairpin (right). (**B**). 16-nt, 18-nt, 19-nt, 20-nt, and 22-nt siRNA duplexes were used for RNA interference (RNAi) experiments. * 16-nt, 20-nt, and 22-nt palindromic small RNAs (cpsRNAs) were not detected in this study. (**C**). SARS-CoV-cpsR-19 was detected from a DNA complemented palindrome TCTTTAACAAGCTTGTTAAAGA in the SARS-CoV genome. All of the 22-nt homologous sequences in the civet betacoronavirus genomes were identical to the DNA complemented palindrome, whereas four genotypes of 22-nt homologous sequences were detected in the bat betacoronavirus genomes, however only one of them was identical to it. (**D**). The phylogenetic tree was built by the Neighbor Joining (NJ) method using the *ORF3b* gene from human betacoronavirus, eight homologous sequences from bat betacoronaviruses, and two homologous sequences from civet betacoronaviruses. The branch's length corresponds to an average number of nucleotide changes per 100 nucleotides.

3.3. Clues to Origins of SARS-CoV-cpsR-19

The previously unknown SARS virus generated widespread panic in 2002 and 2003 when it caused 774 deaths and more than 8000 cases of illness. Scientists immediately suspected that civet cats, which are only distant relatives of house cats, may have been the springboard for the transmission of SARS-CoV to humans [17]. Later, scientists concluded that civets were not the original source of SARS. Further investigation showed that the genetic diversity of coronaviruses in bats increased the possibility of variants crossing the species barrier to cause disease outbreaks in the human population [18]. To investigate the origins of SARS-CoV-cpsR-19, we obtained coronavirus genome sequences that were associated with bats, palm civets, rats, mice, monkeys, dogs, bovines, hedgehogs, giraffes, waterbucks, and equines from the NCBI GenBank database. The results of the sequence analysis showed that the DNA complemented palindrome TCTTTAACAAGCTTGTTAAAGA was only located in the *ORF3b* genes of betacoronaviruses. Next, we blasted the *ORF3b* gene of human betacoronavirus (GenBank: DQ497008.1) with all of the obtained betacoronavirus genomes, except for those that were obtained from experiments with mice and monkeys. The results showed that the *ORF3b* gene from human betacoronavirus had homologous genes from the betacoronaviruses of bats and palm civets (Supplementary file 1) rather than from those of other species. The DNA complemented palindrome also had 22-nt homologous sequences in the bat and civet betacoronavirus genomes. All of the 22-nt homologous sequences in the civet betacoronavirus genomes were identical to the DNA complemented palindrome, whereas four genotypes of 22-nt homologous sequences were detected in the bat betacoronavirus genomes, however only one of them was identical to it (Figure 2C). Four genotypes had no, one, two, and three mismatches with the DNA complemented palindrome and their corresponding *ORF3b* homologous sequences had identities of 96.77%, 96.13%, 87.96%, and 85.16%, respectively. This suggested that one betacoronavirus variant containing the DNA complemented palindrome could have originated from bats and was then passed onto palm civets and finally to humans. This was consistent with the results of the phylogenetic analysis using the *ORF3b* homologous sequences from the bat and civet betacoronavirus genomes (Figure 2D). In the phylogenetic tree, all of the human and civet betacoronaviruses containing the DNA complemented palindrome were grouped into one clade. The nearest relative of the human and civet clade was the bat betacoronavirus (GenBank: JX993988.1) containing the DNA complemented palindrome, and the next nearest relative was the bat betacoronavirus (GenBank: DQ412042.1) containing a homologous 22-nt sequence with one mismatch with the DNA complemented palindrome.

Table 2. DNA complemented palindromes in the severe acute respiratory syndrome coronavirus (SARS-CoV) genome.

DNA Complemented Palindrome	Start	End	Length	GC%	Tm	MFE
GGTAACTATAAAGTTACC	1783	1800	18	33	20	−6.9
AATGTGAGAATCACATT	2779	2795	17	29	18	−4.4
AAGAAACTAAGTTTCTT	3923	3939	17	24	18	−3.5
ATGGTAAGCTTTACCAT	3971	3987	17	35	18	−4.5
AAATGCAAATCTGCATTT	4234	4251	18	28	18	−4.9
ATATGTCTATGACATAT	4949	4965	17	24	18	−4.3
CCTCATGTAAATCATGAGG	5020	5038	19	42	22	−9.0
ATAACAATTGTTAT	5207	5220	14	14	12	−0.9
ACTTCAACAGCTTGAAGT	5241	5258	18	39	18	−4.7
ACTTCAAATTCATTTGAAGT	6256	6275	20	25	20	−5.5
GTACTTTTACTAAAAGTAC	6734	6752	19	26	20	−4.6
ATCTACCAGTGGTAGAT	9189	9205	17	41	20	−6.5
TTACCTTCCAAGGTAA	10,892	10,907	16	38	16	−4.0
CCACTTATTAAGTGG	14,160	14,174	15	40	18	−4.7
CCCATTTAATAAATGGG	14,882	14,898	17	35	20	−5.7
CAGTGACAATGTCACTG	16,463	16,479	17	47	22	−7.7
CACCTTTGAAAAAGGTG	16,760	16,776	17	41	20	−5.6
TGTAAGAGAATTTCTTACA	17,651	17,669	19	26	18	−5.3
TGAATATGACTATGTCATATTCA	17,783	17,805	23	26	26	−9.7
CTACTTTAAGAAAGTAG	20,081	20,097	17	29	18	−3.6
AGCATTCTTGGAATGCT	21,106	21,122	17	41	20	−6.1
TTCCTCTTAAATTAAGAGGAA	21,337	21,357	21	29	22	−9.1
GCATTACTACAGAAGTAATGC	23,599	23,619	21	38	22	−7.9
AGCCCTTTATAAGGGCT	25,480	25,496	17	47	22	−8.7
TCTTTAACAAGCTTGTTAAAGA *	25,962	25,983	22	27	20	−7.2
CAACGGTACTATTACCGTTG	26,406	26,425	20	45	24	−9.0
ACCTTCATGAAGGT	28,048	28,061	14	43	14	−2.6
GCAAATTGCACAATTTGC *	29,028	29,045	18	39	18	−4.0
TAAAATTAATTTTA	29,668	29,681	14	0	NA	NA

In total, 29 DNA complemented palindromes were identified in the SARS-CoV genome (GenBank: DQ497008.1). * In this study, we only detected psRNAs from two of 29 DNA complemented palindromes. Minimum Free Energy (MFE) was calculated using RNAfold (http://rna.tbi.univie.ac.at/cgi-bin/RNAWebSuite/RNAfold.cgi). Tm (melting temperature) of cpsRNA hairpins was calculated by the formula: Tm = 4*(C+G) + 2*(A+T) and only using the nucleic acids in the stems of DNA complemented palindromes.

3.4. Preliminary Studies on Biological Functions of SARS-CoV-cpsR-19

Our previous study showed that the psRNA hsa-tiR-MDL1-16 contained the Transcription Initiation Site (TIS) of the human mitochondrial H-strand and could be involved in mtDNA transcription regulation [9]. This inspired us to speculate that cpsRNAs could also have specific biological functions and we investigated SARS-CoV-cpsR-19 using RNAi and Annexin V/PI staining and detection. 16-nt, 18-nt, 19-nt, 20-nt, and 22-nt segments from the DNA complemented palindrome TCTTTAACAAG<u>CTTGTTAAAGA</u> were used to produce siRNA duplexes by pSIREN-RetroQ plasmid transfection (see Materials and Methods). As a result, the 19-nt and 20-nt segments significantly induced cell apoptosis 2.76- and 1.48-fold 48 h after their transfection, respectively. Particularly, the 19-nt segment significantly induced cell apoptosis 7.94-fold (36.04%/4.54%) 72 h after its transfection into PC-9 cells (Figure 3). Using the 19-nt segment, we also tested cell apoptosis in five other human cell lines and one mouse cell line. The results showed that the 19-nt segment significantly induced cell apoptosis in the A549, MCF-7, and H1299 cell lines, however it did not in the MB231, H520, and L929 (mouse) cell lines. The siRNA duplexes that were induced by the 19-nt segment could silence cell-specific transcripts to induce cell apoptosis through RNAi. These results suggested that SARS-CoV-cpsR-19 had significant biological functions and could play a role in SARS-CoV infection or pathogenesis.

To confirm the results using Annexin V/PI staining and detection, the expression levels of three cell-apoptosis marker genes (*BAX*, *BCL2*, and *CASP3*) were measured by qPCR assays using the same RNAi protocol in the Hela cell line (see Materials and Methods). The results showed 1.08-, 1.65-, 1.60-, 1.24-, and 1.30-fold increases in *BAX/BCL2* ratios which were caused by 16-nt, 18-nt, 19-nt, 20-nt, and 22-nt segments, respectively (Figure 3E). As a low *BAX/BCL2* ratio (<1) and a high *BAX/BCL2* ratio (>1) is against and in favor of cell apoptosis, it suggested that 18-nt and 19-nt segments significantly induced cell apoptosis. The RNAi using 18-nt, 19-nt, and 20-nt segments also caused significantly increased expression of *CASP3* by 1.72, 2.58, and 1.89 fold changes in Hela cells. Therefore, the experimental results by qPCR assays were consistent with those using Annexin V/PI staining and detection. In addition, the expression levels of two novel long non-coding RNAs (lncRNAs) (*MDL1* and *MDL1AS*) from mitochondrial genome were measured to investigate their expression changes that were caused by RNAi using 16-nt, 18-nt, 19-nt, 20-nt, and 22-nt segments. *MDL1* and *MDL1AS* were recently discovered [9] and were predicted to be markers which can indicate the activities of mitochondria and even the whole cells. The RNAi using 16-nt, 18-nt, 19-nt, 20-nt, and 22-nt segments also caused increased relative expression levels of *MDL1* by 1.28, 5.43, 1.58, 9.25, and 1.41 fold changes in Hela cells (Figure 3E). It suggested that qPCR of *MDL1* produce higher sensitivities than that of *BAX/BCL2* and *CASP3* in the detection of cell apoptosis.

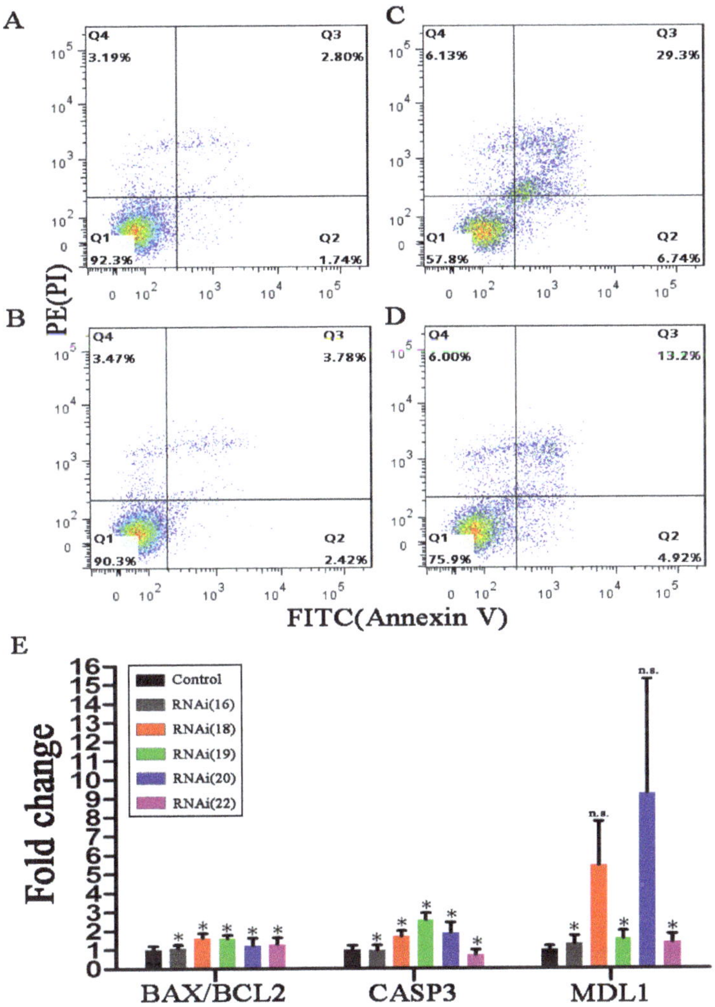

Figure 3. RNAi and cellular experiments for validation. PC-9 cells were divided into six groups named 16, 18, 19, 20, 22, and control for transfection using plasmids containing 16-nt, 18-nt, 19-nt, 20-nt, 22-nt segments, and their controls (Figure 2B). Each group had three replicate samples for plasmid transfection and cell apoptosis measurement. The 19-nt and 20-nt segments significantly induced cell apoptosis, whereas the 16-nt, 18-nt, and 22-nt did not show significantly positive results. The samples in this figure were selected randomly from the control (**A**), 18 (**B**), 19 (**C**), and 20 (**D**) group. (**E**). The experimental results by qPCR assays were consistent with those using Annexin V/PI staining and detection. The primers of *MDL1* were used to amplify both *MDL1* and *MDL1AS*, as they are sense–antisense transcripts from the same loci [9] and are predicted to be markers which can indicate the activities of mitochondria and even the whole cells.

4. Conclusions and Discussion

SARS-CoV-cpsR-19 from the DNA complemented palindrome TCTTTAACAAGCTTGTTAAAGA (DQ497008: 25962-25983) is located in the *ORF3b* gene of the SARS-CoV MA15 genome. In one previous study, this 22-nt DNA palindrome was studied using two mathematics models (M0 and M1) to conclude that this DNA palindrome was quite unlikely to occur by chance [19]. These authors also found that the underrepresentation of 4-nt DNA palindromes existed in all of the coronaviruses and the significant underrepresentation of 6-nt DNA palindromes existed only in SARS but not in the other six coronaviruses. Then, they hypothesized that this avoidance of 6-nt DNA palindromes in the SARS genome would offer a protective effect on the virus, making it comparatively more difficult to be destroyed by some host-defence mechanisms (e.g., restriction enzymes). In this study, we unexpectedly discovered cpsRNAs from the DNA palindrome TCTTTAACAAGCTTGTTAAAGA and reported that SARS-CoV-cpsR-19 could play a role in SARS-CoV infection or pathogenesis. We hypothesized that cpsRNAs, as a novel class of small RNAs, were produced by some host-defence mechanisms (e.g., RNAi) and that they could interact with the immune system of the host cells to help virus survival or infection.

Another previous study reported that another sRNA AGGAACUGGCCCAGAAGCUUC in the SARS-CoV genome, named small viral RNA-N (svRNA-N), contributed to SARS-CoV pathogenesis [20]. SARS-CoV-cpsR-19 was detected in very low abundance using four runs of sRNA-seq data (see Materials and Methods) in this study but was not detected using the sRNA-seq data (NCBI SRA: SRP094035) in that study [20]. One possible reason could be that SARS-CoV-cpsR-19 can form one hairpin and the T_m (melting temperature) of this hairpin is about 20 °C. Therefore, most of SARS-CoV-cpsR-19 cannot be ligated to adapters during the sRNA-seq library preparation process at room temperature (~20 °C). As the optimal secondary structure of svRNA-N (Minimum Free Energy (MFE) = −1.7) is less stable than those of SARS-CoV-cpsR-19 (MFE = −2.6) and all the other possible cpsRNAs (Table 2), svRNA-N is more easily captured during the sRNA-seq library preparation process. SARS-CoV-cpsR-19 induced cell apoptosis, whereas svRNA-N caused pulmonary inflammation. In Vivo experiments revealed that the biogenesis pathway that was responsible for svRNA-N's synthesis was not dependent on Dicer or Drosha as the canonical biogenesis of miRNAs and svRNA-N were able to silence mRNA expression by targeting its 3′ UTR. However, the mechanisms underlying infection or pathogenesis of both SARS-CoV-cpsR-19 and svRNA-N are still unknown.

Author Contributions: S.G. conceived this project. S.G. and W.B. supervised this project. S.G., Z.C., and D.Y. analyzed the data. H.J. and S.L. curated the sequences and prepared all of the figures, tables, and additional files. C.L., Y.H., and W.S. performed the experiments. S.G. drafted the main manuscript. S.G. and J.R. revised the manuscript. All of the authors have read and approved the manuscript.

Acknowledgments: We thank Dr. Qingsong Wang from Tianjin Medical University for her cell lines, Yangguang Han and Jinke Huang from TianJin KeYiJiaXin Technology Co., Ltd. for his professional guidance on the RNAi and cellular experiments. This manuscript has been released as a pre-print at https://www.biorxiv.org/content/early/2017/12/13/185876.

References

1. Chen, Z.; Sun, Y.; Yang, X.; Wu, Z.; Guo, K.; Niu, X.; Wang, Q.; Ruan, J.; Bu, W.; Gao, S. Two featured series of rRNA-derived RNA fragments (rRFs) constitute a novel class of small RNAs. *PLoS ONE* **2017**, *12*, e0176458. [CrossRef] [PubMed]
2. Kreuze, J.F.; Perez, A.; Untiveros, M.; Quispe, D.; Fuentes, S.; Barker, I.; Simon, R. Complete viral genome sequence and discovery of novel viruses by deep sequencing of small RNAs: A generic method for diagnosis, discovery and sequencing of viruses. *Virology* **2009**, *388*, 1–7. [CrossRef] [PubMed]
3. Li, R.; Gao, S.; Hernandez, A.G.; Wechter, W.P.; Fei, Z.; Ling, K. Deep sequencing of small RNAs in tomato for virus and viroid identification and strain differentiation. *PLoS ONE* **2012**, *7*, e37127. [CrossRef] [PubMed]
4. Zheng, Y.; Gao, S.; Chellappan, P.; Li, R.; Marco, G.; Dina, G.; Segundo, F.; Ling, K.; Jan, K.; Fei, Z. VirusDetect: An automated pipeline for efficient virus discovery using deep sequencing of small RNAs. *Virology* **2017**, *500*, 130–138. [CrossRef] [PubMed]
5. Nayak, A.; Tassetto, M.; Kunitomi, M.; Andino, R. *RNA Interference-Mediated Intrinsic Antiviral Immunity in Invertebrates*; Springer: Berlin/Heidelberg, Germany, 2013; Volume 371, pp. 183–200.
6. Wang, F.; Sun, Y.; Ruan, J.; Chen, R.; Chen, X.; Chen, C.; Kreuze, J.F.; Fei, Z.; Zhu, X.; Gao, S. Using small RNA deep sequencing to detect human viruses. *BioMed Res. Int.* **2016**, *2016*, 2596782. [CrossRef] [PubMed]
7. Niu, X.; Sun, Y.; Chen, Z.; Li, R.; Padmanabhan, C.; Ruan, J.; Kreuze, J.F.; Ling, K.; Fei, Z.; Gao, S. Using small RNA-seq data to detect siRNA duplexes induced by plant viruses. *Genes* **2017**, *8*, 163. [CrossRef] [PubMed]
8. Roberts, A.; Deming, D.; Paddock, C.D.; Cheng, A.; Yount, B.; Vogel, L.; Herman, B.D.; Sheahan, T.; Heise, M.; Genrich, G.L. A mouse-adapted SARS-coronavirus causes disease and mortality in BALB/c mice. *PLoS Pathog.* **2007**, *3*, e5. [CrossRef] [PubMed]

9. Gao, S.; Tian, X.; Chang, H.; Sun, Y.; Wu, Z.; Cheng, Z.; Dong, P.; Zhao, Q.; Ruan, J.; Bu, W. Two novel lncRNAs discovered in human mitochondrial DNA using PacBio full-length transcriptome data. *Mitochondrion* **2017**, *36*. [CrossRef] [PubMed]
10. Elbashir, S.M.; Harborth, J.; Weber, K.; Tuschl, T. Analysis of gene function in somatic mammalian cells using small interfering RNAs. *Methods* **2002**, *26*, 199–213. [CrossRef]
11. Peng, X.; Gralinski, L.; Ferris, M.T.; Frieman, M.B.; Thomas, M.J.; Proll, S.; Korth, M.J.; Tisoncik, J.R.; Heise, M.; Luo, S. Integrative deep sequencing of the mouse lung transcriptome reveals differential expression of diverse classes of small RNAs in response to respiratory virus infection. *Mbio* **2011**, *2*, e00198-11. [CrossRef] [PubMed]
12. Zhang, M.; Zhan, F.; Sun, H.; Gong, X.; Fei, Z.; Gao, S. Fastq_clean: An optimized pipeline to clean the Illumina sequencing data with quality control. In Proceedings of the 2014 IEEE International Conference on Bioinformatics and Biomedicine (BIBM), Belfast, UK, 2–5 November 2014.
13. Langmead, B.; Trapnell, C.; Pop, M.; Salzberg, S.L. Ultrafast and memory-efficient alignment of short DNA sequences to the human genome. *Genome Biol.* **2009**, *10*, R25. [CrossRef] [PubMed]
14. Gao, S.; Ou, J.; Xiao, K. *R Language and Bioconductor in Bioinformatics Applications, Chinese ed.*; Tianjin Science and Technology Translation Publishing Co., Ltd.: Tianjin, China, 2014.
15. Clustal Omega. Available online: http://www.ebi.ac.uk/Tools/msa/clustalo/ (accessed on 18th May 2014).
16. Srivastava, S.K.; Robins, H.S. Palindromic nucleotide analysis in human T cell receptor rearrangements. *PLoS ONE* **2012**, *7*, e52250. [CrossRef] [PubMed]
17. Enserink, M. Clues to the animal origins of SARS. *Science* **2003**, *300*, 1351. [CrossRef] [PubMed]
18. Li, W.; Shi, Z.; Yu, M.; Ren, W.; Smith, C.; Epstein, J.H.; Wang, H.; Crameri, G.; Hu, Z.; Zhang, H. Bats are natural reservoirs of SARS-like coronaviruses. *Science* **2005**, *310*, 676–679. [CrossRef] [PubMed]
19. Chew, D.S.H.; Choi, K.P.; Heidner, H.; Leung, M.Y. Palindromes in SARS and other coronaviruses. *Informs J. Comput.* **2004**, *16*, 331–340. [CrossRef] [PubMed]
20. Morales, L.; Oliveros, J.C.; Fernandez-Delgado, R.; Tenoever, B.R.; Enjuanes, L.; Sola, I. SARS-CoV-encoded small RNAs contribute to infection-associated lung pathology. *Cell Host Microbe* **2017**, *21*, 344–355. [CrossRef] [PubMed]

8

An Exonic Switch Regulates Differential Accession of microRNAs to the Cd34 Transcript in Atherosclerosis Progression

Miguel Hueso [1,*], Josep M. Cruzado [1], Joan Torras [1] and Estanis Navarro [2,*]

1 Department of Nephrology, Hospital Universitari Bellvitge and Bellvitge Research Institute (IDIBELL), L'Hospitalet de Llobregat, 08907 Barcelona, Spain; jmcruzado@bellvitgehospital.cat (J.M.C.); 15268jta@comb.cat (J.T.)

2 Independent Researcher, Esplugues de Llobregat, 08950 Barcelona, Spain

* Correspondence: mhueso@idibell.cat (M.H.); estanis.navarro@gmail.com (E.N.)

Abstract: Background: $CD34^+$ Endothelial Progenitor Cells (EPCs) play an important role in the recovery of injured endothelium and contribute to atherosclerosis (ATH) pathogenesis. Previously we described a potential atherogenic role for miR-125 that we aimed to confirm in this work. Methods: Microarray hybridization, TaqMan Low Density Array (TLDA) cards, qPCR, and immunohistochemistry (IHC) were used to analyze expression of the miRNAs, proteins and transcripts here studied. Results: Here we have demonstrated an increase of resident CD34-positive cells in the aortic tissue of human and mice during ATH progression, as well as the presence of clusters of CD34-positive cells in the intima and adventitia of human ATH aortas. We introduce miR-351, which share the seed sequence with miR-125, as a potential effector of CD34. We show a splicing event at an internal/cryptic splice site at exon 8 of the murine *Cd34* gene (exonic-switch) that would regulate the differential accession of miRNAs (including miR-125) to the coding region or to the 3'UTR of *Cd34*. Conclusions: We introduce new potential mediators of ATH progression (CD34 cell-clusters, miR-351), and propose a new mechanism of miRNA action, linked to a cryptic splicing site in the target-host gene, that would regulate the differential accession of miRNAs to their cognate binding sites.

Keywords: atherosclerosis; miR-125b; CD34 cell-clusters; CD34 cryptic splicing; ORF miRNA target; exon switch

---

1. Background

Atherosclerosis (ATH) is a chronic inflammatory disease initiated at the vascular endothelium which shows a complex pathogenesis [1]. Endothelium damage displays a focal distribution, particularly at sites of disturbed blood flow, suggesting the contribution of vascular repair mechanisms in ATH progression. Recent studies suggest that endothelial progenitor cells (EPCs) do have an important role in the recovery and repair of the injured endothelium [2], since mobilized EPCs are able to migrate to damaged sites where they would differentiate into mature endothelial cells (ECs) in situ [3]. EPCs come from three major sources; i.e., peripheral blood and bone marrow, the vascular wall, and resident monocytes and macrophages [4,5]. In patients with ATH it has been observed a decrease in the number of circulant EPCs [6], although it is not well know if this is the result from a decreased production or from an increased EPC cell death [7].

Bone marrow-derived EPCs express the progenitor receptor CD34 and are released into the peripheral blood for vascular repair/angiogenesis, since these have reparative potential of endothelial dysfunction [8]. CD34, a membrane glycoprotein, is a critical component of the group of surface receptors that regulate migration and engraftment of progenitor cells to target tissues [9]. *CD34* is expressed as two variant forms after alternative splicing of the so-called "exon-X" at the *CD34* gene [10]. These two forms differ in their cytoplasmic domain, which is almost totally lacking in the variant-1, a fact that abrogates phosphorylation by protein kinase C (PKC) at two sites in the Cterminal cytoplasmic domain [11]. Interestingly, bioinformatic analyses have detected a cluster of miRNA binding sites at the putative 3'UTR of the human and murine *CD34* (Hueso, M., this work) that could be involved in the regulation of the expression balance among both isoforms.

MicroRNAs (miRNAs) are very small RNAs (over 22 nucleotides long) with critical roles in the regulation of gene expression and whose changes in expression have been related to the onset and progression of different diseases (see [12] for a recent review). As of March 2018, the human miRNAome included 1917 mature miRNAs [13]). Aberrant miRNA expression profiles have been described during the progression of ATH, cardiovascular disease (CVD) [14] or renal fibrosis [15]. Thus, plaque progression and rupture were linked to the expression of miR-23a-5p, miR-210 and miR-222 [16,17] while, on the contrary, miR-19b and miR-33a/miR-33b had protective roles on plaque stability [18,19]. Furthermore, plasma miR-144 and miR-33 levels were seen to be increased in coronary artery disease (CAD) patients [20,21], miR-155 was seen to inhibit transformation of macrophages into foam cells by targeting CEH expression [22], and miR-181b was found to be overexpressed in human atherosclerotic plaques and abdominal aortic aneurysms, where it downregulated expression of the tissue inhibitor of MMP-3, and elastin [23]. Lastly, miR-296 has been described as a positive regulator of ATH onset and progression by promoting neovascularization and favoring M1 macrophage polarization [24].

We have previously described the relationship of mir-125b expression with ATH progression. Aortic tissue from ApoE-deficient mice fed with a western diet showed a significant overexpression of miR-125b, but not of the highly related miR-125a. This overexpression was reversed by treating mice with a siRNA against the immune mediator CD40 [25]. MiR-125b belongs to a conserved family (composed by the highly related miR-125a, miR-125b-1 and miR-125b-2) with key roles in cellular differentiation, proliferation and apoptosis [26]. Furthermore, there is experimental evidence of the involvement of the miR-125 family with pathogenesis of ATH. In this sense, it has been described that up-regulation of miR-125a contributed to the differentiation of monocytes towards the inflammatory phenotype [27] through the repression of the TNFAIP3 a negative regulator of NF-kB signaling [28], and that miR-125a was upregulated in oxLDL activated macrophages [29]. Lastly, members of the miR-125 family have been shown to target transcripts whose product could be also involved in mechanisms promoting ATH onset and progression, as STAT3 [30], FOXP3 [31], VEGF [32], MARK1 [33], and BCL2, BCL2L12 and MCL1 [34] among others.

We are interested in the mechanisms of ATH progression, and more specifically in those which involve genes that are implicated in the function of stem or progenitor cells, such as *CD34* or *Lin28a/b*. *CD34* was previously detected as a target of miR-125 in miRNA by bioinformatic analysis (M. Hueso et al., unpublished results) and here we have deepened the molecular relationship among miR-125 and *CD34* in the context of ATH progression. The most significant finding of this work is the description of a new molecular mechanism of miRNA action by which an splicing event at a internal/cryptic splice site of the murine *Cd34* gene would regulate the differential accession of a number of micro-RNAs (including miR-125) to the coding region or to the 3'UTR of the two isoforms of the *Cd34* transcript [10], likely disturbing the expression balance of both CD34 protein isoforms.

2. Methods

2.1. Ethics Statement

Here, we have used authorized autopsy material from the Pathology Department at the Hospital Universitari de Bellvitge (HUB), L'Hospitalet, Barcelona, Spain. Confidential information from patients was protected following national normatives. This study was performed conforming the declaration of Helsinki, and with the approval by the Clinical Research Ethics Committee of HUB (PR163/13).

2.2. Patients and Samples

The characteristics of patients and samples studied in this work have been reported previously [25,35]. Briefly, abdominal aortas were obtained from deceased patients from the HUB, and included atherosclerotic plaques, incipient atherosclerotic plaques, and normal abdominal aortas without injury.

2.3. Mice

In this work we used tissue samples from 25 female ApoE$^{-/-}$ mice (Homocigous*Apoe*^{tm1Unc} in the C57BL/6 background, Jackson Laboratory, Charles River, Wilmington, MA, USA) that were stored in a previous study [25]. Briefly, 5 mice were considered as controls and sacrificed at 8 weeks (8W). Other 20 mice were i.p. injected with 50 µg of siRNA against CD40 (siCD40), or of a scrambled siRNA (SC) as control, twice weekly. Mice were euthanized by isoflurane anesthesia after 14 or 24 weeks of treatment (SC/14w, n = 5; siCD40/14W, n = 5; SC/24W, n = 5; siCD40/24w, n = 5). Mice were fed with a western diet enriched in cholesterol (0.2% w/w), which provided 42% of calories as fat (TD.88137; Harlan-Tekland, Madison, WI, USA). We isolated the ascendant aortas from 10 ApoE$^{-/-}$ mice treated with the anti-siCD40 siRNA and from 10 ApoE$^{-/-}$ controls, 5 mice treated with the scrambled siRNA (SC) and 5 mice treated with the vehicle (veh) for 24 weeks. This study followed the EU guidelines on animal care and the protocols were approved by the Ethics Committee for Animal Research of the University of Barcelona-HUB. The effect of the anti-CD40 treatment had been previously confirmed by checking the downregulation in the expression of CD40 mRNA and the lowering in the counting of CD40-positive cells (see [25] and references therein).

2.4. Reactives

Biotinylated secondary antibodies and vectastain were from Vector (Burlingame, CA, USA). Expression of the isoforms of CD34 (Mn01310773 and Mn00519283) was measured with Taqman Gene Expression Assays (Life technologies, Thermo Fisher Scientific Inc., Waltham, MA, USA). For the anti-CD34 staining, we used an anti-CD34 antibody (RabMab EP373Y; ab81289) from Abcam (Cambridge, UK) following standard procedures (see [25] for details).

2.5. Histological and Immunohistochemical CD34 Analysis

Preparation of aortic tissue for protein/RNA expression analysis has been described in detail elsewhere [25,35]. Briefly, aortic sections (aortic arch, descending thoracic aorta, abdominal aorta above renal arteries and abdominal aorta below renal arteries) were split for RNA extraction or for IHC after fixing with 4% PFA. ATH lesions were assessed in sequential, 5 µm thick, sections of the aortic roots. stained with hematoxylin/eosin (H/E) or with a specific antibody by standard procedures. A semiquantitative evaluation of the immunostaining scored the total number of antigen-positive cells with regard of the number of total cells (nuclear H/E staining). Lastly, the ProgResCapturePro 2.7.7 software (Jenoptik, Jena, Germany) was used for morphometric image analysis.

2.6. qPCR and miRNA Expression Data

Primary murine miRNA expression data was extracted from the original Taqman Low Density Array card (TLDA) experiment on murine aortas, already described [25], by using the ExpressionSuite Software version 1.1 (Applied Biosystems-Thermo Fisher Scientific, Waltham, MA, USA). RNA extraction from aortas and cDNA conversion have been described previously [25]. Expression of human *CD34* was measured by using Taqman probes in the conditions stated by the manufacturer. The ΔCt/ΔΔCt method was used to compare results with the basal conditions samples

(8w mice). For mRNA expression normalization, the housekeeping GAPDH was used, while miRNA expression was normalized by comparison with RNU48 (human) or U6-snRNA (mouse) also from the TLDA data.

2.7. Data Mining and Bioinformatic Analysis

The genomic structures of the miR-125a/miR-125b-1 and miR-125b-2 clusters were obtained from the Ensembl genome browser, release 94 [36]. MiRNA sequences were retrieved from the Mirbase [13]. The genomic structure of the murine *Cd34* gene was obtained by "Blasting" the complete murine genome (Annotation release 106) with the cDNA sequences for the two *Cd34* isoforms at the NCBI server. Prediction of Cd34 miRNA binding sites was made at the "Targetscan" for mouse, release 7.2 [37].

2.8. Statistics

All data are expressed as mean ±SD. The Kruskal-Wallis test (SPSS 20.0 software, SPSS Inc. Chicago, IL, USA) was used to compare miRNA-expression among basal (8W), control SC-siRNA and anti-CD40 siRNA-treated mice. The Mann-Whitney test was used to compare CD34 expression among siRNA control/treated mice A or in aortas from patients with or without CKD. A value of $p < 0.05$ was considered as statistically significant.

3. Results

We are interested in the mechanisms of ATH progression, and more specifically on those that depend on the function of different lineages of stem cells, such as hemopoietic progenitors, or endothelial progenitor cells, characterized by the expression of the CD34 marker. In this context we have recently described an animal model of ATH progression and reversion in which ApoE-deficient mice were fed with a fat-enriched chow to facilitate ATH development and treated for 14 or 24 weeks with an anti-CD40 siRNA to reverse disease progression, or with an irrelevant, scrambled sequence siRNA as control [25]. This disease model was characterized by the upregulation of miR-125b, a member of the miR-125 family, with ATH progression, which was reversed by the treatment with the anti-CD40 specific siRNA [25]. Interestingly, another highly related member of the family, miR-125a, which has been proposed by bioinformatic analysis to target CD34 did not show up in our previous analysis [25]. Here we address the relationship among the members of the miR-125 family and CD34, and describe a new regulatory mechanism by which a cryptic splice event at the CD34 gene could regulate the accessibility of miR-125a (and other miRNAs) to the 3'UTR or to the coding region of the CD34 transcript.

3.1. MiR-125: A Complex miRNA Family

miR-125b is encoded by two different genes in human and mice (miR-125b-1 and miR-125b-2), co-clustered with genes encoding members of the miR-99 (miR-99a and miR-99b, miR-100) and let-7 (let-7a-2, let-7c-1 and let-7e) families (see Figure 1A). Furthermore, miR-125b is also highly homologous to miR-125a with which it shares the seed sequence (Figure 1B). Bioinformatic analysis predicted specific promoters for human miR-125b-1/2, but expression of miR-125a seemed to be co-regulated with that of the members of its cluster, miR-99b and let-7e [38]. Given the high homology among miR-125a and miR-125b, and the fact that both shared the same seed sequence but had apparently different roles during ATH progression, we re-evaluated the original data from the TLDA experiment to study changes in miR-125a expression during ATH progression. Figure 1C shows the individual, absolute Ct values obtained for each one of the miR-125a, miR-125b and the housekeeping gene U6-snRNA in the different experimental conditions tested. Furthermore, Table 1 shows the same data expressed as the ΔCt values after normalization to U6 (miR-125-U6). Both clearly show that expression of miR-125a and miR-125b was very similar in the basal, initial condition at 8 weeks (9.43 ± 0.5 cycles vs. 10.47 ± 0.6 cycles; for miR-125a and miR-125b, respectively; $n = 2$), diverged after 24W of ATH progression (6.7 ± 0.6 vs. 3.97 ± 0.6 cycles, respectively; $n = 3$) and returned to almost basal levels after treatment with the anti-CD40 siRNA (7.16 ± 1.6 vs. 6.26 ± −2.0 cycles, respectively; $n = 3$). Given the small differences in the expression of miR-125a and miR-125b, we considered that miR-125a could also have a role, though perhaps less relevant that miR-125b, in ATH progression.

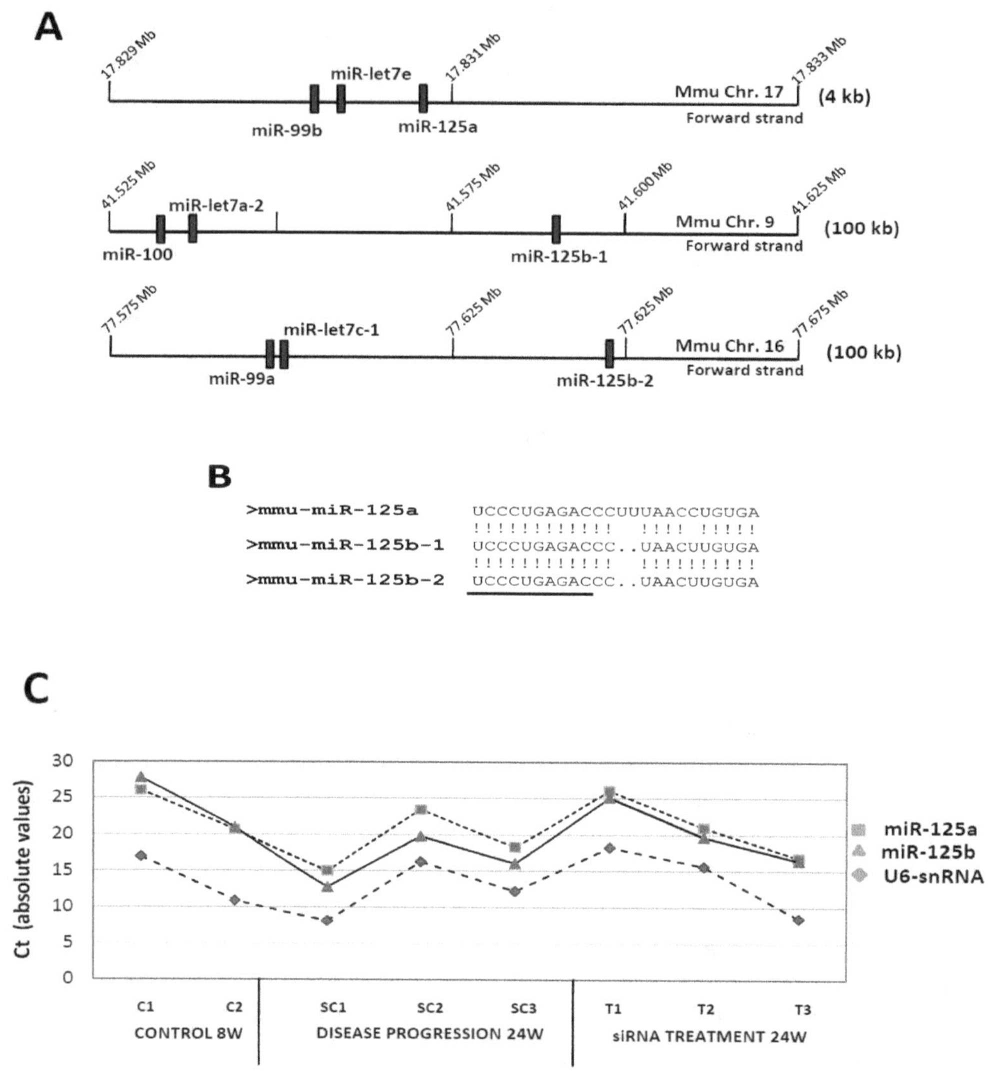

Figure 1. Structure and expression of the murine miR-125a/miR-125b-1/miR-125b-2 gene clusters in ATH progression. (**A**) Graphic diagram of the three miRNA gene clusters. miRNA gene locations (relative positions) were taken from the Ensembl genome browser [36]. Ensemble code for miR-125a was ENSMUSG00000065479, for miR-125b-1 was ENSMUSG00000093354, and for miR-125b-2 was ENSMUSG00000065472. Only miRNA genes and no other coding/non-coding genes are shown. Graphics (genomic regions and gene boxes) are not drawn to scale, and scales are different for each cluster (see numbering for absolute positions in the chromosome). Numbers to the right show the amount of genomic sequence covered by each graph. (**B**) Homology among miR-125a/miR-125b-1/miR-125b-2. Shown are the sequences of the mature miRNAs. (!) stands for a homologous nucleotide, while (.) stands for a deleted nucleotide. Underlined, the seed sequence. (**C**) Absolute expression (without normalization) of murine miR-125a, miR-125b and U6-snRNA (as normalizator) in control, basal samples (8W) in mice treated with the control SC-siRNA (24W, ATH progression) and in mice treated with the specific anti-CD40 siRNA (siRNA treatment, 24W). Shown are the Cts for each individual sample. Expression data was extracted from the original Taqman Low Density Array card (TLDA) experiment [25].

Table 1. Expression of the individual members of the murine miR-125a/miR-125b-1/miR-125b-2 gene clusters in ATH progression and after treatment with the specific anti-CD40 siRNA. The different miRNAs were tested by Taqman Low Density Array card (TLDAs) in the different experimental conditions stated [25], and individual expression values were normalized to U6 expression (ΔCt Gene-U6 snRNA) and expressed as mean ±SD of the different samples. C8W, control mice at 8 weeks. SC24W, mice treated with the scrambled control siRNA for 24 weeks (control group for ATH progression, see Materials and Methods). T24W, mice treated with the anti-CD40 siRNA for 24 weeks (treatment group). (*) Only one sample could be analyzed.

	miRNAExpression								
	Cluster 99b/let7e/125a			Cluster 100/let7a-2/125b-1			Cluster 99a/let7c-1/125b-2		
	99b	let7e	125a	100	let7a-2	125b-1	99a	let7c-1	125b-2
C8W (n = 2)	10.95 ± 0.2	8.22 ± 0.2	9.43 ± 0.5	10.64 ± 0.5	13.47 *	10.47 ± 0.6	10.6 ± 0.5	10.8 ± 0.6	10.47 ± 0.6
SC24W (n = 3)	4.51 ± 1.3	3.35 ± 1.8	6.7 ± 0.6	5.57 ± 0.8	7.20 ± 1.8	3.97 ± 0.6	6.32 ± 0.3	3.03 ± 2.2	3.97 ± 0.6
T24W (n = 3)	6.27 ± 2.8	4.99 ± 2.8	7.16 ± 1.6	6.15 ± 1.8	7.5 ± 2.8	6.26 ± 2.0	6.98 ± 2.1	5.74 ± 4.1	6.26 ± 2.0

3.2. *CD34-Positive CellsAre Clustered in Human Aortic Lesions*

We had previous evidences of the involvement of CD34 cells in the process of ATH progression (M. Hueso, unpublished results), so that we performed a more complete study of CD34-positive cells during ATH progression. We first analyzed human abdominal aortas from deceased patients by IHQ to detect CD34-positive cells. Figure 2 shows representative views of the human neointimae (Figure 2A,B) and adventitiae (Figure 2D) from aortic sections with ATH plaques, as well as a representative section from a region with no plaques (Figure 2C), all of them stained with the anti-CD34 antibody. It can be clearly seen the presence of clusters of CD34-positive cells in the lesions (Figure 2A,B,D), which at a larger magnification showed a fibroblastoid morphology (arrows at Figure 2B). Interestingly, the non-lesion tissue showed a different distribution of the CD34-positive cells which were non detected in the intima, but scattered in muscle layer (Figure 2C for a representative field). We could not analyze the perivascular adipose tissue (PVAT) because it was lost during preparation of the human samples.

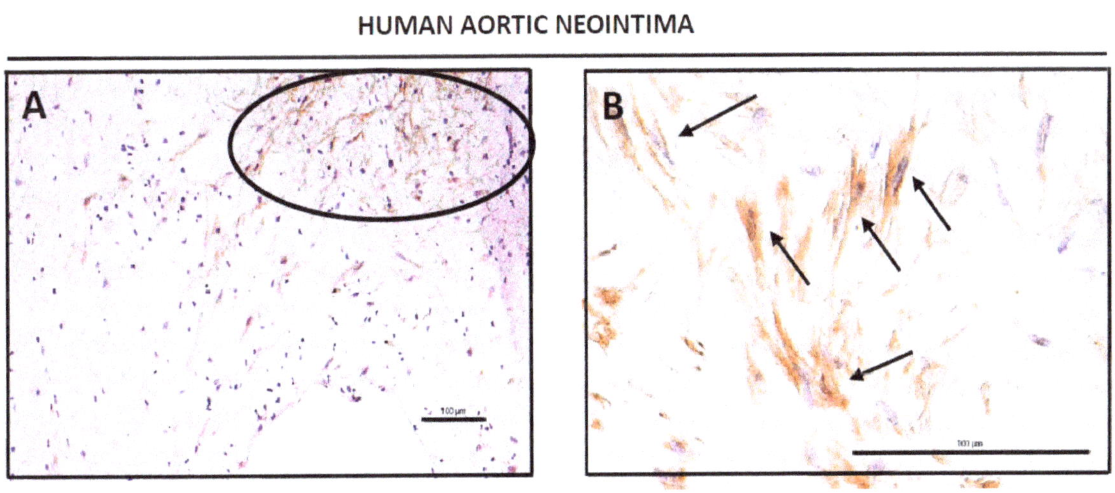

Figure 2. *Cont.*

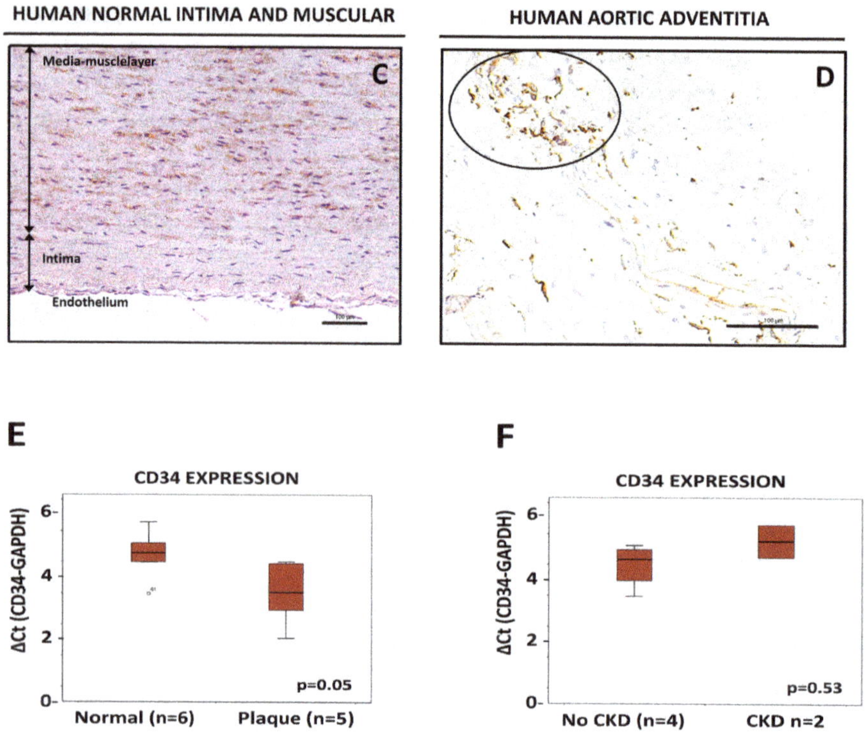

Figure 2. Detection of CD34-positive cells in human aortic tissue. Whole human arteries were isolated, sliced, prepared for IHQ analysis and stained for CD34 as described in Material and Methods. (**A**,**B**) Shown are representative images of two different neointimas, stained for anti-CD34, at different magnifications (20× in 3A, 40× in 3B). The circle in (**A**) shows a cluster of CD34-positive cells, and arrows in (**B**) show±± individual fibroblastoid cells positive for CD34 staining. Tissues were counterstained with hematoxilin-eosin. Bars are 100 µm. (**C**) Representative image of a human aortic tissue from a normal section (non-lesion) showing the intima and the muscular layer (vertical arrows) and stained for anti-CD34. Tissue was counterstained with hematoxilin-eosin. Bar is 100 µm. (**D**) Representative image of a human adventitia stained for anti-CD34. The circle in (**D**) shows a cluster of CD34-positive cells. Tissue was counterstained with hematoxilin-eosin. Bar is 100 µm. (**E**) Expression of *CD34* mRNA measured by qPCR in human ATH plaques compared with normal abdominal aortas. (**F**) Expression of *CD34* mRNA in aortic tissue with normal vascular walls from CKD patients when compared with non-CKD patients.

We next dissected human ATH samples into plaque and normal tissue, which were tested by qPCR for mRNA levels of *CD34*. Figure 2E show that *CD34* mRNA levels in the plaque were significantly higher than in normal aortic tissue (ΔCt = 3.5 ± 1.03 cycles in plaque vs. 4.7 ± 0.76 cycles in normal tissue; $p = 0.05$). Since ATH is a frequent complication of Chronic Kidney Disease, we also tested *CD34* mRNA expression in human normal aortas of CDK and non-CKD patients, but only a trend to a lower expression of *CD34* in CKD samples was found (ΔCt = 4.47 ± 0.71 cycles in non-CKD vs. 5.26 ± 0.75 cycles in CKD; $p = 0.53$, see Figure 2F).

We took advantage of the mouse model of ApoE-deficient mice treated with an anti-CD40 siRNA to study the presence of CD34-positive cells in aortic tissues of animals subjected to different treatments. As in humans, we also found clusters of CD34-positive cells in the intima and adventitia of the murine samples (M. Hueso, unpublished result). Furthermore, we had the opportunity of studying the PVAT in these murine samples. As seen in Figure 3A,B individual CD34 cells were found in the perivascular aortic fat (PVAT) of mice treated with the control, scrambled siRNA (Figure 3A) or with the anti-CD40 siRNA (Figure 3B). When different fields were analyzed and counted, it arose that the number of CD34-positive cells was lower in the PVAT of mice treated with the CD40-siRNA when compared with SC-siRNA treated animals (from 14 ± 9% CD34+ cells per field in SC-animals vs. 12 ± 9% CD34+ cells per field in siCD40-treated animals, see Figure 3C), as well as in the adventitia (18 ± 6% CD34+ cells per field vs. 9 ± 3% respectively; $p < 0.008$, see Figure 3C) or in the intima (11 ± 1% CD34+ cells per field vs. 3 ± 6% respectively; $p < 0.002$, see Figure 3C) suggesting that ATH lesions were associated with an increase in the number of tissue resident CD34 cells.

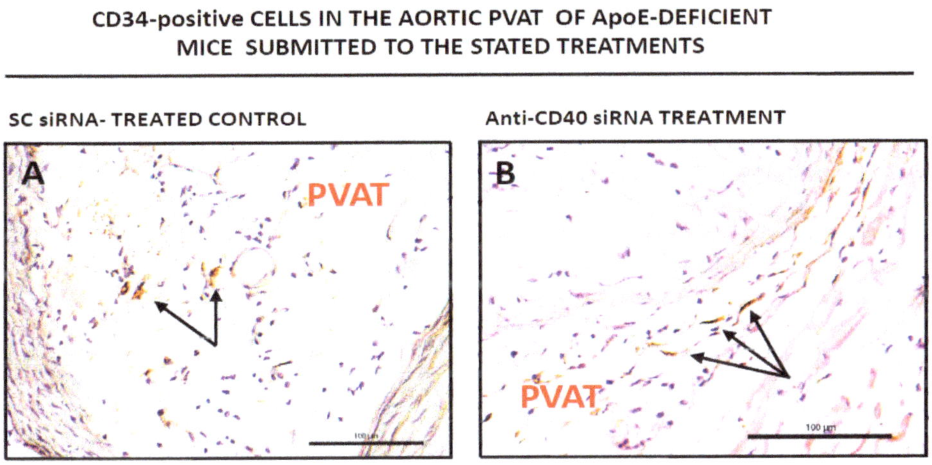

C

% OF CD34-positive CELLS IN AORTIC TISSUES OF ApoE-DEFICIENT MICE SUBMITTED TO THE STATED TREATMENTS

	Treated with SC-control (24W)	Treated with αCD40-siRNA (24W)
PVAT (p=0.5 ; n.s.)	14±9	12±9
ADVENTITIA (p<0.008)	18±6	9±3
INTIMA (p=0.002)	11±1	3±6

Figure 3. Detection of CD34-positive cells in the murine aortic perivascular tissue. Aortas from ApoE-deficient mice treated with a scrambled (SC) siRNA for 24 weeks as control, or with an anti-CD40 specific siRNA also for 24 weeks, were isolated, sliced, prepared for IHQ analysis and stained for CD34 as described in Material and Methods. (**A**,**B**) Show representative fields with individual CD34-positive cells (arrows) in the SC-siRNA control (**A**) or in the anti-CD40 specific siRNA (**B**). Tissues were counterstained with hematoxilin-eosin. Bars are 100 µm. (**C**) Quantification of CD34-positive cells in the PVAT, intima and adventitia of aortas of ApoE-deficient mice submitted to the stated treatments. A total of 10 fields were examined by two different researchers and the result expressed as % of CD34-positive cells with regard of the total number of eosin-positive nuclei in the field.

3.3. An Exonic Switch Regulates Differential Accession of miR-125, and Other miRNAs, to the CDS or to the 3'UTR of the Cd34 Transcript

We examined the relationship of *Cd34* with miRNAs, among them the miR-125a, during ATH progression. The murine *Cd34* gene is composed by 8 exons and is submitted to a complex splicing regulation. Exon 8 includes an internal/cryptic splice site (I/CSS, Figure 4A) and two in-frame stop-codons (TGA1 and TGA2). Activation of the usual-I/CSS splice sites lead to the expression of two different *Cd34* mRNAs. The transcript variant 1, TV1, links exon 7 to the entire exon 8 and incorporates a premature stop-codon (TGA1 at Figure 4B), which originates a short, truncated form of CD34 lacking almost the entire intracellular domain (Figure 4C). The second *Cd34* transcript, TV2, links exon 7 to the I/CSS (Figure 4A) and incorporates a partial exon 8 which lacks the premature stop-codon and terminates at the TGA2 (Figure 4B) to originate a "full-length" CD34 protein (Figure 4C). In this way, the long transcript (including the entire exon 8) originates a short protein by activating a premature STOP codon, while the short transcript (incorporating a partial exon 8) originates the long CD34 protein (Figure 4B).

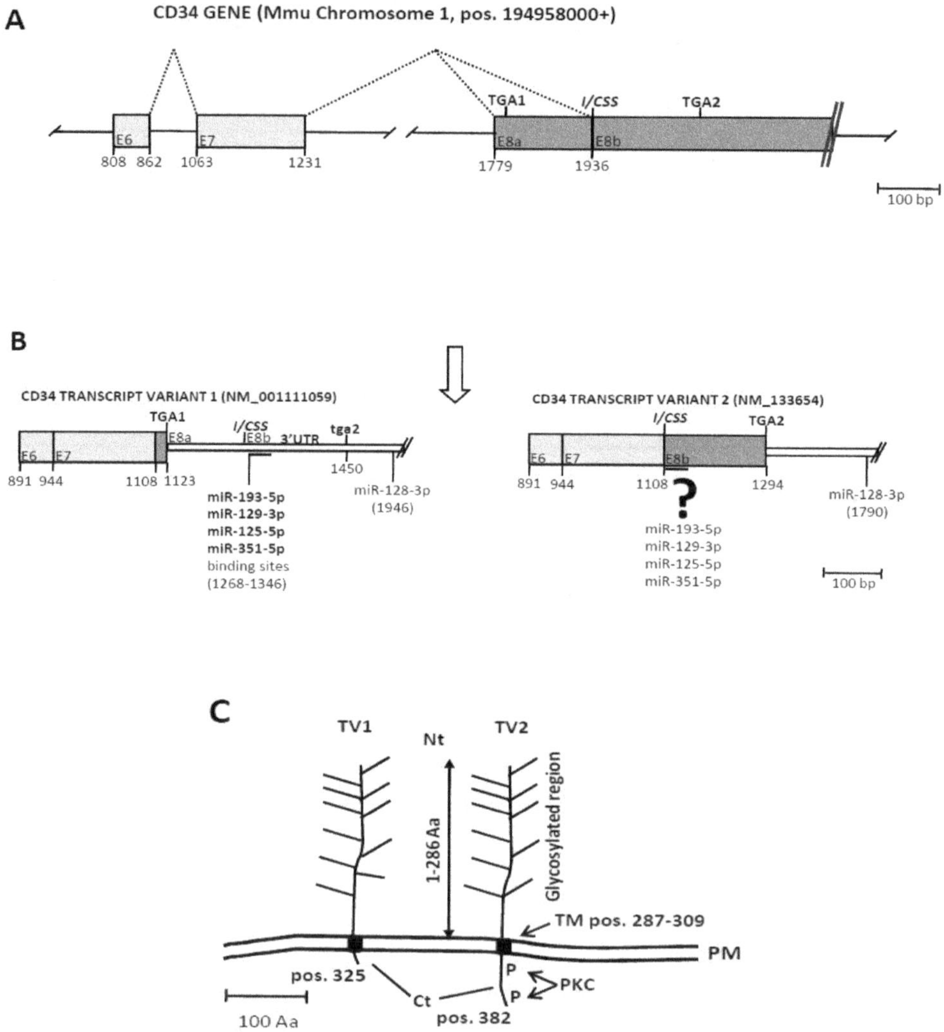

Figure 4. Structure and expression of the murine *Cd34* genetic loci and its interaction with miR-125 and other miRNAs. (**A**) Diagram of the murine *Cd34* gene. Exons are represented by colored boxes, intergenic regions by solid lines and splicing events by doted lines. Shown are exons 6,7 and 8 only. Former exon X is here shown as part "a" of exon 8 (see text). Shown are also the two stop codons (TGA1, TGA2) at exon 8, as well as the internal/cryptic splice site (*I/CSS*) at the virtual juncture of exon 8a/b. Numbering refers to the position in the NCBI Reference Sequence: NC_000067.6. Drawn to scale (1cm = 100bp). (**B**) Diagram showing the two *Cd34* transcripts originated by the internal/cryptic splice event at exon 8. Coding regions are expressed as colored boxes while the 3'UTRs are shown as white bars. Also shown are the two stop codons (the active one in capital letters), the internal/cryptic splice site and the miRNA binding sites predicted by the Targetscanmouse at the 3'UTR of *Cd34* transcript variant 1. Binding sites at the equivalent positions at the coding region of transcript variant 2 are shown under the question mark. Numbering refers to the Genbank accession numbers of the transcripts. Drawn to scale (1cm = 100bp). Ensembl code for *Cd34*-TV1: ENSMUST00000110815. Ensembl code for *Cd34*-TV2: ENSMUST00000016638. (**C**) Diagram showing the two different proteins encoded by CD34 transcripts variant 1 and 2 (TV1, TV2). Shown are the glycosylated Nterminal extracellular domain (Nt), the transmembrane domain (TM) and the different Cterminal domains (Ct) of both protein isoforms. Protein kinase C phosphorylation sites are labelled in TV2. Numbering refers to the Genbank accession numbers of the transcripts (see B). Drawn to scale (1cm = 100Aa). PM stands for plasma membrane.

Bioinformatic analysis with mouse "Targetscan" of the *Cd34* transcript allowed the detection of binding sites for miR-193, miR-129, miR-125, and miR-351clustered into 70 bp of sequence immediately downstream of the I/CSS. Interestingly, these binding sites would be located at the 3'UTR of *Cd34*-TV1 but at the coding region of *Cd34*-TV2 (Figures 4B and 5A), depending on the splice event that took place, so that a differential splicing would direct the binding of miR-125 (and the other miRNAs) to alternative locations in the *Cd34* transcript, the 3'UTR of *Cd34*-TV1 or the coding region of *Cd34*-TV2 (Figure 5A). Interestingly, two of the miRNAs that targeted the exon 8b of *Cd34*, miR-125 and miR-351 had an almost identical seed region (Figure 5A). Furthermore, another binding site for miR-128-3p was present at the far 3'UTR of both *Cd34* isoforms, but its location was not affected by the splicing event (see Figure 4B).

A

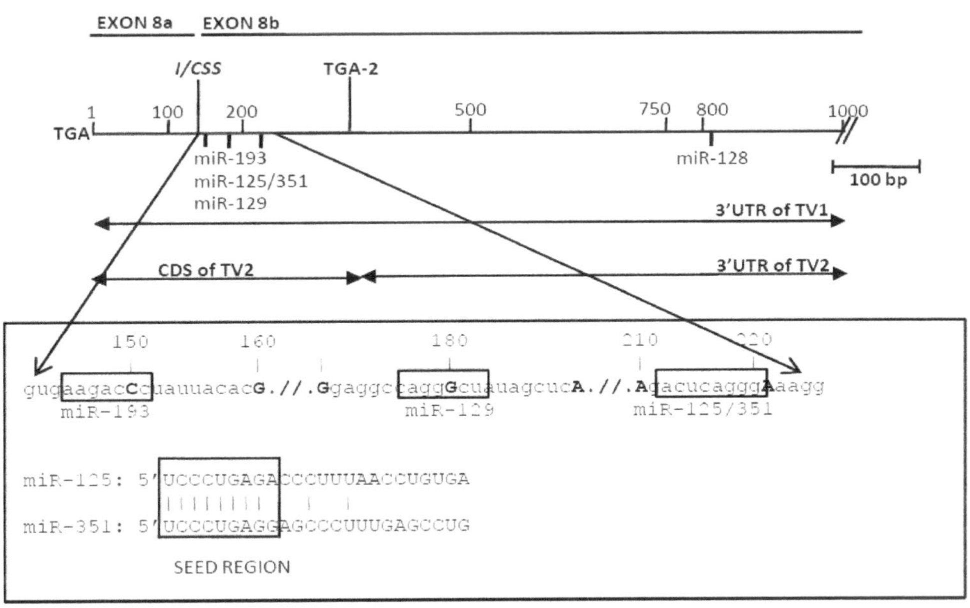

B

Expression of miRNAs targeting CD34-TV1 at different conditions					
	miR-193	**miR-129**	**miR-125**	**miR-351**	**miR-128**
C8W (n=3)	4.59± 2.10	17.88±0.48	9.43±0.5	10.07*	17.14*
SC24W (n=3)	2.77±1.83	9.7±0.48	6.07±0.6	14.28±2.02	12.04±1.66
siCD40-24W (n=3)	8.74±4.67	9.65±2.27	7.16±1.6	16.02±3.76	13.63±3.36

Figure 5. Detailed structure of the *Cd34* region harboring the predicted miRNA binding sites. (**A**) Diagram of 1 kb of sequence at the 3'end of *Cd34*. Shown is the position of the two stop codons (TGA1 and TGA2), the internal/cryptic splice site (I/CSS) and the position of the miRNA binding sites. Expanded is the sequence of the 70 bp, which concentrate the miRNA binding sites. Boxed, the seed sequences of the miRNAs and the comparison among those of miR-125 and miR-351. (**B**) Expression of the miRNAs targeting *Cd34* at different experimental conditions. Data shown after normalization to U6-snRNA. Expression data extracted from the original Taqman Low Density Array card (TLDA) experiment [25]. (*) Only one sample could be analyzed.

Analysis of the expression of the *Cd34*-targeting miRNAs (Figure 5B) showed that miR-193, miR-129, miR-125 and miR-128 shared the same expression pattern characterized by an overexpression in SC24W (i.e., disease progression) that was reversed by the anti-CD40 treatment. On the other hand, miR-351 was the only one whose expression decreased in SC24W, with its values not recovering after si-CD40 treatment.

Lastly, we measured expression *Cd34* expression by qPCR in the SC24W and siCD40-24W samples from our disease model, but unfortunately we did not have (basal) control due to sample exhaustion. We could test expression of *Cd34*−TV1 but not of TV2 since there was not a specific commercial Taqman probe for this isoform, so instead we used a probe that amplified both isoforms. The results confirmed our previous results for CD34 mRNA and protein (as number of cells expressing CD34): in the sense that *Cd34*-TV1 and *Cd34*−(TV1 + TV2) were more expressed in disease progression SC24W (TV1: 6.24 $\pm$ 0.62 cycles; (TV1 + TV2): 4.73 $\pm$ 0.75 cycles) than in the treatment group siCD4/24W (TV1: 7.88 $\pm$ 0.37 cycles; (TV1 + TV2): 6.42 $\pm$ 0.50 cycles). Furthermore, the difference among *Cd34*−TV1 and *Cd34*−(TV1 + TV2) indicated that both, TV1 and TV2 were expressed.

4. Discussion

ATH progression has an hematopoietic component that includes expansion of the hematopoietic stem-cell compartment [39], and we are interested in the role of the stem-cell genes *Cd34* and *Lin28a/b* in the development of atherosclerotic lesions. Originally characterized as an hemopoietic progenitor cell marker [40], CD34 was subsequently detected in vascular endothelial cells [41] and neovascularized tissues [42]. CD34 has been shown to be important in the development of ATH lesions, mainly in neo/re-vascularization events, with CD34-positive cells being detected more frequently in inflammatory-erosive plaques, when compared with plaques of the lipid or degenerative-necrotic plaques [43]. It is thus evident that elucidating the mechanisms of *CD34* expression and CD34-cell differentiation would have a positive impact on CVD disease.

We have previously demonstrated the involvement of miR-125b in the progression of ATH [25], but our data was inconclusive about the role of the other highly related member of the family, miR-125a (Figure 1A,B). It is interesting to notice that *Cd34*, *Lin28a/b* and the miR-125 family shared a molecular link, since data mining of miRNA targets showed that *Cd34* and *Lin28a/b* were predicted targets of miR-125a/b. Furthermore, miR-125a was recently described as a regulator of hemopoietic progenitor cell stemness [44]. In this context, we have addressed the study of the relationship among *Cd34* and miR-125(a,b) with the aim to generate new tools for the study of ATH progression.

We firstly substantiated the involvement of miR-125a in ATH progression by reanalyzing our previous miRNA data to compare expression of the highly related miR-125a and miR-125b. As seen in Figure 1C, while expression of both (in absolute Ct values) was very similar in the control samples, miR-125a was slightly less expressed than miR-125b during ATH progression (SC1-SC3), thus justifying why miR-125a did not show up in our earlier analysis of ATH-progression related miRNA genes. Nevertheless, after treatment with the specific anti-CD40 siRNA both miRNAs showed similar expression levels. The difference in expression levels was confirmed after normalization with the U6-snRNA control (Table 1), and could be likely due to the fact that miR-125b was encoded by two different genes (b-1 and b-2) while miR-125a was encoded by a single gene (Figure 1A).

We next studied CD34 expression in human ATH plaques, as well as in aortic tissues from the mice. CD34-positive cells could be detected as clusters of cells in the neointimas and adventitias of human (Figure 2) and mouse (Figure 3) diseased aortas. Furthermore, individual CD34-positive cells could also be detected in the murine perivascular adipose tissue (PVAT) (Figure 3). Interestingly, in aortic tissue from non diseased aortas the distribution of CD34-positive cells was totally different, being mostly detected as scattered cells in the media layer (Figure 2C). One possibility is that this differential distribution reflected distinct migratory or homing abilities of CD34-positive cells in the context of normal or ATH-diseased aortic tissue.

We next quantified CD34 expression/positive cells by two methods in our two disease models, by PCR (human, Figure 2E) and by directly counting CD34-positive cells (mouse, Figure 3C). Both approaches showed an increase in CD34 in ATH lesions. Increased levels of CD34 mRNA were detected in human ATH plaques when compared with normal aortic tissue (Figure 2E) and more cells were detected in SC24W mice (disease progression) than in siCD40 mice (Figure 3). This is very interesting because most works on CD34-positive PCs and ATH progression has been made with blood-circulating cells and there are scant data on lesion resident cells. Furthermore, a number of authors showed a direct correlation among newly formed blood vessels (as CD34-positive cells) at the neointima and ATH progression [45], that statin treated patients had reduced intraplaque angiogenesis [46], or that a catechin supplementation in the diet of apoE deficient mice resulted in a reduction of ATH lesions and a significant reduction in CD34 expression [47], in line with the results here shown. On the other hand ATH progression has been related with reduced levels of circulating EPCs [48], while a diminution in circulating CD34+/CD45(dim)/VEGFR2- or CD34+/CD45(dim)/CD133+/VEGFR2-cells was shown to highlight patients with coronary endothelial dysfunction [49].

Thus, it is clear that ATH progresses with a simultaneous increase in *CD34* expression and in the number of CD34-positive resident cells, a fact that would be apparently in full contradiction with a role for the miR-125 family in the regulation of *CD34* mRNA stability, since these are also overexpressed in ATH progression. Nevertheless, the end of the coding domain and 3'UTR region of the *CD34* gene is a very complex genomic region, and upon a detailed analysis two features arose that could help to understand the contradictory scenario. Firstly, this region is very poor in predicted miRNA binding sites (after Targetscan), with only five of them, four clustered in a short, 70 bp region. This must be a distinctive feature of the CD34 gene since, on the other hand, only one miRNA (miR-665) has been experimentally confirmed as targeting CD34 [50], although this was not predicted by Targetscan. Furthermore, two of these miRNAs, miR-125 and miR-351, shared the same seed sequence, but only this last showed an expression profile (downregulated in ATH progression) compatible with being a CD34 efector (Figure 5B), thus casting serious doubts on the role of the miR-125 family in the regulation of CD34 stability. It is thus possible that only miR-351, but not miR-125, be involved in CD34 regulation, and that data-mining and bioinformatic analysis were misled by the similarities among their seed sequences. In this sense, miR-351 has been characterized as a developmental regulator that represses differentiation of mesenchymal stem cells (to osteoblasts) by targeting the vitamin D receptor [51], or suppresses angiogenesis [52], roles compatible with its downregulation during ATH progression in our animal model

The second feature to consider when dealing with miRNAs targeting *Cd34* is the structure of its 3'end region. Originally, the *Cd34* gene was shown to be structured in 8 exons [53], and a subsequent work showed that an alternative splicing event at exons X and 8 generated two transcript variants which differed in the Cterminal domain of the protein, generating an isoform almost lacking this domain and another one with a full-length Cterminal domain [10]. This last one included two PKC phosphorylation cytosplamic sites [54], suggesting that the truncated form was unable for signaling. Nevertheless, we have carefully reanalyzed the genomic sequence data of *Cd34* and concluded that the splicing event at exon 8 was more compatible with a splicing at an internal/cryptic splice site than with an alternative splicing as previously described, since the originally defined alternative exon X is in fact part of exon 8 and there is no intronic sequence among them.

This cryptic splicing (exon switch) is intriguing because it would have a compelling consequence on the miRNA binding sites of *Cd34*, since these would be located at the beginning of the 3'UTR sequence at the *Cd34*-TV1, but at the end of the coding region in *Cd34*-TV2 (Figures 4 and 5). This is very interesting because recent work suggested that miRNAs targeting coding regions would cause translational inhibition rather than degradation of the transcript (see [55] and references therein). In this way *Cd34*-TV2 could be also considered as a "miRNA sponge" that would decrease microRNA availability and relieve the repression of target RNAs through sequestering miRNAs away from parental mRNA. Should this be the case, this could have an impact on the balance of CD34 isoform expression, as well as in the function of miR-125s since these could bind to sites in different transcript contexts (3′UTR and CDS) and likely compete with miR-351 for the binding site. This complex miRNA binding region at the 3′ end of the *Cd34* transcript is worth studying "in vitro" with advanced tools of molecular biology to describe its real impact on the dynamics of the *Cd34* transcript and protein as well as on the cells expressing this membrane marker and their role in human disease.

In conclusion, here we have demonstrated an increase in tissue resident CD34-positive cells in the aortic tissue of human and mice during ATH progression. Secondly, although we have failed to provide a sound functional relationship among miR-125 and *Cd34*, we have detected miR-351 as a possible, alternative *Cd34* effector. Lastly, we describe a novel regulatory mechanism of miRNA function by which a cryptic splice event at the *CD34* gene could regulate the accessibility of miR-125a to the 3′UTR or to the coding region of the *CD34* transcript, likely producing different effects on its translation or its stability. Work is in progress to confirm this hypothesis.

Author Contributions: Formal analysis, M.H. and E.N.; Funding acquisition, M.H., J.M.C. and J.T.; Investigation, M.H., J.M.C., J.T. and E.N.; Methodology., M.H. and E.N.; Project administration, J.M.C. and J.T.; Resources, J.M.C. and J.T.; Supervision, M.H., J.M.C., J.T. and E.N.; Writing-original draft, M.H. and E.N.; Writing-review & editing, M.H., J.M.C., J.T. and E.N.

Acknowledgments: We thank the CERCA program/Generalitat de Catalunya, Spain, for institutional support.

References

1. Hueso, M.; Cruzado, J.M.; Torras, J.; Navarro, E. ALUminating the Path of Atherosclerosis Progression: Chaos Theory Suggests a Role for Alu Repeats in the Development of Atherosclerotic Vascular Disease. *Int. J. Mol. Sci.* **2018**, *19*, 1734. [CrossRef] [PubMed]
2. Andreou, I.; Tousoulis, D.; Tentolouris, C.; Antoniades, C.; Stefanadis, C. Potential role of endothelial progenitor cells in the pathophysiology of heart failure: Clinical implications and perspectives. *Atherosclerosis* **2006**, *189*, 247–254. [CrossRef]
3. Urbich, C.; Dimmeler, S. Endothelial progenitor cells: Characterization and role in vascular biology. *Circ. Res.* **2004**, *95*, 343–353. [CrossRef]
4. Du, F.; Zhou, J.; Gong, R.; Huang, X.; Pansuria, M.; Virtue, A.; Li, X.; Wang, H.; Yang, X.F. Endothelial progenitor cells in atherosclerosis. *Front. Biosci.* **2012**, *17*, 2327–2349. [CrossRef]
5. Rohde, E.; Malischnik, C.; Thaler, D.; Maierhofer, T.; Linkesch, W.; Lanzer, G.; Guelly, C.; Strunk, D. Blood monocytes mimic endothelial progenitor cells. *Stem Cells* **2006**, *24*, 357–367. [CrossRef] [PubMed]
6. Hill, J.M.; Zalos, G.; Halcox, J.P.; Schenke, W.H.; Waclawiw, M.A.; Quyyumi, A.A.; Finkel, T. Circulating endothelial progenitor cells, vascular function, and cardiovascular risk. *N. Engl. J. Med.* **2003**, *348*, 593–600. [CrossRef] [PubMed]
7. Verma, S.; Kuliszewski, M.A.; Li, S.H.; Szmitko, P.E.; Zucco, L.; Wang, C.H.; Badiwala, M.V.; Mickle, D.A.; Weisel, R.D.; Fedak, P.W.; et al. C-reactive protein attenuates endothelial progenitor cell survival, differentiation, and function: Further evidence of a mechanistic link between C-reactive protein and cardiovascular disease. *Circulation* **2004**, *109*, 2058–2067. [CrossRef] [PubMed]
8. Riesinger, L.; Saemisch, M.; Nickmann, M.; Methe, H. CD34(+) circulating cells display signs of immune activation in patients with acute coronary syndrome. *Heart Vessel.* **2018**. [CrossRef]
9. Leung, K.T.; Chan, K.Y.; Ng, P.C.; Lau, T.K.; Chiu, W.M.; Tsang, K.S.; Li, C.K.; Kong, C.K.; Li, K. The tetraspanin CD9 regulates migration, adhesion, and homing of human cord blood CD34+ hematopoietic stem and progenitor cells. *Blood* **2011**, *117*, 1840–1850. [CrossRef]
10. Suda, J.; Sudo, T.; Ito, M.; Ohno, N.; Yamaguchi, Y.; Suda, T. Two types of murine CD34 mRNA generated by alternative splicing. *Blood* **1992**, *79*, 2288–2295. [PubMed]
11. Lanza, F.; Healy, L.; Sutherland, D.R. Structural and functional features of the CD34 antigen: An update. *J. Biol. Regul. Homeost. Agents* **2001**, *15*, 1–13.

12. Lekka, E.; Hall, J. Noncoding RNAs in disease. *FEBS Lett.* **2018**, *592*, 2884–2900. [CrossRef]
13. Kozomara, A.; Griffiths-Jones, S. miRBase: Annotating high confidence microRNAs using deep sequencing data. *Nucleic Acids Res.* **2014**, *42*, D68–D73. [CrossRef] [PubMed]
14. Jackson, A.O.; Regine, M.A.; Subrata, C.; Long, S. Molecular mechanisms and genetic regulation in atherosclerosis. *Int. J. Cardiol. Heart Vasc.* **2018**, *21*, 36–44. [CrossRef] [PubMed]
15. Chung, A.C.; Lan, H.Y. MicroRNAs in renal fibrosis. *Front. Physiol.* **2015**, *6*, 50. [CrossRef] [PubMed]
16. Yang, S.; Ye, Z.M.; Chen, S.; Luo, X.Y.; Chen, S.L.; Mao, L.; Li, Y.; Jin, H.; Yu, C.; Xiang, F.X.; et al. MicroRNA-23a-5p promotes atherosclerotic plaque progression and vulnerability by repressing ATP-binding cassette transporter A1/G1 in macrophages. *J. Mol. Cell. Cardiol.* **2018**, *123*, 139–149. [CrossRef] [PubMed]
17. Rink, C.; Khanna, S. MicroRNA in ischemic stroke etiology and pathology. *Physiol. Genom.* **2011**, *43*, 521–528. [CrossRef] [PubMed]
18. Li, S.; Geng, Q.; Chen, H.; Zhang, J.; Cao, C.; Zhang, F.; Song, J.; Liu, C.; Liang, W. The potential inhibitory effects of miR19b on vulnerable plaque formation via the suppression of STAT3 transcriptional activity. *Int. J. Mol. Med.* **2018**, *41*, 859–867. [CrossRef]
19. Price, N.L.; Rotllan, N.; Canfran-Duque, A.; Zhang, X.; Pati, P.; Arias, N.; Moen, J.; Mayr, M.; Ford, D.A.; Baldan, A.; et al. Genetic Dissection of the Impact of miR-33a and miR-33b during the Progression of Atherosclerosis. *Cell Rep.* **2017**, *21*, 1317–1330. [CrossRef]
20. Chen, B.; Luo, L.; Wei, X.; Gong, D.; Jin, L. Altered Plasma miR-144 as a Novel Biomarker for Coronary Artery Disease. *Ann. Clin. Lab. Sci.* **2018**, *48*, 440–445.
21. Reddy, L.L.; Shah, S.A.; Ponde, C.K.; Rajani, R.M.; Ashavaid, T.F. Circulating miRNA-33: A potential biomarker in patients with Coronary Artery Disease (CAD). *Biomarkers* **2018**. [CrossRef] [PubMed]
22. Zhang, F.; Zhao, J.; Sun, D.; Wei, N. MiR-155 inhibits transformation of macrophages into foam cells via regulating CEH expression. *Biomed. Pharmacother.* **2018**, *104*, 645–651. [CrossRef] [PubMed]
23. Di Gregoli, K.; Mohamad Anuar, N.N.; Bianco, R.; White, S.J.; Newby, A.C.; George, S.J.; Johnson, J.L. MicroRNA-181b Controls Atherosclerosis and Aneurysms Through Regulation of TIMP-3 and Elastin. *Circ. Res.* **2017**, *120*, 49–65. [CrossRef] [PubMed]
24. Li, H.; Ouyang, X.P.; Jiang, T.; Zheng, X.L.; He, P.P.; Zhao, G.J. MicroRNA-296: A promising target in the pathogenesis of atherosclerosis? *Mol. Med.* **2018**, *24*, 12. [CrossRef] [PubMed]
25. Hueso, M.; De Ramon, L.; Navarro, E.; Ripoll, E.; Cruzado, J.M.; Grinyo, J.M.; Torras, J. Silencing of CD40 in vivo reduces progression of experimental atherogenesis through an NF-kappaB/miR-125b axis and reveals new potential mediators in the pathogenesis of atherosclerosis. *Atherosclerosis* **2016**, *255*, 80–89. [CrossRef] [PubMed]
26. Sun, Y.M.; Lin, K.Y.; Chen, Y.Q. Diverse functions of miR-125 family in different cell contexts. *J. Hematol. Oncol.* **2013**, *6*, 6. [CrossRef] [PubMed]
27. Graff, J.W.; Dickson, A.M.; Clay, G.; McCaffrey, A.P.; Wilson, M.E. Identifying functional microRNAs in macrophages with polarized phenotypes. *J. Biol. Chem.* **2012**, *287*, 21816–21825. [CrossRef]
28. Kim, S.W.; Ramasamy, K.; Bouamar, H.; Lin, A.P.; Jiang, D.; Aguiar, R.C. MicroRNAs miR-125a and miR-125b constitutively activate the NF-kappaB pathway by targeting the tumor necrosis factor alpha-induced protein 3 (TNFAIP3, A20). *Proc. Natl. Acad. Sci. USA* **2012**, *109*, 7865–7870. [CrossRef]
29. Chen, T.; Huang, Z.; Wang, L.; Wang, Y.; Wu, F.; Meng, S.; Wang, C. MicroRNA-125a-5p partly regulates the inflammatory response, lipid uptake, and ORP9 expression in oxLDL-stimulated monocyte/macrophages. *Cardiovasc. Res.* **2009**, *83*, 131–139. [CrossRef]
30. Xiu, L.; Xing, Q.; Mao, J.; Sun, H.; Teng, W.; Shan, Z. miRNA-125b-5p Suppresses Hypothyroidism Development by Targeting Signal Transducer and Activator of Transcription 3. *Med. Sci. Monit.* **2018**, *24*, 5041–5049. [CrossRef]
31. Wang, S.; Wu, J.; Ren, J.; Vlantis, A.C.; Li, M.Y.; Liu, S.Y.W.; Ng, E.K.W.; Chan, A.B.W.; Luo, D.C.; Liu, Z.; et al. MicroRNA-125b Interacts with Foxp3 to Induce Autophagy in Thyroid Cancer. *Mol. Ther.* **2018**, *26*, 2295–2303. [CrossRef] [PubMed]
32. Wu, Q.B.; Chen, J.; Zhu, J.W.; Yin, X.; You, H.Y.; Lin, Y.R.; Zhu, H.Q. MicroRNA125 inhibits RKO colorectal cancer cell growth by targeting VEGF. *Int. J. Mol. Med.* **2018**, *42*, 665–673. [CrossRef] [PubMed]
33. Natalia, M.A.; Alejandro, G.T.; Virginia, T.J.; Alvarez-Salas, L.M. MARK1 is a Novel Target for miR-125a-5p: Implications for Cell Migration in Cervical Tumor Cells. *MicroRNA* **2018**, *7*, 54–61. [CrossRef] [PubMed]

34. Tong, Z.; Liu, N.; Lin, L.; Guo, X.; Yang, D.; Zhang, Q. miR-125a-5p inhibits cell proliferation and induces apoptosis in colon cancer via targeting BCL2, BCL2L12 and MCL1. *Biomed. Pharmacother.* **2015**, *75*, 129–136. [CrossRef] [PubMed]
35. Hueso, M.; Torras, J.; Carrera, M.; Vidal, A.; Navarro, E.; Grinyo, J. Chronic Kidney Disease is associated with an increase of Intimal Dendritic cells in a comparative autopsy study. *J. Inflamm.* **2015**, *12*, 26. [CrossRef] [PubMed]
36. Cunningham, F.; Achuthan, P.; Akanni, W.; Allen, J.; Amode, M.R.; Armean, I.M.; Bennett, R.; Bhai, J.; Billis, K.; Boddu, S.; et al. Ensembl 2019. *Nucleic Acids Res.* **2018**. [CrossRef] [PubMed]
37. Agarwal, V.; Bell, G.W.; Nam, J.W.; Bartel, D.P. Predicting effective microRNA target sites in mammalian mRNAs. *eLife* **2015**, *4*. [CrossRef] [PubMed]
38. Shaham, L.; Binder, V.; Gefen, N.; Borkhardt, A.; Izraeli, S. MiR-125 in normal and malignant hematopoiesis. *Leukemia* **2012**, *26*, 2011–2018. [CrossRef]
39. Van der Valk, F.M.; Kuijk, C.; Verweij, S.L.; Stiekema, L.C.A.; Kaiser, Y.; Zeerleder, S.; Nahrendorf, M.; Voermans, C.; Stroes, E.S.G. Increased haematopoietic activity in patients with atherosclerosis. *Eur. Heart J.* **2017**, *38*, 425–432. [CrossRef]
40. Sutherland, D.R.; Watt, S.M.; Dowden, G.; Karhi, K.; Baker, M.A.; Greaves, M.F.; Smart, J.E. Structural and partial amino acid sequence analysis of the human hemopoietic progenitor cell antigen CD34. *Leukemia* **1988**, *2*, 793–803.
41. Fina, L.; Molgaard, H.V.; Robertson, D.; Bradley, N.J.; Monaghan, P.; Delia, D.; Sutherland, D.R.; Baker, M.A.; Greaves, M.F. Expression of the CD34 gene in vascular endothelial cells. *Blood* **1990**, *75*, 2417–2426. [PubMed]
42. Schlingemann, R.O.; Rietveld, F.J.; de Waal, R.M.; Bradley, N.J.; Skene, A.I.; Davies, A.J.; Greaves, M.F.; Denekamp, J.; Ruiter, D.J. Leukocyte antigen CD34 is expressed by a subset of cultured endothelial cells and on endothelial abluminal microprocesses in the tumor stroma. *Lab. Investig.* **1990**, *62*, 690–696. [PubMed]
43. Murashov, I.S.; Volkov, A.M.; Kazanskaya, G.M.; Kliver, E.E.; Chernyavsky, A.M.; Nikityuk, D.B.; Lushnikova, E.L. Immunohistochemical Features of Different Types of Unstable Atherosclerotic Plaques of Coronary Arteries. *Bull. Exp. Biol. Med.* **2018**, *166*, 102–106. [CrossRef] [PubMed]
44. Wojtowicz, E.E.; Lechman, E.R.; Hermans, K.G.; Schoof, E.M.; Wienholds, E.; Isserlin, R.; van Veelen, P.A.; Broekhuis, M.J.; Janssen, G.M.; Trotman-Grant, A.; et al. Ectopic miR-125a Expression Induces Long-Term Repopulating Stem Cell Capacity in Mouse and Human Hematopoietic Progenitors. *Cell Stem Cell* **2016**, *19*, 383–396. [CrossRef] [PubMed]
45. Nakano, T.; Nakashima, Y.; Yonemitsu, Y.; Sumiyoshi, S.; Chen, Y.X.; Akishima, Y.; Ishii, T.; Iida, M.; Sueishi, K. Angiogenesis and lymphangiogenesis and expression of lymphangiogenic factors in the atherosclerotic intima of human coronary arteries. *Hum. Pathol.* **2005**, *36*, 330–340. [CrossRef] [PubMed]
46. Koutouzis, M.; Nomikos, A.; Nikolidakis, S.; Tzavara, V.; Andrikopoulos, V.; Nikolaou, N.; Barbatis, C.; Kyriakides, Z.S. Statin treated patients have reduced intraplaque angiogenesis in carotid endarterectomy specimens. *Atherosclerosis* **2007**, *192*, 457–463. [CrossRef] [PubMed]
47. Auclair, S.; Milenkovic, D.; Besson, C.; Chauvet, S.; Gueux, E.; Morand, C.; Mazur, A.; Scalbert, A. Catechin reduces atherosclerotic lesion development in apo E-deficient mice: A transcriptomic study. *Atherosclerosis* **2009**, *204*, e21–e27. [CrossRef]
48. Schmidt-Lucke, C.; Rossig, L.; Fichtlscherer, S.; Vasa, M.; Britten, M.; Kamper, U.; Dimmeler, S.; Zeiher, A.M. Reduced number of circulating endothelial progenitor cells predicts future cardiovascular events: Proof of concept for the clinical importance of endogenous vascular repair. *Circulation* **2005**, *111*, 2981–2987. [CrossRef]
49. Boilson, B.A.; Kiernan, T.J.; Harbuzariu, A.; Nelson, R.E.; Lerman, A.; Simari, R.D. Circulating CD34+ cell subsets in patients with coronary endothelial dysfunction. *Nat. Clin.Pract. Cardiovasc. Med.* **2008**, *5*, 489–496. [CrossRef]
50. Fan, J.; Li, H.; Nie, X.; Yin, Z.; Zhao, Y.; Zhang, X.; Yuan, S.; Li, Y.; Chen, C.; Wang, D.W. MiR-665 aggravates heart failure via suppressing CD34-mediated coronary microvessel angiogenesis. *Aging* **2018**, *10*, 2459–2479. [CrossRef]
51. Hou, Q.; Huang, Y.; Luo, Y.; Wang, B.; Liu, Y.; Deng, R.; Zhang, S.; Liu, F.; Chen, D. MiR-351 negatively regulates osteoblast differentiation of MSCs induced by (+)-cholesten-3-one through targeting VDR. *Am. J. Transl. Res.* **2017**, *9*, 4963–4973. [PubMed]

52. Zhang, Y.; Liu, Y.; Zhang, H.; Wang, M.; Zhang, J. Mmu-miR-351 attenuates the survival of cardiac arterial endothelial cells through targeting STAT3 in the atherosclerotic mice. *Biochem. Biophys. Res. Commun.* **2015**, *468*, 300–305. [CrossRef] [PubMed]
53. Brown, J.; Greaves, M.F.; Molgaard, H.V. The gene encoding the stem cell antigen, CD34, is conserved in mouse and expressed in haemopoietic progenitor cell lines, brain, and embryonic fibroblasts. *Int. Immunol.* **1991**, *3*, 175–184. [CrossRef] [PubMed]
54. Fackler, M.J.; Civin, C.I.; Sutherland, D.R.; Baker, M.A.; May, W.S. Activated protein kinase C directly phosphorylates the CD34 antigen on hematopoietic cells. *J. Biol. Chem.* **1990**, *265*, 11056–11061. [PubMed]
55. Brummer, A.; Hausser, J. MicroRNA binding sites in the coding region of mRNAs: Extending the repertoire of post-transcriptional gene regulation. *BioEssays* **2014**, *36*, 617–626. [CrossRef]

9

Host–MicroRNA–Microbiota Interactions in Colorectal Cancer

Ce Yuan [1,2], Clifford J. Steer [3,4] and Subbaya Subramanian [1,2,*]

1 Bioinformatics and Computational Biology Program, University of Minnesota, Minneapolis, MN 55455, USA; yuanx236@umn.edu
2 Department of Surgery, University of Minnesota Medical School, Minneapolis, MN 55455, USA
3 Department of Medicine, University of Minnesota Medical School, Minneapolis, MN 55455, USA; steer001@umn.edu
4 Department of Genetics, Cell Biology, and Development, University of Minnesota Medical School, Minneapolis, MN 55455, USA
* Correspondence: subree@umn.edu

Abstract: Changes in gut microbiota composition have consistently been observed in patients with colorectal cancer (CRC). Yet, it is not entirely clear how the gut microbiota interacts with tumor cells. We know that tumor cells undergo a drastic change in energy metabolism, mediated by microRNAs (miRNAs), and that tumor-derived miRNAs affect the stromal and immune cell fractions of the tumor microenvironment. Recent studies suggest that host intestinal miRNAs can also affect the growth and composition of the gut microbiota. Our previous CRC studies showed a high-level of interconnectedness between host miRNAs and their microbiota. Considering all the evidence to date, we postulate that the altered nutrient composition and miRNA expression in the CRC microenvironment selectively exerts pressure on the surrounding microbiota, leading to alterations in its composition. In this review article, we present our current understanding of the role of miRNAs in mediating host–microbiota interactions in CRC.

Keywords: colorectal cancer; microRNAs; gut microbiota; metabolic interactions

---

1. Introduction

An average human intestine contains more than 100 trillion bacteria (collectively known as the gut microbiota) [1]. In recent decades, a number of studies have suggested that the gut microbiota is crucial to human health and to the development of diseases, including colorectal cancer (CRC) [2–7]. Those studies determined that altered microbiota composition and function (dysbiosis) is a common signature of CRC. Bacterial candidates such as *Fusobacterium nucleatum* and *Bacteroides fragilis* are consistently enriched in tumor tissues, and included in that signature. Specific factors in those bacteria, including FadA and Fap2 protein from *F. nucleatum* and *B. fragilis* toxins, that play a role in CRC pathobiology have been identified [8–16]. However, our knowledge of the vast majority of other bacteria associated with the CRC microenvironment is limited. Moreover, we are just beginning to understand the complex interactions between host and microbiota in CRC, as well as other clinical disorders including neurodegenerative diseases [17].

In healthy humans, a key factor associated with microbiota variations is host genetics [18–21]. In a study of healthy twins, Goodrich et al. found that host genetics drive microbiota composition and can also affect the host metabolic phenotype [19]. Several other studies have found an association between the abundance of *Bifidobacterium* species and the presence of single-nucleotide polymorphisms (SNPs) in close proximity to the host lactase gene locus [18,22]. This association suggests that the *Bifidobacterium* species conceivably assists the host in metabolizing lactose.

A recent CRC study found that loss-of-function mutations in the mitogen-activated protein kinase (MAPK) and Wnt signaling pathways are associated with specific sets of microbiota profiles [2]. Furthermore, mutations in the tumor-suppressor adenomatous polyposis coli (APC) gene are also associated with a distinct inter-microbiota association network [2]. These findings suggest that a common genetic factor might orchestrate the dynamic host–microbiota interaction(s) and functional relationship(s). Indeed, other recent studies have provided experimental evidence that microRNAs (miRNAs) can influence the survival and composition of gut bacteria [3,23,24]. Moreover, miRNAs have important intermediate roles in regulating CRC transformation and progression via the action of signaling pathways, including MAPK, Wnt, and APC [25,26]. MicroRNAs are small noncoding RNAs (about 22 nt) that play an important role in regulating and fine-tuning gene expression [27]. In mammalian cells, miRNAs regulate gene expression through posttranscriptional modifications in two distinct, albeit paired, mechanisms. First, if the miRNA has an extensive complementary binding site in the messenger RNA (mRNA) target, then it will guide the RNA-induced silencing complex (RISC) to cleave the mRNA, thus inhibiting translation. Second, if the miRNA only partially binds to the 3′ untranslated region (3′UTR) of the mRNA, then the miRNA-RISC will act to repress mRNA translation [28]. Both mechanisms lead to the decreased translation of mRNAs, which alters their respective downstream functions. Because miRNAs can act upon mRNA targets with limited complementarity, each miRNA can target a wide range of mRNAs in mammalian cells and each mRNA can be targeted by numerous miRNAs. More than 30% of human genes are estimated to have conserved binding sites in the 3′UTR [29]. Clearly, given this vast and enormously complex regulatory network, miRNAs are immensely important in regulating critical cellular processes. We are only now beginning to understand the sophisticated cross-talk of miRNAs, not only with each other, but with the myriad of target mRNAs.

Based on mounting evidence, we postulate that the altered nutrient composition and miRNA expression in the CRC microenvironment selectively influences the surrounding microbiota, leading to alterations in its composition. In this review, we present our current understanding of the role of miRNAs in mediating host–microbiota interactions in CRC (Figure 1). After highlighting the evidence pointing to their central role, we reflect on the future direction of this rapidly evolving field.

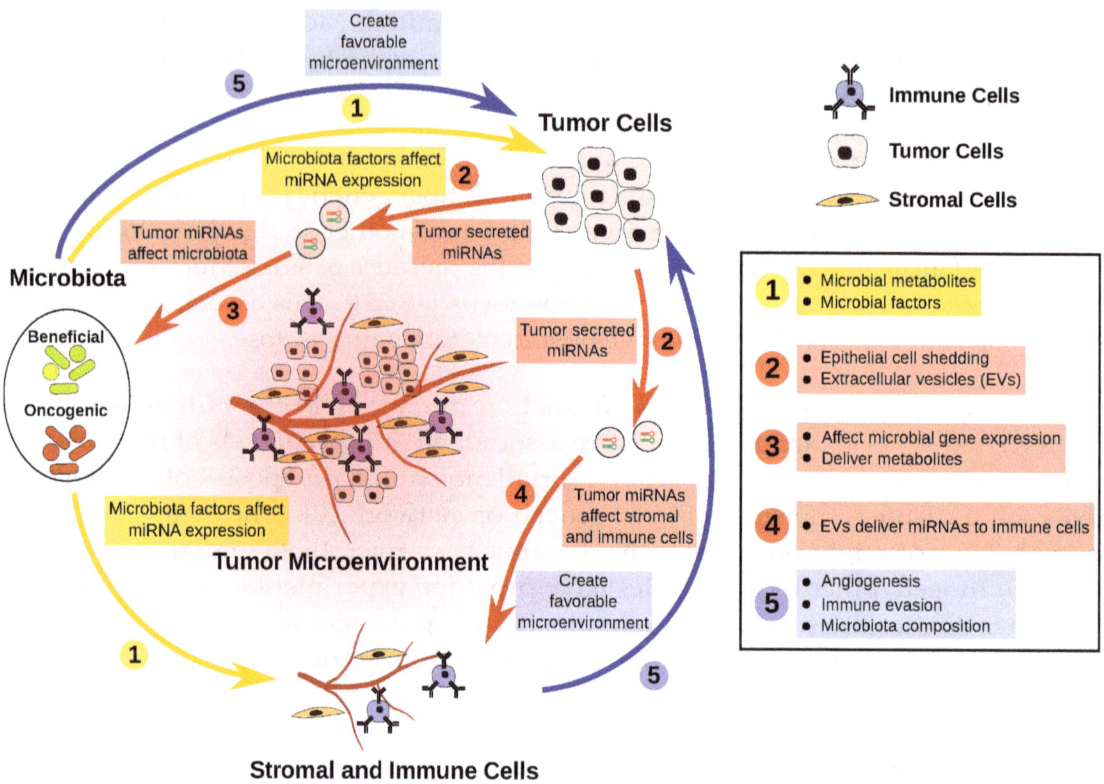

Figure 1. Host–microRNA–microbiota interactions in colorectal cancer. Microbiota composition has a functional effect on the cancer cells, via stromal and tumor infiltrating immune cells by regulating various cellular process (**1**). Microbial-metabolites and other secreted factors affect miRNA/gene expression profiles in cells present in the tumor microenvironment. In turn, tumor cells affect the microbiota composition of the stromal and tumor infiltrating immune cells through shedding of epithelial cells and/or secreting extracellular vesicles (EVs) containing miRNAs (**2**). The tumor-miRNAs alter the microbiota composition by affecting the gene expression of the microbiota and by delivering cancer-secreted metabolites (**3**). The tumor-derived miRNAs also have a role in regulating stromal and tumor infiltrating immune cells by affecting gene expression through miRNAs delivered in EVs (**4**). Such interactions will finally create a favorable microenvironment for tumor cells that include angiogenesis, immune evasion, and microbiota composition (**5**).

2. Microbiota and Colorectal Cancer

In the healthy intestine, the microbiota maintains a stable structure and actively participates in energy harvesting and nutrient production from undigested food [30,31]. However, this balance is disrupted in patients with CRC. Current evidence suggests that the microbiota regulates host functions via both metabolites and secreted factors.

2.1. *Microbial Metabolites*

In a normal colon, the microbiota produces a vast number of metabolites. Some of them, including vitamin K, biotin, and short-chain fatty acids (SCFAs), are essential for maintaining homeostasis in the colon microenvironment [31]. In fact, the major energy source (~70%) required by colon epithelium is butyrate, which is produced by the microbiota through fermentation of complex carbohydrates. Without the microbiota, the colon epithelium undergoes autophagy and fails to maintain its normal structure and function [32]. Similarly, mice lacking a microbiota (i.e., germ-free mice or those treated by broad-spectrum antibiotics) develop significantly fewer tumors in the colon [33–35]. However, in humans, using broad-spectrum antibiotics to treat CRC is not feasible, because of the risk of introducing harmful and highly resistant secondary infections such as *Clostridium difficile*.

In our current understanding, a few main classes of bacterial metabolites play a key role in the pathogenesis of CRC and the immune microenvironment. These metabolites include SCFAs, polyamines, secondary bile acids, and phytochemicals. Their role in CRC has been extensively reviewed and documented [31,36–38]. We have recently also explored the role of miRNAs in mediating the effect of microbial metabolites on CRC and its microenvironment [39].

2.2. Microbial Factors

Early studies consistently found a greater population of *F. nucleatum* in microbiota samples in patients with CRC than in healthy controls [40,41]. This bacterium is commonly found in the human oral microbiota and is frequently associated with gum diseases; it is, however, not commonly present in the gut microbiota. Through the Fap2 virulence factor, it uniquely binds with the D-galactose-β(1-3)-N-acetyl-D-galactosamine (Gal-GalNAc) carbohydrate moiety expressed on the tumor surface of CRCs [14]. Once it localizes to the CRC microenvironment, it targets the Wnt/β-catenin signaling pathway by binding, via association with the FadA virulence factor, to the E-cadherin protein on the cell surface [16]. The Wnt/β-catenin signaling pathway is critical during tumor initiation, tumor migration, and metabolic reprogramming [42–45]. The role of the Wnt/β-catenin signaling pathway in CRC has been previously reviewed [45].

Another bacterial protein targeting the same Wnt/β-catenin signaling pathway is the Bacteroides fragilis toxin (bft) produced by *B. fragilis* [46]. The bft virulence factor is able to bind to the E-cadherin protein, similar to that of FadA, but additionally cleaves the protein, which can alter the intestinal tight-junction function [47]. The Wnt/β-catenin pathway is a major signaling pathway that controls the expression of many important tumor-related genes, including MYC. The transcription factor MYC, transactivates miRNAs, such as the miR-17-92 cluster, that are highly expressed in CRC [48–51].

Additionally, *F. nucleatum* can also induce CRC cell proliferation by upregulating miR-21, via activation of the nuclear factor kappa-light-chain-enhancer of activated B cells (NF-κB) pathway via toll-like receptor 4 (TLR4) signaling [52]. The *Escherichia coli* bacterium harboring the pks genomic island also plays an important role in CRC. When CRC cells come in contact with the colibactin genotoxin produced by *E. coli*, the cells undergo cellular senescence [53,54]. This process is mediated by the cellular upregulation of miR-20a-5p, which results in the downregulation of sentrin-specific protease 1 (SENP1). This process then alters p53 small ubiquitin-like modifier (SUMO)ylation, which has been shown to affect the growth and metastasis of tumor cells [55].

In addition to factors that are virulent, many bacteria also produce beneficial factors that can reduce inflammation and modulate the immune system. In germ-free mice, early studies found impaired intestinal immune systems, which were amenable to treatment [56]. Specifically, the *B. fragilis* polysaccharide A (PSA) is one such immunomodulatory factor that maintains the proper function of CD4+ T cells [57]. Several other polysaccharides produced by *B. fragilis* are also beneficial in maintaining proper immune function. Immunization with *B. fragilis* polysaccharides, or the adoptive transfer of T cells specific to *B. fragilis*, can even boost the treatment effect of anti-cytotoxic T-lymphocyte antigen 4 (CTLA-4) immunotherapy [58]. The seemingly conflicting role of *B. fragilis* within gut bacteria is only the tip of the iceberg in current microbiota research and the fine and highly complex balance between functions.

In light of all this evidence, we created the first system-level map of interactions between host miRNAs and the microbiota [3]. Our comprehensive map helped us analyze correlations between host miRNA expression levels and mucosa-associated microbiota profiles, specifically in patients with CRC.

3. MicroRNAs and Colorectal Cancer

Previous studies have identified numerous aberrant miRNA expression patterns in CRC [25,59–62]. Specifically, the miR-17-92 cluster, miR-21, miR-182, and miR-503 are consistently overexpressed in tumor (vs. normal) tissues [3,26,48,49,59,63–71]. Any alteration(s) in expression levels of these miRNAs could, in turn, affect a wide array of downstream gene targets. Together, these miRNAs regulate all aspects of tumor pathobiology, including (i) altering tumor metabolism; (ii) promoting cell proliferation; (iii) stimulating angiogenesis; (iv) down-regulating tumor-suppressor genes; (v) promoting evasion of immune surveillance; and (vi) creating a favorable tumor microenvironment that promotes invasion and metastasis.

Our laboratory previously reported that, during the adenoma to adenocarcinoma transition, miR-182 and miR-503 were sequentially overexpressed and targeted the tumor-suppressor *FBXW7* gene [69]. Other researchers have observed, during CRC transformation, an increased expression of the miR-17-92 cluster and miR-21 [48,72]. In CRC adenocarcinoma, members of the miR-17-92 cluster target transforming growth factor-beta (TGF-β), which in turn stimulates angiogenesis in the tumor microenvironment, thus promoting tumor growth [70]. Additionally, miR-19, a member of the miR-17-92 cluster, downregulates expression of the tumor-suppressor phosphatase and tensin homolog (PTEN), thereby activating the protein kinase B (AKT)/mammalian target of rapamycin (mTOR) pathway in tumor cells [73]. The AKT/mTOR pathway is the main metabolic sensing pathway, responsible for regulating glucose transport into cells [74]. Since glucose is the main fuel source of CRC cells, an activated AKT/mTOR pathway promotes tumor cell proliferation [75].

The tumor-suppressor *PDCD4* gene, which is commonly downregulated in CRC, is a target of miR-21 [67]. Inhibiting the *PDCD4* gene can lead to an increase in the metastasis potential of tumor cells. Another important pathway commonly altered in CRC tumors is the Wnt/β-catenin pathway [25,26]. Dozens of miRNAs have been shown to extensively regulate the genes involved in the Wnt/β-catenin pathway [25].

The complex microenvironment of the CRC tumor also involves stromal cell and immune cell fractions, which can be regulated by cancer-derived miRNAs [76–78]. Studies have found that the miR-17-92 cluster, commonly overexpressed in CRC cells, is also upregulated in CRC stromal cells [68,72,79]. Strikingly, these miRNAs are not only endogenously produced by stromal cells, but also packaged in the microvesicles of tumor cells, and then delivered to stromal cells [80,81]. Similar intracellular regulation mediated by miRNAs is also found in immune cell fractions [82]. Additionally, endogenous miRNA dysregulation is prevalent in CRC immune cell fractions, usually as a downstream effect of tumor-secreted factors such as cytokines and chemokines [83–85]. Collectively, this evidence suggests that miRNAs are important in regulating tumor cells, in addition to maintaining the tumor microenvironment. It is clear that the relationship between miRNAs and CRC is multifaceted, interrelated, and highly complex.

4. Host Regulation of Microbiota Mediated by MicroRNAs

In reestablishing germ-free mice with a normal microbiota, studies have found altered intestinal miRNA profiles, suggesting that the microbiota regulates host miRNA expression [86,87]. Moreover, the responses of intestinal cells to facilitating the microbiota process depends on the cell type, and intestinal epithelial stem cells are especially sensitive to microbiota reestablishment [87].

Because miRNAs are highly stable, several studies in the clinical arena were able to detect higher levels of miR-21 and miR-92a, among other miRNAs, in the fecal samples of patients with CRC [65,88,89]. This finding facilitated in developing a noninvasive CRC screening method and delineating the potential role of miRNAs in interacting with the trillions of microbes in the human gut.

Intestinal miRNAs develop from two main sources, including the host and the food [23,24]. The intestinal epithelial cells are the main contributors of host-derived miRNAs, either via shedding of cells or excretion of exosomes. Evidence has shown that miRNAs from food can be absorbed by the host and can affect host gene expression [90–92]. But certain food-sourced miRNAs remain stable in the digestive tract and reach the intestines [93,94]. This evidence suggests that miRNAs can mediate cross-species regulation. The idea remains nascent, so insight into how miRNAs mediate host–microbiota interactions is still limited. Liu et al. first demonstrated such regulation, showing that miRNAs present in the feces can regulate gene expression and growth of bacteria [23]. Specifically, they found that mice lacking the Dicer gene, which enables mature miRNA processing, had different microbiota profiles than wild-type mice. More importantly, the study reported that hsa-miR-515-5p promoted the growth of *F. nucleatum* in vitro by targeting the 16S ribosomal RNA (rRNA) gene.

Notably, however, hsa-miR-515-5p shows very low expression levels in CRC tumors, so they are not significantly different from normal tissue. Thus, interactions between hsa-miR-515-5p and *F. nucleatum* might not be significant in CRC pathogenesis. However, more importantly, this study found that fecal miRNA transplantation restores fecal microbiota composition in mice with Dicer gene knockout. Several recent studies found that fecal microbiota transplantation (FMT) offers a potential therapeutic benefit that enables an immunotherapeutic response [35,95–98]. Based on growing evidence, it is plausible that fecal miRNAs play an important role in modulating the CRC microbiota as well as immunotherapy responses.

Recently, Teng et al. demonstrated that miRNAs encapsulated in plant-derived exosome-like nanoparticles (ELNs) can enter bacteria and alter bacterial genes [24]. The process for bacterial uptake of ELNs is determined primarily by the lipid composition of the outer membrane. They found that ELNs enriched with phosphatidylcholine were preferentially taken up by the *Ruminococcus* species, whereas ELNs enriched with phosphatidic acid (PA) were primarily taken up by *Lactobacillus rhamnosus*. After the ELNs are taken up by specific bacteria, the miRNA contents are released into bacterial cells. Teng et al. also found that mdo-miR7267-3p encapsulated in the PA-enriched ELNs targets the *Lactobacillus* monooxygenase ycnE, which then increases its production of indole-3-carboxaldehyde (I3A). The I3A metabolite then promotes interleukin-22 (IL-22) production and helps repair damaged colon mucosa [99].

There is developing evidence to support the notion that host or exogenous miRNAs might be biologically active in bacteria, thereby affecting bacterial gene expression. Although small RNAs similar to miRNAs exist in bacteria and function similarly to miRNAs, it remains unknown as to how miRNAs function in bacteria [100]. Several studies have reported that exogenous miRNAs from plant or animal sources can be taken up by human cells and exert biological functions [90–94,101–103]. Additional studies are required to ascertain whether miRNAs can indeed affect bacteria and to delineate the precise mechanism(s).

5. Metabolic Changes in Colorectal Cancer and Microbiota Mediated by MicroRNAs

The prevailing "driver-passenger" model suggests that dysbiosis in the CRC microbiota is initially caused by colonization of driver bacteria. This is followed by a gradual change in the tumor microenvironment, an increase in the number of driver bacteria, and secondary colonization of passenger bacteria that benefit from the changed environment [104]. That model, together with other studies, suggest that a gradual metabolic change in the tumor microenvironment during cancer progression could be the cause of dysbiosis [31]. Again, we explored that issue in our recent review of the role of miRNAs in mediating the effect of microbial metabolites on CRC and its microenvironment [39].

One of the hallmarks of tumor growth is their increased use of glycolysis as a main energy source, known universally as the Warburg effect [105]. Because the normal colon uses butyrate as its major energy source, any change in that source preferred by proliferating tumor cells will undoubtedly profoundly alter the nutrient composition of the tumor microenvironment [106,107]. Indeed, several studies have found altered metabolite levels in CRC tissues and stools [107–110]. A significantly lower glucose level and higher levels of lactate and fatty acids have been found in CRC tumor tissues, as compared with adjacent normal tissues. In stool samples from patients with CRC, a higher level of amino acids and a lower level of fatty acids have also been observed [108]. Interestingly, the CRC microbiota has shown reduced carbohydrate metabolism and an increase in the biosynthesis of amino acids and fatty acids [41]. In CRC, the switch in the nutrient source preferred by proliferating tumor cells appears to alter the nutrient composition in the tumor microenvironment. At the same time, the nutrient metabolism of the tumor microbiota seems to complement the nutrient needs of the tumor. This could be due to factors associated with the tumor nutrient microenvironment, and by the miRNAs excreted by tumor cells, on the surrounding microbiota [3]. Given the role of miRNAs in mediating such metabolic changes, we believe that miRNAs play a central, if not critical role, in mediating host–microbiota metabolic interactions in CRC.

6. Conclusions and Perspectives

With thousands of bacterial species living in the human digestive tract, it is becoming quite evident that they profoundly affect human health. Our review of the recent literature regarding CRC underscores a complex metabolic interplay between the host and its microbiota, mediated in part by miRNAs. Based on the current literature, we offer five major points in host–microbiota interactions mediated by miRNAs (Figure 1):

1. The CRC microbiota has reduced representation of beneficial bacteria. These bacteria produce metabolites and other factors that can potentially slow CRC progression, in part via the modulation of miRNAs that regulate tumor cells.
2. Dysregulation of miRNAs in tumor cells can affect the survival, or the gene expression, of certain bacteria in the microbiota.
3. Dysregulated miRNAs in tumor cells can be packaged and delivered to both stromal and immune cell fractions, creating a more favorable microenvironment for tumor cells.
4. Overrepresentation of oncogenic bacteria in the CRC microbiota can modulate tumor cells, as well as the tumor microenvironment, through miRNA modulation, thereby resulting in a more favorable condition for tumor growth.
5. This negative feedback loop perpetuates CRC progression.

Potential methods to break such a negative feedback loop include:

1. Interfering with host-mediated microbiota modulation by designing strategies to deliver anti-miRNAs to block the effect of host-miRNAs on the microbiota.
2. Modulating the microbiota through miRNAs that promote the growth of beneficial bacteria while suppressing the growth of oncogenic bacteria, in conjunction with chemotherapy or immunotherapy.

Based on both experimental and computational data, we conclude that miRNAs mediate and critically influence host–microbiota interactions. Clearly, miRNAs are a major part of a complex web of highly dynamic interactions. Other factors, such as nutrient availability in the CRC microenvironment, could also play an important role. In the future, it will be imperative to use a combination of approaches to comprehensively survey the CRC microenvironment, in order to discover all potential players in mediating such interactions.

Acknowledgments: S.S. is supported by research grants funded by the NIH R03CA219129 and Chainbreaker Grant; C.Y., by the MnDrive—University of Minnesota Informatics Institute graduate fellowship. Due to space restrictions, we cannot cite many other significant contributions made by numerous researchers and laboratories in this potentially important and rapidly progressing field. We thank Mary Knatterud for assisting in manuscript editing.

References

1. Turnbaugh, P.J.; Ley, R.E.; Hamady, M.; Fraser-Liggett, C.M.; Knight, R.; Gordon, J.I. The human microbiome project. *Nature* **2007**, *449*, 804–810. [CrossRef]
2. Burns, M.B.; Montassier, E.; Abrahante, J.; Priya, S.; Niccum, D.E.; Khoruts, A.; Starr, T.K.; Knights, D.; Blekhman, R. Colorectal cancer mutational profiles correlate with defined microbial communities in the tumor microenvironment. *PLoS Genet.* **2018**, *14*, e1007376. [CrossRef]
3. Yuan, C.; Burns, M.B.; Subramanian, S.; Blekhman, R. Interaction between host microRNAs and the gut microbiota in colorectal cancer. *mSystems* **2018**, *3*, e00205-17. [CrossRef] [PubMed]

4. Dejea, C.M.; Wick, E.C.; Hechenbleikner, E.M.; White, J.R.; Mark Welch, J.L.; Rossetti, B.J.; Peterson, S.N.; Snesrud, E.C.; Borisy, G.G.; Lazarev, M.; et al. Microbiota organization is a distinct feature of proximal colorectal cancers. *Proc. Natl. Acad. Sci. USA* **2014**, *111*, 18321–18326. [CrossRef]
5. Kostic, A.D.; Gevers, D.; Pedamallu, C.S.; Michaud, M.; Duke, F.; Earl, A.M.; Ojesina, A.I.; Jung, J.; Bass, A.J.; Tabernero, J.; et al. Genomic analysis identifies association of Fusobacterium with colorectal carcinoma. *Genome Res.* **2012**, *22*, 292–298. [CrossRef]
6. Shah, M.S.; DeSantis, T.; Yamal, J.-M.; Weir, T.; Ryan, E.P.; Cope, J.L.; Hollister, E.B. Re-purposing 16S rRNA gene sequence data from within case paired tumor biopsy and tumor-adjacent biopsy or fecal samples to identify microbial markers for colorectal cancer. *PLoS ONE* **2018**, *13*, e0207002. [CrossRef]
7. García-Castillo, V.; Sanhueza, E.; McNerney, E.; Onate, S.A.; García, A. Microbiota dysbiosis: A new piece in the understanding of the carcinogenesis puzzle. *J. Med. Microbiol.* **2016**, *65*, 1347–1362. [CrossRef] [PubMed]
8. Boleij, A.; Hechenbleikner, E.M.; Goodwin, A.C.; Badani, R.; Stein, E.M.; Lazarev, M.G.; Ellis, B.; Carroll, K.C.; Albesiano, E.; Wick, E.C.; et al. The Bacteroides fragilis toxin gene is prevalent in the colon mucosa of colorectal cancer patients. *Clin. Infect. Dis.* **2015**, *60*, 208–215. [CrossRef] [PubMed]
9. Toprak, N.U.; Yagci, A.; Gulluoglu, B.M.; Akin, M.L.; Demirkalem, P.; Celenk, T.; Soyletir, G. A possible role of *Bacteroides fragilis* enterotoxin in the aetiology of colorectal cancer. *Clin. Microbiol. Infect.* **2006**, *12*, 782–786. [CrossRef]
10. Wu, S.; Shin, J.; Zhang, G.; Cohen, M.; Franco, A.; Sears, C.L. The Bacteroides fragilis toxin binds to a specific intestinal epithelial cell receptor. *Infect. Immun.* **2006**, *74*, 5382–5390. [CrossRef]
11. Kostic, A.D.; Chun, E.; Robertson, L.; Glickman, J.N.; Gallini, C.A.; Michaud, M.; Clancy, T.E.; Chung, D.C.; Lochhead, P.; Hold, G.L.; et al. *Fusobacterium nucleatum* potentiates intestinal tumorigenesis and modulates the tumor-immune microenvironment. *Cell Host Microbe* **2013**, *14*, 207–215. [CrossRef]
12. Mima, K.; Cao, Y.; Chan, A.T.; Qian, Z.R.; Nowak, J.A.; Masugi, Y.; Shi, Y.; Song, M.; da Silva, A.; Gu, M.; et al. Fusobacterium nucleatum in colorectal carcinoma tissue according to tumor location. *Clin. Transl. Gastroenterol.* **2016**, *7*, e200. [CrossRef]
13. Mima, K.; Nishihara, R.; Qian, Z.R.; Cao, Y.; Sukawa, Y.; Nowak, J.A.; Yang, J.; Dou, R.; Masugi, Y.; Song, M.; et al. Fusobacterium nucleatum in colorectal carcinoma tissue and patient prognosis. *Gut* **2016**, *65*, 1973–1980. [CrossRef]
14. Abed, J.; Emgård, J.E.M.; Zamir, G.; Faroja, M.; Almogy, G.; Grenov, A.; Sol, A.; Naor, R.; Pikarsky, E.; Atlan, K.A.; et al. Fap2 mediates Fusobacterium nucleatum colorectal adenocarcinoma enrichment by binding to tumor-expressed Gal-GalNAc. *Cell Host Microbe* **2016**, *20*, 215–225. [CrossRef]
15. Rubinstein, M.R.; Baik, J.E.; Lagana, S.M.; Han, R.P.; Raab, W.J.; Sahoo, D.; Dalerba, P.; Wang, T.C.; Han, Y.W. Fusobacterium nucleatum promotes colorectal cancer by inducing Wnt/β-catenin modulator Annexin A1. *EMBO Rep.* **2019**. [CrossRef]
16. Rubinstein, M.R.; Wang, X.; Liu, W.; Hao, Y.; Cai, G.; Han, Y.W. Fusobacterium nucleatum promotes colorectal carcinogenesis by modulating E-cadherin/β-catenin signaling via its FadA adhesin. *Cell Host Microbe* **2013**, *14*, 195–206. [CrossRef]
17. Schroeder, B.O.; Bäckhed, F. Signals from the gut microbiota to distant organs in physiology and disease. *Nat. Med.* **2016**, *22*, 1079–1089. [CrossRef]
18. Blekhman, R.; Goodrich, J.K.; Huang, K.; Sun, Q.; Bukowski, R.; Bell, J.T.; Spector, T.D.; Keinan, A.; Ley, R.E.; Gevers, D.; et al. Host genetic variation impacts microbiome composition across human body sites. *Genome Biol.* **2015**, *16*, 191. [CrossRef] [PubMed]
19. Goodrich, J.K.; Waters, J.L.; Poole, A.C.; Sutter, J.L.; Koren, O.; Blekhman, R.; Beaumont, M.; Van Treuren, W.; Knight, R.; Bell, J.T.; et al. Human genetics shape the gut microbiome. *Cell* **2014**, *159*, 789–799. [CrossRef]
20. Bongers, G.; Pacer, M.E.; Geraldino, T.H.; Chen, L.; He, Z.; Hashimoto, D.; Furtado, G.C.; Ochando, J.; Kelley, K.A.; Clemente, J.C.; et al. Interplay of host microbiota, genetic perturbations, and inflammation promotes local development of intestinal neoplasms in mice. *J. Exp. Med.* **2014**, *211*, 457–472. [CrossRef]
21. Davenport, E.R.; Cusanovich, D.A.; Michelini, K.; Barreiro, L.B.; Ober, C.; Gilad, Y. Genome-wide association studies of the human gut microbiota. *PLoS ONE* **2015**, *10*, e0140301. [CrossRef]
22. Goodrich, J.K.; Davenport, E.R.; Beaumont, M.; Jackson, M.A.; Knight, R.; Ober, C.; Spector, T.D.; Bell, J.T.; Clark, A.G.; Ley, R.E. Genetic determinants of the gut microbiome in UK twins. *Cell Host Microbe* **2016**, *19*, 731–743. [CrossRef] [PubMed]

23. Liu, S.; da Cunha, A.P.; Rezende, R.M.; Cialic, R.; Wei, Z.; Bry, L.; Comstock, L.E.; Gandhi, R.; Weiner, H.L. The host shapes the gut microbiota via fecal microrna. *Cell Host Microbe* **2016**, *19*, 32–43. [CrossRef] [PubMed]
24. Teng, Y.; Ren, Y.; Sayed, M.; Hu, X.; Lei, C.; Kumar, A.; Hutchins, E.; Mu, J.; Deng, Z.; Luo, C.; et al. Plant-derived exosomal microRNAs shape the gut microbiota. *Cell Host Microbe* **2018**, *24*, 637–652.e8. [CrossRef] [PubMed]
25. Slattery, M.L.; Mullany, L.E.; Sakoda, L.C.; Samowitz, W.S.; Wolff, R.K.; Stevens, J.R.; Herrick, J.S. Expression of Wnt-signaling pathway genes and their associations with miRNAs in colorectal cancer. *Oncotarget* **2018**, *9*, 6075–6085. [CrossRef] [PubMed]
26. Li, Y.; Lauriola, M.; Kim, D.; Francesconi, M.; D'Uva, G.; Shibata, D.; Malafa, M.P.; Yeatman, T.J.; Coppola, D.; Solmi, R.; et al. Adenomatous polyposis coli (APC) regulates miR17-92 cluster through β-catenin pathway in colorectal cancer. *Oncogene* **2016**, *35*, 4558–4568. [CrossRef] [PubMed]
27. Bartel, D.P. MicroRNAs: Genomics, biogenesis, mechanism, and function. *Cell* **2004**, *116*, 281–297. [CrossRef]
28. Fabian, M.R.; Sonenberg, N. The mechanics of miRNA-mediated gene silencing: A look under the hood of miRISC. *Nat. Struct. Mol. Biol.* **2012**, *19*, 586–593. [CrossRef]
29. Lewis, B.P.; Burge, C.B.; Bartel, D.P. Conserved seed pairing, often flanked by adenosines, indicates that thousands of human genes are microRNA targets. *Cell* **2005**, *120*, 15–20. [CrossRef]
30. The Human Microbiome Project Consortium. Structure, function and diversity of the healthy human microbiome. *Nature* **2012**, *486*, 207–214. [CrossRef] [PubMed]
31. Louis, P.; Hold, G.L.; Flint, H.J. The gut microbiota, bacterial metabolites and colorectal cancer. *Nat. Rev. Microbiol.* **2014**, *12*, 661–672. [CrossRef]
32. Donohoe, D.R.; Garge, N.; Zhang, X.; Sun, W.; O'Connell, T.M.; Bunger, M.K.; Bultman, S.J. The microbiome and butyrate regulate energy metabolism and autophagy in the mammalian colon. *Cell Metab.* **2011**, *13*, 517–526. [CrossRef]
33. Dove, W.F.; Clipson, L.; Gould, K.A.; Luongo, C.; Marshall, D.J.; Moser, A.R.; Newton, M.A.; Jacoby, R.F. Intestinal neoplasia in the ApcMin mouse: Independence from the microbial and natural killer (beige locus) status. *Cancer Res.* **1997**, *57*, 812–814. [PubMed]
34. Zackular, J.P.; Baxter, N.T.; Chen, G.Y.; Schloss, P.D. Manipulation of the gut microbiota reveals role in colon tumorigenesis. *mSphere* **2016**, *1*. [CrossRef] [PubMed]
35. Vivarelli, S.; Salemi, R.; Candido, S.; Falzone, L.; Santagati, M.; Stefani, S.; Torino, F.; Banna, G.L.; Tonini, G.; Libra, M. Gut microbiota and cancer: From pathogenesis to therapy. *Cancers* **2019**, *11*, 38. [CrossRef]
36. Blacher, E.; Levy, M.; Tatirovsky, E.; Elinav, E. Microbiome-modulated metabolites at the interface of host immunity. *J. Immunol.* **2017**, *198*, 572–580. [CrossRef]
37. Johnson, C.H.; Spilker, M.E.; Goetz, L.; Peterson, S.N.; Siuzdak, G. Metabolite and microbiome interplay in cancer immunotherapy. *Cancer Res.* **2016**, *76*, 6146–6152. [CrossRef]
38. O'Keefe, S.J.D. Diet, microorganisms and their metabolites, and colon cancer. *Nat. Rev. Gastroenterol. Hepatol.* **2016**, *13*, 691–706. [CrossRef]
39. Yuan, C.; Subramanian, S. MicroRNA mediated tumor-microbiota metabolic interactions in colorectal cancer. *DNA Cell Biol.* **2019**. [CrossRef] [PubMed]
40. Castellarin, M.; Warren, R.L.; Freeman, J.D.; Dreolini, L.; Krzywinski, M.; Strauss, J.; Barnes, R.; Watson, P.; Allen-Vercoe, E.; Moore, R.A.; et al. *Fusobacterium nucleatum* infection is prevalent in human colorectal carcinoma. *Genome Res.* **2012**, *22*, 299–306. [CrossRef]
41. Burns, M.B.; Lynch, J.; Starr, T.K.; Knights, D.; Blekhman, R. Virulence genes are a signature of the microbiome in the colorectal tumor microenvironment. *Genome Med.* **2015**, *7*, 55. [CrossRef] [PubMed]
42. Pate, K.T.; Stringari, C.; Sprowl-Tanio, S.; Wang, K.; TeSlaa, T.; Hoverter, N.P.; McQuade, M.M.; Garner, C.; Digman, M.A.; Teitell, M.A.; et al. Wnt signaling directs a metabolic program of glycolysis and angiogenesis in colon cancer. *EMBO J.* **2014**, *33*, 1454–1473. [CrossRef]
43. Konsavage, W.M.; Kyler, S.L.; Rennoll, S.A.; Jin, G.; Yochum, G.S. Wnt/β-catenin signaling regulates Yes-associated protein (YAP) gene expression in colorectal carcinoma cells. *J. Biol. Chem.* **2012**, *287*, 11730–11739. [CrossRef]
44. Strillacci, A.; Valerii, M.C.; Sansone, P.; Caggiano, C.; Sgromo, A.; Vittori, L.; Fiorentino, M.; Poggioli, G.; Rizzello, F.; Campieri, M.; et al. Loss of miR-101 expression promotes Wnt/β-catenin signalling pathway activation and malignancy in colon cancer cells. *J. Pathol.* **2013**, *229*, 379–389. [CrossRef]
45. Bienz, M.; Clevers, H. Linking colorectal cancer to Wnt signaling. *Cell* **2000**, *103*, 311–320. [CrossRef]

46. Wu, S.; Morin, P.J.; Maouyo, D.; Sears, C.L. Bacteroides fragilis enterotoxin induces c-Myc expression and cellular proliferation. *Gastroenterology* **2003**, *124*, 392–400. [CrossRef] [PubMed]
47. Wu, S.; Lim, K.C.; Huang, J.; Saidi, R.F.; Sears, C.L. Bacteroides fragilis enterotoxin cleaves the zonula adherens protein, E-cadherin. *Proc. Natl. Acad. Sci. USA* **1998**, *95*, 14979–14984. [CrossRef]
48. Diosdado, B.; van de Wiel, M.A.; Terhaar Sive Droste, J.S.; Mongera, S.; Postma, C.; Meijerink, W.J.H.J.; Carvalho, B.; Meijer, G.A. MiR-17-92 cluster is associated with 13q gain and c-Myc expression during colorectal adenoma to adenocarcinoma progression. *Br. J. Cancer* **2009**, *101*, 707–714. [CrossRef]
49. Mogilyansky, E.; Rigoutsos, I. The miR-17/92 cluster: A comprehensive update on its genomics, genetics, functions and increasingly important and numerous roles in health and disease. *Cell Death Differ.* **2013**, *20*, 1603–1614. [CrossRef]
50. O'Donnell, K.A.; Wentzel, E.A.; Zeller, K.I.; Dang, C.V.; Mendell, J.T. c-Myc-regulated microRNAs modulate E2F1 expression. *Nature* **2005**, *435*, 839–843. [CrossRef]
51. Dang, C.V. c-Myc target genes involved in cell growth, apoptosis, and metabolism. *Mol. Cell Biol.* **1999**, *19*, 1–11. [CrossRef] [PubMed]
52. Yang, Y.; Weng, W.; Peng, J.; Hong, L.; Yang, L.; Toiyama, Y.; Gao, R.; Liu, M.; Yin, M.; Pan, C.; et al. *Fusobacterium nucleatum* increases proliferation of colorectal cancer cells and tumor development in mice by activating Toll-Like receptor 4 signaling to nuclear factor-κB, and up-regulating expression of microRNA-21. *Gastroenterology* **2017**, *152*, 851–866.e24. [CrossRef]
53. Dalmasso, G.; Cougnoux, A.; Delmas, J.; Darfeuille-Michaud, A.; Bonnet, R. The bacterial genotoxin colibactin promotes colon tumor growth by modifying the tumor microenvironment. *Gut Microbes* **2014**, *5*, 675–680. [CrossRef] [PubMed]
54. Cougnoux, A.; Dalmasso, G.; Martinez, R.; Buc, E.; Delmas, J.; Gibold, L.; Sauvanet, P.; Darcha, C.; Déchelotte, P.; Bonnet, M.; et al. Bacterial genotoxin colibactin promotes colon tumour growth by inducing a senescence-associated secretory phenotype. *Gut* **2014**, *63*, 1932–1942. [CrossRef] [PubMed]
55. Baek, S.H. A novel link between SUMO modification and cancer metastasis. *Cell Cycle* **2006**, *5*, 1492–1495. [CrossRef]
56. Gordon, H.A. Morphological and physiological characterization of germfree life. *Ann. N.Y. Acad. Sci.* **1959**, *78*, 208–220. [CrossRef] [PubMed]
57. Mazmanian, S.K.; Liu, C.H.; Tzianabos, A.O.; Kasper, D.L. An immunomodulatory molecule of symbiotic bacteria directs maturation of the host immune system. *Cell* **2005**, *122*, 107–118. [CrossRef]
58. Vétizou, M.; Pitt, J.M.; Daillère, R.; Lepage, P.; Waldschmitt, N.; Flament, C.; Rusakiewicz, S.; Routy, B.; Roberti, M.P.; Duong, C.P.M.; et al. Anticancer immunotherapy by CTLA-4 blockade relies on the gut microbiota. *Science* **2015**, *350*, 1079–1084. [CrossRef] [PubMed]
59. Sarver, A.L.; French, A.J.; Borralho, P.M.; Thayanithy, V.; Oberg, A.L.; Silverstein, K.A.T.; Morlan, B.W.; Riska, S.M.; Boardman, L.A.; Cunningham, J.M.; et al. Human colon cancer profiles show differential microRNA expression depending on mismatch repair status and are characteristic of undifferentiated proliferative states. *BMC Cancer* **2009**, *9*, 401. [CrossRef]
60. Sarver, A.L.; Sarver, A.E.; Yuan, C.; Subramanian, S. OMCD: Oncomir cancer database. *BMC Cancer* **2018**, *18*, 1223. [CrossRef] [PubMed]
61. Wong, N.W.; Chen, Y.; Chen, S.; Wang, X. OncomiR: An online resource for exploring pan-cancer microRNA dysregulation. *Bioinformatics* **2018**, *34*, 713–715. [CrossRef]
62. Falzone, L.; Scola, L.; Zanghì, A.; Biondi, A.; Di Cataldo, A.; Libra, M.; Candido, S. Integrated analysis of colorectal cancer microRNA datasets: Identification of microRNAs associated with tumor development. *Aging* **2018**, *10*, 1000–1014. [CrossRef] [PubMed]
63. Koga, Y.; Yasunaga, M.; Takahashi, A.; Kuroda, J.; Moriya, Y.; Akasu, T.; Fujita, S.; Yamamoto, S.; Baba, H.; Matsumura, Y. MicroRNA expression profiling of exfoliated colonocytes isolated from feces for colorectal cancer screening. *Cancer Prev. Res. (Phila Pa)* **2010**, *3*, 1435–1442. [CrossRef] [PubMed]
64. Fang, L.; Li, H.; Wang, L.; Hu, J.; Jin, T.; Wang, J.; Yang, B.B. MicroRNA-17-5p promotes chemotherapeutic drug resistance and tumour metastasis of colorectal cancer by repressing PTEN expression. *Oncotarget* **2014**, *5*, 2974–2987. [CrossRef] [PubMed]
65. Wu, C.W.; Ng, S.S.M.; Dong, Y.J.; Ng, S.C.; Leung, W.W.; Lee, C.W.; Wong, Y.N.; Chan, F.K.L.; Yu, J.; Sung, J.J.Y. Detection of miR-92a and miR-21 in stool samples as potential screening biomarkers for colorectal cancer and polyps. *Gut* **2012**, *61*, 739–745. [CrossRef] [PubMed]

66. Kulda, V.; Pesta, M.; Topolcan, O.; Liska, V.; Treska, V.; Sutnar, A.; Rupert, K.; Ludvikova, M.; Babuska, V.; Holubec, L.; et al. Relevance of miR-21 and miR-143 expression in tissue samples of colorectal carcinoma and its liver metastases. *Cancer Genet. Cytogenet.* **2010**, *200*, 154–160. [CrossRef]
67. Asangani, I.A.; Rasheed, S.A.K.; Nikolova, D.A.; Leupold, J.H.; Colburn, N.H.; Post, S.; Allgayer, H. MicroRNA-21 (miR-21) post-transcriptionally downregulates tumor suppressor Pdcd4 and stimulates invasion, intravasation and metastasis in colorectal cancer. *Oncogene* **2008**, *27*, 2128–2136. [CrossRef] [PubMed]
68. Bullock, M.D.; Pickard, K.M.; Nielsen, B.S.; Sayan, A.E.; Jenei, V.; Mellone, M.; Mitter, R.; Primrose, J.N.; Thomas, G.J.; Packham, G.K.; et al. Pleiotropic actions of miR-21 highlight the critical role of deregulated stromal microRNAs during colorectal cancer progression. *Cell Death Dis.* **2013**, *4*, e684. [CrossRef]
69. Li, L.; Sarver, A.L.; Khatri, R.; Hajeri, P.B.; Kamenev, I.; French, A.J.; Thibodeau, S.N.; Steer, C.J.; Subramanian, S. Sequential expression of miR-182 and miR-503 cooperatively targets FBXW7, contributing to the malignant transformation of colon adenoma to adenocarcinoma. *J. Pathol.* **2014**, *234*, 488–501. [CrossRef]
70. Dews, M.; Fox, J.L.; Hultine, S.; Sundaram, P.; Wang, W.; Liu, Y.Y.; Furth, E.; Enders, G.H.; El-Deiry, W.; Schelter, J.M.; et al. The myc-miR-17~92 axis blunts TGFβ signaling and production of multiple TGF β-dependent antiangiogenic factors. *Cancer Res.* **2010**, *70*, 8233–8246. [CrossRef]
71. Hu, S.; Liu, L.; Chang, E.B.; Wang, J.-Y.; Raufman, J.-P. Butyrate inhibits pro-proliferative miR-92a by diminishing c-Myc-induced miR-17-92a cluster transcription in human colon cancer cells. *Mol. Cancer* **2015**, *14*, 180. [CrossRef] [PubMed]
72. Yamamichi, N.; Shimomura, R.; Inada, K.; Sakurai, K.; Haraguchi, T.; Ozaki, Y.; Fujita, S.; Mizutani, T.; Furukawa, C.; Fujishiro, M.; et al. Locked nucleic acid in situ hybridization analysis of miR-21 expression during colorectal cancer development. *Clin. Cancer Res.* **2009**, *15*, 4009–4016. [CrossRef] [PubMed]
73. Grillari, J.; Hackl, M.; Grillari-Voglauer, R. miR-17-92 cluster: Ups and downs in cancer and aging. *Biogerontology* **2010**, *11*, 501–506. [CrossRef] [PubMed]
74. Makinoshima, H.; Takita, M.; Saruwatari, K.; Umemura, S.; Obata, Y.; Ishii, G.; Matsumoto, S.; Sugiyama, E.; Ochiai, A.; Abe, R.; et al. Signaling through the phosphatidylinositol 3-kinase (PI3K)/mammalian target of rapamycin (mTOR) axis is responsible for aerobic glycolysis mediated by glucose transporter in epidermal growth factor receptor (EGFR)-mutated lung adenocarcinoma. *J. Biol. Chem.* **2015**, *290*, 17495–17504. [CrossRef] [PubMed]
75. Weijenberg, M.P.; Hughes, L.A.E.; Bours, M.J.L.; Simons, C.C.J.M.; van Engeland, M.; van den Brandt, P.A. The mTOR pathway and the role of energy balance throughout life in colorectal cancer etiology and prognosis: Unravelling mechanisms through a multidimensional molecular epidemiologic approach. *Curr. Nutr. Rep.* **2013**, *2*, 19–26. [CrossRef] [PubMed]
76. Kohlhapp, F.J.; Mitra, A.K.; Lengyel, E.; Peter, M.E. MicroRNAs as mediators and communicators between cancer cells and the tumor microenvironment. *Oncogene* **2015**, *34*, 5857–5868. [CrossRef]
77. Kosaka, N.; Yoshioka, Y.; Hagiwara, K.; Tominaga, N.; Katsuda, T.; Ochiya, T. Trash or Treasure: Extracellular microRNAs and cell-to-cell communication. *Front. Genet.* **2013**, *4*, 173. [CrossRef]
78. Runtsch, M.C.; Round, J.L.; O'Connell, R.M. MicroRNAs and the regulation of intestinal homeostasis. *Front. Genet.* **2014**, *5*, 347. [CrossRef]
79. Nielsen, B.S.; Jørgensen, S.; Fog, J.U.; Søkilde, R.; Christensen, I.J.; Hansen, U.; Brünner, N.; Baker, A.; Møller, S.; Nielsen, H.J. High levels of microRNA-21 in the stroma of colorectal cancers predict short disease-free survival in stage II colon cancer patients. *Clin. Exp. Metastasis* **2011**, *28*, 27–38. [CrossRef]
80. Dews, M.; Homayouni, A.; Yu, D.; Murphy, D.; Sevignani, C.; Wentzel, E.; Furth, E.E.; Lee, W.M.; Enders, G.H.; Mendell, J.T.; et al. Augmentation of tumor angiogenesis by a Myc-activated microRNA cluster. *Nat. Genet.* **2006**, *38*, 1060–1065. [CrossRef]
81. Zhuang, G.; Wu, X.; Jiang, Z.; Kasman, I.; Yao, J.; Guan, Y.; Oeh, J.; Modrusan, Z.; Bais, C.; Sampath, D.; et al. Tumour-secreted miR-9 promotes endothelial cell migration and angiogenesis by activating the JAK-STAT pathway. *EMBO J.* **2012**, *31*, 3513–3523. [CrossRef]
82. Fanini, F.; Fabbri, M. Cancer-derived exosomic microRNAs shape the immune system within the tumor microenvironment: State of the art. *Semin. Cell Dev. Biol.* **2017**, *67*, 23–28. [CrossRef] [PubMed]
83. Sonda, N.; Simonato, F.; Peranzoni, E.; Calì, B.; Bortoluzzi, S.; Bisognin, A.; Wang, E.; Marincola, F.M.; Naldini, L.; Gentner, B.; et al. miR-142-3p prevents macrophage differentiation during cancer-induced myelopoiesis. *Immunity* **2013**, *38*, 1236–1249. [CrossRef] [PubMed]

84. Zhang, M.; Liu, Q.; Mi, S.; Liang, X.; Zhang, Z.; Su, X.; Liu, J.; Chen, Y.; Wang, M.; Zhang, Y.; et al. Both miR-17-5p and miR-20a alleviate suppressive potential of myeloid-derived suppressor cells by modulating STAT3 expression. *J. Immunol.* **2011**, *186*, 4716–4724. [CrossRef]
85. Mei, S.; Xin, J.; Liu, Y.; Zhang, Y.; Liang, X.; Su, X.; Yan, H.; Huang, Y.; Yang, R. MicroRNA-200c promotes suppressive potential of myeloid-derived suppressor cells by modulating PTEN and FOG2 expression. *PLoS ONE* **2015**, *10*, e0135867. [CrossRef]
86. Dalmasso, G.; Nguyen, H.T.T.; Yan, Y.; Laroui, H.; Charania, M.A.; Ayyadurai, S.; Sitaraman, S.V.; Merlin, D. Microbiota modulate host gene expression via microRNAs. *PLoS ONE* **2011**, *6*, e19293. [CrossRef] [PubMed]
87. Peck, B.C.E.; Mah, A.T.; Pitman, W.A.; Ding, S.; Lund, P.K.; Sethupathy, P. Functional transcriptomics in diverse intestinal epithelial cell types reveals robust microrna sensitivity in intestinal stem cells to microbial status. *J. Biol. Chem.* **2017**, *292*, 2586–2600. [CrossRef]
88. Phua, L.C.; Chue, X.P.; Koh, P.K.; Cheah, P.Y.; Chan, E.C.Y.; Ho, H.K. Global fecal microRNA profiling in the identification of biomarkers for colorectal cancer screening among Asians. *Oncol. Rep.* **2014**, *32*, 97–104. [CrossRef]
89. Rotelli, M.T.; Di Lena, M.; Cavallini, A.; Lippolis, C.; Bonfrate, L.; Chetta, N.; Portincasa, P.; Altomare, D.F. Fecal microRNA profile in patients with colorectal carcinoma before and after curative surgery. *Int. J. Colorectal Dis.* **2015**, *30*, 891–898. [CrossRef]
90. Baier, S.R.; Nguyen, C.; Xie, F.; Wood, J.R.; Zempleni, J. MicroRNAs are absorbed in biologically meaningful amounts from nutritionally relevant doses of cow milk and affect gene expression in peripheral blood mononuclear cells, HEK-293 kidney cell cultures, and mouse livers. *J. Nutr.* **2014**, *144*, 1495–1500. [CrossRef] [PubMed]
91. Izumi, H.; Tsuda, M.; Sato, Y.; Kosaka, N.; Ochiya, T.; Iwamoto, H.; Namba, K.; Takeda, Y. Bovine milk exosomes contain microRNA and mRNA and are taken up by human macrophages. *J. Dairy Sci.* **2015**, *98*, 2920–2933. [CrossRef]
92. Zhang, L.; Hou, D.; Chen, X.; Li, D.; Zhu, L.; Zhang, Y.; Li, J.; Bian, Z.; Liang, X.; Cai, X.; et al. Exogenous plant MIR168a specifically targets mammalian LDLRAP1: Evidence of cross-kingdom regulation by microRNA. *Cell Res.* **2012**, *22*, 107–126. [CrossRef]
93. Mu, J.; Zhuang, X.; Wang, Q.; Jiang, H.; Deng, Z.-B.; Wang, B.; Zhang, L.; Kakar, S.; Jun, Y.; Miller, D.; et al. Interspecies communication between plant and mouse gut host cells through edible plant derived exosome-like nanoparticles. *Mol. Nutr. Food Res.* **2014**, *58*, 1561–1573. [CrossRef]
94. Wolf, T.; Baier, S.R.; Zempleni, J. The intestinal transport of bovine milk exosomes is mediated by endocytosis in human colon carcinoma Caco-2 cells and rat small intestinal IEC-6 cells. *J. Nutr.* **2015**, *145*, 2201–2206. [CrossRef]
95. Gopalakrishnan, V.; Spencer, C.N.; Nezi, L.; Reuben, A.; Andrews, M.C.; Karpinets, T.V.; Prieto, P.A.; Vicente, D.; Hoffman, K.; Wei, S.C.; et al. Gut microbiome modulates response to anti-PD-1 immunotherapy in melanoma patients. *Science* **2018**, *359*, 97–103. [CrossRef]
96. Matson, V.; Fessler, J.; Bao, R.; Chongsuwat, T.; Zha, Y.; Alegre, M.-L.; Luke, J.J.; Gajewski, T.F. The commensal microbiome is associated with anti-PD-1 efficacy in metastatic melanoma patients. *Science* **2018**, *359*, 104–108. [CrossRef]
97. Sivan, A.; Corrales, L.; Hubert, N.; Williams, J.B.; Aquino-Michaels, K.; Earley, Z.M.; Benyamin, F.W.; Lei, Y.M.; Jabri, B.; Alegre, M.-L.; et al. Commensal Bifidobacterium promotes antitumor immunity and facilitates anti-PD-L1 efficacy. *Science* **2015**, *350*, 1084–1089. [CrossRef]
98. Gopalakrishnan, V.; Helmink, B.A.; Spencer, C.N.; Reuben, A.; Wargo, J.A. The influence of the gut microbiome on cancer, immunity, and cancer immunotherapy. *Cancer Cell* **2018**, *33*, 570–580. [CrossRef]
99. Zelante, T.; Iannitti, R.G.; Cunha, C.; De Luca, A.; Giovannini, G.; Pieraccini, G.; Zecchi, R.; D'Angelo, C.; Massi-Benedetti, C.; Fallarino, F.; et al. Tryptophan catabolites from microbiota engage aryl hydrocarbon receptor and balance mucosal reactivity via interleukin-22. *Immunity* **2013**, *39*, 372–385. [CrossRef]
100. Gottesman, S.; Storz, G. Bacterial small RNA regulators: Versatile roles and rapidly evolving variations. *Cold Spring Harb. Perspect. Biol.* **2011**, *3*. [CrossRef]
101. Lässer, C.; Alikhani, V.S.; Ekström, K.; Eldh, M.; Paredes, P.T.; Bossios, A.; Sjöstrand, M.; Gabrielsson, S.; Lötvall, J.; Valadi, H. Human saliva, plasma and breast milk exosomes contain RNA: Uptake by macrophages. *J. Transl. Med.* **2011**, *9*, 9. [CrossRef] [PubMed]

102. Deng, Z.; Rong, Y.; Teng, Y.; Mu, J.; Zhuang, X.; Tseng, M.; Samykutty, A.; Zhang, L.; Yan, J.; Miller, D.; et al. Broccoli-derived nanoparticle inhibits mouse colitis by activating dendritic cell AMP-activated protein kinase. *Mol. Ther.* **2017**, *25*, 1641–1654. [CrossRef]
103. Ju, S.; Mu, J.; Dokland, T.; Zhuang, X.; Wang, Q.; Jiang, H.; Xiang, X.; Deng, Z.-B.; Wang, B.; Zhang, L.; et al. Grape exosome-like nanoparticles induce intestinal stem cells and protect mice from DSS-induced colitis. *Mol. Ther.* **2013**, *21*, 1345–1357. [CrossRef]
104. Tjalsma, H.; Boleij, A.; Marchesi, J.R.; Dutilh, B.E. A bacterial driver-passenger model for colorectal cancer: Beyond the usual suspects. *Nat. Rev. Microbiol.* **2012**, *10*, 575–582. [CrossRef]
105. Warburg, O. On the origin of cancer cells. *Science* **1956**, *123*, 309–314. [CrossRef]
106. Bishop, K.S.; Xu, H.; Marlow, G. Epigenetic regulation of gene expression induced by butyrate in colorectal cancer: Involvement of microRNA. *Genet. Epigenet.* **2017**, *9*, 1179237X17729900. [CrossRef]
107. Hu, S.; Dong, T.S.; Dalal, S.R.; Wu, F.; Bissonnette, M.; Kwon, J.H.; Chang, E.B. The microbe-derived short chain fatty acid butyrate targets miRNA-dependent p21 gene expression in human colon cancer. *PLoS ONE* **2011**, *6*, e16221. [CrossRef]
108. Weir, T.L.; Manter, D.K.; Sheflin, A.M.; Barnett, B.A.; Heuberger, A.L.; Ryan, E.P. Stool microbiome and metabolome differences between colorectal cancer patients and healthy adults. *PLoS ONE* **2013**, *8*, e70803. [CrossRef]
109. Hirayama, A.; Kami, K.; Sugimoto, M.; Sugawara, M.; Toki, N.; Onozuka, H.; Kinoshita, T.; Saito, N.; Ochiai, A.; Tomita, M.; et al. Quantitative metabolome profiling of colon and stomach cancer microenvironment by capillary electrophoresis time-of-flight mass spectrometry. *Cancer Res.* **2009**, *69*, 4918–4925. [CrossRef]
110. Brown, D.G.; Rao, S.; Weir, T.L.; O'Malia, J.; Bazan, M.; Brown, R.J.; Ryan, E.P. Metabolomics and metabolic pathway networks from human colorectal cancers, adjacent mucosa, and stool. *Cancer Metab.* **2016**, *4*, 11. [CrossRef]

10

A Two-Cohort RNA-seq Study Reveals Changes in Endometrial and Blood miRNome in Fertile and Infertile Women

Kadri Rekker [1,2,*,†], Signe Altmäe [2,3,†], Marina Suhorutshenko [1,2], Maire Peters [1,2], Juan F. Martinez-Blanch [4], Francisco M. Codoñer [4], Felipe Vilella [5,6], Carlos Simón [5,6,7], Andres Salumets [1,2,8,9] and Agne Velthut-Meikas [2,10]

1 Institute of Clinical Medicine, Department of Obstetrics and Gynecology, University of Tartu, 50406 Tartu, Estonia; marina.suhorutsenko@ut.ee (M.S.); maire.peters@ut.ee (M.P.); andres.salumets@ccht.ee (A.S.)
2 Competence Centre on Health Technologies, 50410 Tartu, Estonia; signealtmae@ugr.es (S.A.); agnevelthut@gmail.com (A.V.-M.)
3 Department of Biochemistry and Molecular Biology, Faculty of Sciences, University of Granada, 18071 Granada, Spain
4 Lifesequencing SL, 46980 Valencia, Spain; juan.martinez@lifesequencing.com (J.F.M.-B.); francisco.codoner@adm.com (F.M.C.)
5 Igenomix Foundation/INCLIVA, 46010 Valencia, Spain; felipe.vilella@igenomix.com (F.V.); carlos.simon@igenomix.com (C.S.)
6 Igenomix SL, 46980 Valencia, Spain
7 Department of Obstetrics and Gynecology, Valencia University, 46010 Valencia, Spain
8 Institute of Biomedicine and Translational Medicine, Department of Biomedicine, University of Tartu, 50412 Tartu, Estonia
9 Department of Obstetrics and Gynecology, University of Helsinki and Helsinki University Hospital, FI-00014 Helsinki, Finland
10 Department of Chemistry and Biotechnology, Tallinn University of Technology, 12618 Tallinn, Estonia
* Correspondence: kadri.rekker@gmail.com
† These authors contributed equally to this work.

Abstract: The endometrium undergoes extensive changes to prepare for embryo implantation and microRNAs (miRNAs) have been described as playing a significant role in the regulation of endometrial receptivity. However, there is no consensus about the miRNAs involved in mid-secretory endometrial functions. We analysed the complete endometrial miRNome from early secretory (pre-receptive) and mid-secretory (receptive) phases from fertile women and from patients with recurrent implantation failure (RIF) to reveal differentially expressed (DE) miRNAs in the mid-secretory endometrium. Furthermore, we investigated whether the overall changes during early to mid-secretory phase transition and with RIF condition could be reflected in blood miRNA profiles. In total, 116 endometrial and 114 matched blood samples collected from two different population cohorts were subjected to small RNA sequencing. Among fertile women, 91 DE miRNAs were identified in the mid-secretory vs. early secretory endometrium, while no differences were found in the corresponding blood samples. The comparison of mid-secretory phase samples between fertile and infertile women revealed 21 DE miRNAs from the endometrium and one from blood samples. Among discovered novel miRNAs, chr2_4401 was validated and showed up-regulation in the mid-secretory endometrium. Besides novel findings, we confirmed the involvement of miR-30 and miR-200 family members in mid-secretory endometrial functions.

Keywords: endometrial receptivity; infertility; microRNA; recurrent implantation failure; small RNA-seq

1. Introduction

The establishment of a receptive endometrium is essential for successful embryo implantation. The endometrium is receptive within a limited time frame during the mid-secretory cycle phase and aberrations in the processes involved in transition to this stage can lead to infertility [1,2]. Indeed, impaired endometrial receptivity is suspected to play a major role in female infertility among women suffering from recurrent implantation failure (RIF). Gaining insights into the complex mechanisms controlling changes within the endometrium is crucial to understanding not only embryo implantation but also endometrial dysfunction that can lead to infertility. Although hundreds of simultaneously up- and down-regulated genes have been implicated in the processes of acquiring endometrial receptivity [3,4] and the development of RIF [5–10], the precise molecular mechanisms regulating gene expression necessary for endometrial function in fertility and infertility-associated diseases are not well understood.

MicroRNAs (miRNAs), non-coding RNAs (ncRNAs) of ~22 nucleotides in length, act as post-transcriptional regulators of gene expression by inhibiting the stability or repressing the translation of their target messenger RNA (mRNA) molecules [11]. To date, more than 2500 annotated miRNAs in the human genome are known, and as each miRNA may regulate hundreds of genes, it is estimated that miRNAs collectively adjust the expression of over one third of genes in the human genome [11,12]. Hence, miRNAs orchestrate a large variety of processes, including the cyclic changes in the female reproductive tract [13] and cellular processes involved in implantation, such as cellular differentiation, proliferation and apoptosis [14].

The involvement of miRNAs in the mid-secretory endometrial functions has been demonstrated [15–17] and aberrant miRNA profiles in RIF have been identified in a few previous microarray-based [18–21] and two RNA-sequencing (RNA-seq) based studies [22,23]. However, as the analysis settings and platforms have been different and only a small number of samples have been analysed, there is no consensus about the list of miRNAs involved in endometrial receptivity and in endometrial dysfunction(s), advocating further investigations.

Therefore, we set out to analyse the complete spectrum of endometrial miRNAs–miRNome from paired early secretory (ES) and mid-secretory (MS) phase samples collected from two independent population cohorts using the comprehensive RNA-seq technology. Additionally, MS endometrial samples from RIF patients were included in order to identify the dysregulated miRNAs in infertility. Furthermore, we investigated whether the overall cyclic changes in the female body during ES to MS transition and with RIF condition could also be reflected in blood miRNA profiles.

2. Materials and Methods

2.1. Study Participants

The study was approved by the Research Ethics Committee of the University of Tartu, Estonia (No 221/M-31), and the Ethical Clinical Research Committee of IVI Clinic, Valencia, Spain (No 1201-C-094-CS). Informed consent was signed by all women who entered the study and all research was performed in accordance with relevant guidelines and regulations.

The study participants were recruited and samples were collected by two independent research teams from two countries: Estonia (EST) and Spain (ESP). All protocols were standardized across the study centres with exceptions highlighted in the "Sample collection" paragraph.

The healthy fertile group consisted in total of 39 women: EST $n = 22$, age 30.2 $\pm$ 3.3 years (mean $\pm$ standard deviation), body mass index (BMI) 23.1 $\pm$ 4.2 kg/m^2; and ESP $n = 17$, age 29.4 $\pm$ 3.6 years, BMI 23.2 $\pm$ 2.8 kg/m^2; with self-reported regular menstrual cycles. All women had at least one live-born child (1.5 $\pm$ 1.0 for EST cohort; 1.2 $\pm$ 0.4 for ESP cohort) within the last 10 years (5.3 $\pm$ 2.6 years, data available for the EST cohort) and no previous infertility records.

RIF patient group consisted in total of 38 women: EST $n = 21$, age 35.1 $\pm$ 3.9 years, BMI 22.3 $\pm$ 2.3 kg/m^2; ESP $n = 17$, age 36.7 $\pm$ 3.3 years, BMI 25.2 $\pm$ 5.3 kg/m^2. All RIF patients had undergone at least three (4.0 ± 1.5 for EST cohort; 3.0 ± 1.7 for ESP cohort) unsuccessful in vitro fertilisation (IVF) treatment cycles with embryo transfers. No donor eggs were used in the treatments. RIF patients were not undergoing hormonal stimulation for IVF at the time of sample collection. Among these patients, tubal and male factor were the main reasons for infertility along with endometriosis and unexplained infertility. Two participants were suffering from secondary infertility with the experience of childbirth in average 6.5 ± 0.5 years ago. None of the participants had received hormonal treatments for at least three months prior to the time of sample collection.

2.2. Sample Collection

All women performed urine-based ovulation tests to determine the luteinizing hormone (LH) surge using commercial kits (BabyTime® hLH urine cassette, Pharmanova, Beit Shemesh, Israel). The day with positive outcome of the ovulation test is referred to as LH + 0. Healthy women donated endometrium and blood samples during two time-points of the same menstrual cycle: LH + 1 to LH + 3 corresponding to the ES phase and LH + 7 to LH + 9 corresponding to the MS phase. Women in the RIF group provided samples at MS phase (LH + 7 to LH + 9).

Endometrial tissue was obtained using Pipelle catheter (Laboratoire CCD, Paris, France). The biopsy was divided into two parts: one part was rinsed in sterile phosphate-buffered saline (PBS) and frozen at −80 °C in RNA*later* solution (Thermo Fisher Scientific, Waltham, MA, USA) in EST or without any additives in ESP, the other portion was placed in 10% buffered formalin for histological evaluation. Histological dating was conducted to confirm the endometrial phase. In addition, the receptivity status of all endometrial biopsies was assessed and confirmed by analysing a set of 57 receptivity biomarkers as part of our previous studies [4,24].

Blood samples were collected and processed differently by the two research centres. In EST, whole blood samples were collected into PAXgene Blood RNA Tubes (Qiagen, Hilden, Germany), incubated at room temperature for 2 h and then frozen until further use. In ESP, blood samples were collected with tubes containing K_2-EDTA anticoagulant, the buffy coat fractions containing leukocytes and platelets were separated by a centrifugation procedure with Histopaque-1077 (Sigma-Aldrich, St. Louis, MO, USA), and were frozen until further use.

2.3. RNA Extraction from Endometrial Tissue

Up to 30 mg of the endometrial tissue was processed using miRNeasy Mini and RNeasy MinElute kits (Qiagen), following the manufacturer's protocol for isolating small RNA (<200 nucleotides) separately from large RNA molecules. DNase I treatment was performed on column using RNase-Free DNase Set (Qiagen). Purified RNA quantity was determined with Bioanalyzer 2100 Small RNA kit (Agilent Technologies, Santa Clara, CA, USA).

2.4. RNA Extraction from Blood Samples

In EST, RNA was extracted from the whole blood samples using PAXgene Blood miRNA Kit (Qiagen) according to manufacturer's instructions. In ESP, total RNA was isolated from buffy coat (leukocytes and platelets) using the miRNeasy Mini Kit and then cleaned with RNA Cleanup kit (Qiagen). On-column DNase I treatment was performed with RNase-Free DNase Set (Qiagen) for all samples.

2.5. Small RNA Sequencing

A total of 116 endometrial samples (EST n = 65; ESP n = 51) and 114 blood samples (EST n = 65; ESP n = 49) were subjected to small RNA sequencing in the current study (Figure 1). Small RNA libraries were constructed following the TruSeq Small RNA Library Preparation Guide (Illumina, San Diego, CA, USA). As input, 1 µg of total RNA or small RNA fraction was used; 12-plexing of samples was performed by inserting Illumina index sequences into each library. Final purification step was performed by manually selecting libraries corresponding to the length of inserted miRNAs (area between 145–160 bp) using gel electrophoresis (6% Novex TBE gels). Libraries were quantified and validated with Agilent 2100 Bioanalyzer (Agilent Technologies), normalized, pooled, used for cluster generation on the cBot and sequenced on HiSeq 2000/2500 with a configuration of 50 cycles Single Reads following manufacturer's instructions (Illumina) at Lifesequencing S.L., Valencia, Spain (ESP samples) or at Estonian Genome Center Core Facility, Tartu, Estonia (EST samples).

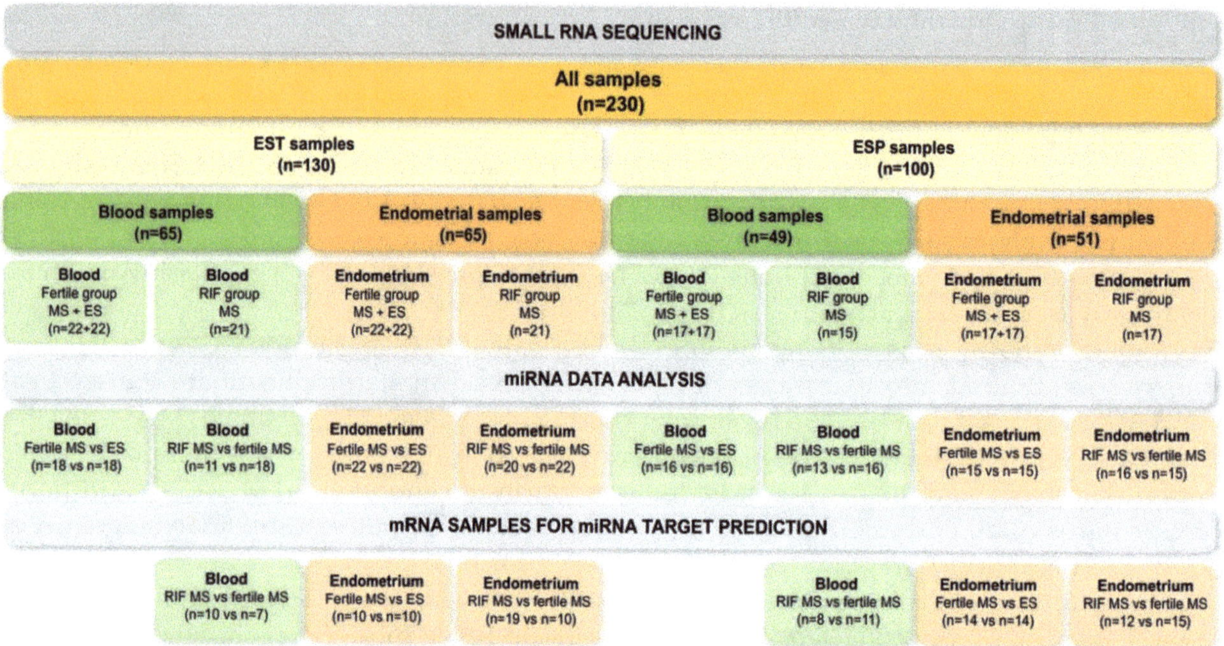

Figure 1. Number of samples included to small RNA sequencing, to final microRNA (miRNA) analysis and to messenger RNA (mRNA) sequencing analysis. EST—Estonian samples; ESP—Spanish samples; RIF—recurrent implantation failure; ES—early secretory phase; MS—mid-secretory phase.

2.6. SMALL RNA Sequencing Data Analysis

Small RNA sequencing data were deposited into the Gene Expression Omnibus (GEO accession number GSE108966). Quality control and adapter trimming of raw sequences was performed using Trimmomatic version 0.36 [25]. All sequences were scanned with 4-base long sliding window and reads were dropped if the average quality of base pair was below 30 according to the Phred +33 quality score system or if total read length was less than 17.

Identification of miRNAs from sequencing reads was performed by the miRDeep2 algorithm [26]. Firstly, mapper.pl algorithm was used to map the trimmed high-quality sequences to the UCSC human genome version hg19. Common reads were collapsed and passed as input to miRDeep2.pl algorithm together with mature and stem-loop sequences of human miRNAs from miRBase version 21 [27]. Standard miRDeep2 settings were used for the analysis.

Raw read counts for each miRNA from miRDeep2 analysis were used as input for differential expression analysis in edgeR package version 3.16.5 [28,29]. Analysis was run on R software environment version 3.3.1. Samples with library size below 500,000 reads and those performing as clear outliers according to the multidimensional scaling (MDS) plot were omitted. The remaining number of patient samples used for each comparison is displayed in Figure 1. Count per million (CPM) of at least 1 in 75% of samples in the smaller group was set as a threshold for each miRNA to be included in the analysis for each separate comparison. The remaining raw library sizes were scaled according to the method of weighted trimmed mean of M-values [30].

For comparing samples from ES and MS phases, a generalized linear model (GLM) likelihood ratio test with paired design was performed. For comparison of samples from fertile group and from patients with RIF, GLM test with age adjustment was implemented as age between those two groups was statistically different (*t*-test, $p < 0.05$). EST and ESP datasets were analysed separately and miRNAs that were detected as differentially expressed (DE) in both sample sets were included for further analysis.

2.7. Novel miRNA Identification

Novel miRNAs were predicted from 92 blood samples and 110 endometrial samples (Table 1). The reads predicted as potential candidate miRNAs by miRDeep2 were subjected to BLAST in order to discriminate the sequences corresponding to other human ncRNAs such as rRNA, snRNA, tRNA, lncRNA etc. Mature miRNA sequences of *Pan troglodytes* were included as input for evolutionary conservation analysis, which provides more solid confirmation for finding novel unannotated miRNAs from high-throughput sequencing experiments [31]. Sequences were considered as potentially novel miRNAs if they met the following criteria: (a) miRDeep2 score ≥ 5, (b) significant randfold *p*-value, (c) detected in at least 3 samples within a sample/biofluid type (endometrium or blood), (d) detected from two different datasets (EST + ESP) or detected from different sample types (endometrium and blood). miRDB was used for novel miRNA target prediction [32] and Ingenuity Pathway Analysis (IPA, 2017, Qiagen Bioinformatics Redwood City, CA, USA) was performed on predicted targets.

2.8. mRNA Data Analysis for miRNA Target Prediction

For miRNA target gene prediction purposes, mRNA data from our previously published RNA-seq study were retrieved from Gene Expression Omnibus (GSE98386) [4,24]. The mRNA dataset comprised of a subset of the same samples that were used for miRNA analysis in the current study (Figure 1). Raw mRNA sequencing read quality was inspected with FastQC v.0.11.3 (http://www.bioinformatics.babraham.ac.uk/projects/fastqc/) before and after the preprocessing step. Adapter removal and trimming was performed with Trimmomatic tool v. 0.32 [25]. Quality threshold was set to 30 according to Phred-33 scoring system and reads <36 nucleotides were omitted. FASTQ Quality Filter from FASTX-Toolkit v. 0.0.13 (http://hannonlab.cshl.edu/fastx_toolkit/index.html) was used as a second level of quality control, where the average quality threshold of 30 for every 50 nucleotides was set. Good quality reads were mapped to the human reference genome hg19 using TopHat v. 2.0.11 [33]. Read counts for every gene in Ensembl v.75 annotation file were generated with HTSeq v.0.6.1 [34]. Differential gene expression analysis was performed by using edgeR package according to the same parameters as described in miRNA data analysis. List of DE mRNAs that were common for EST and ESP samples were passed to miRNA-mRNA interaction analysis.

2.9. Integrated Analysis of miRNA and mRNA Data

Differentially expressed miRNAs and mRNAs in endometrial samples from ES vs. MS phases from fertile women and from MS phase of RIF patients were further analysed using the software IPA. The tool microRNA Target Filter was applied which allows prioritization of experimentally validated and predicted mRNA targets from TargetScan, TarBase, miRecords, and the Ingenuity Knowledge Base. Expression pairing of our miRNA and mRNA RNA-seq data was performed using the confidence filter of 'experimentally observed' and 'high confidence', and the expression pairing filter of opposite expression directions was applied (pairs of up-regulated miRNA with down-regulated mRNA, and down-regulated miRNA with up-regulated mRNA). The involvement of detected miRNA target genes in different canonical pathways was analysed. The overall level of gene expression in the canonical pathways was analysed by GOplot package version 1.0.2 in R statistical environment. The z-score refers to the level of under- or overrepresented mRNAs in each pathway and is calculated as $z = (\text{number of up-regulated genes} - \text{number of down-regulated genes})/\sqrt{\text{count}}$ [35].

miRNA–mRNA interactions were not analysed on data from blood samples because no DE miRNAs between blood samples corresponding to MS vs. ES phase were identified in either of the two cohorts of fertile women (Figure 2), and no consensus mRNAs were identified between EST and ESP cohorts when comparing blood mRNAs from MS cycle phase of fertile vs. RIF women.

2.10. Novel miRNA Validation Using Quantitative Real-Time Polymerase Chain Reaction

Custom TaqMan Small RNA assay (Thermo Fisher Scientific) was used for novel miRNA (chr2_4401, sequence: gaacacugaaguuaauggcug) validation with eight paired ES and MS endometrial samples from fertile women and eight MS endometrial samples from RIF women. The average level of miR-151a-5p and miR-196b-5p was used as reference for normalization. These reference miRNAs were chosen due to their stable expression levels according to the current small RNA sequencing data. Additionally, five paired endometrial stromal ($CD13^+$) and epithelial cell ($CD9^+$) samples (two from ES and three from MS phase) isolated by fluorescence-activated cell sorting (FACS) were used to determine the cell type specificity of the novel miRNA. Cell sorting was performed according to our previous publication [36]. cDNA synthesis was conducted with TaqMan MicroRNA Reverse Transcription Kit (Thermo Fisher Scientific) and quantitative real-time polymerase chain reaction (qRT-PCR) was performed with TaqMan Universal PCR Master Mix, No AmpErase UNG (Thermo Fisher Scientific). Real-time experiments were performed in duplicate. Relative miRNA expression levels were compared between the studied groups by paired (ES vs. MS) or unpaired (fertile vs. RIF) two-tailed t-test (Excel, Microsoft Corporation, Redmond, WA, USA) and p-value ≤ 0.05 was considered as significant. Fold change (FC) was calculated using the $2^{-\Delta\Delta Ct}$ method [37].

2.11. Data Availability

The datasets analysed during the current study are available in the Gene Expression Omnibus repository (GEO), accession numbers GSE108966 and GSE98386.

3. Results

3.1. miRNAs in the Endometrium

In this study, 65 endometrial samples from the Estonian cohort (EST) and 51 samples from independent Spanish validation cohort (ESP) were included for miRNA data analysis (Figure 1). In total, 615 miRNAs were detected from endometrial samples among the EST cohort and 624 miRNAs among ESP cohort (CPM of at least 1 in 75% of samples within a sample group of ES or MS of fertile group, or RIF). Most abundant miRNAs in endometrial tissues from both cohorts were miR-10b-5p, miR-10a-5p and miR-27b-3p (Table S1A,B).

miRNAs were considered as DE if they showed significantly altered levels (FDR < 0.05) also in validation cohort (both in EST and ESP). From fertile women, 91 DE endometrial miRNAs were confirmed from MS vs. ES (Figure 2A) of which 49 were down- and 42 up-regulated in MS endometrium (Table S2A).

The comparison of MS endometrial samples from RIF patients vs. fertile women in EST and ESP validation datasets revealed 21 DE miRNAs (Figure 2B), out of which eight miRNAs were more abundantly expressed in RIF patients and 13 miRNAs in fertile women (Table S2B).

miR-424-5p was the only DE miRNA present in both comparisons—between the ES and MS endometria of fertile women (Table S2A) as well as between fertile women and RIF patients (Table S2B). Interestingly, while down-regulated in the MS endometrium of fertile women (average FC = −1.87 between the two cohorts), it was up-regulated in RIF patients' samples from the same phase (average FC = 1.74). When predicting miR-424-5p target genes from our mRNA dataset (Figure 1, see Methods section), we detected 85 targets (Table S3A) with involvement in different canonical pathways important in MS endometrial functions such as glucocorticoid, insulin receptor, axonal guidance and interleukins signalling (Table S4A). The miRNA-mRNA target analysis among endometrial samples from RIF women, based on our experimental datasets, identified miR-424-5p to target solely Serine/Threonine Kinase 2 (*SGK2*) gene (Table S3B).

3.2. *miRNAs in Blood*

Altogether, 65 blood samples from EST and 49 samples from ESP validation cohort were included for miRNA data analysis (Figure 1). In total, 305 miRNAs were detected from blood samples among EST cohort and 710 miRNAs among the independent ESP cohort (Table S1C,D). miRNAs were isolated from whole blood in EST samples and from the buffy coat fraction in ESP samples. Possibly due to the differences in sample treatment protocols between the centres, the most abundant miRNAs in blood varied between the two cohorts: miR-486-5p, miR-92a-3p and miR-451a demonstrated the highest read counts among EST blood samples, while miR-26a-5p, miR-191-5p and miR-181a-5p exhibited the highest expression levels among ESP samples.

No DE miRNAs were detected between blood samples corresponding to ES and MS cycle phase in fertile women either in the EST dataset or in the validation ESP dataset (Figure 2C). The comparison of blood samples corresponding to MS cycle phase from fertile and RIF women revealed that miR-30a-5p was significantly up-regulated among RIF patients in EST cohort (FC = 3.0, FDR = 0.01); and the difference was also confirmed in ESP cohort (FC = 1.9, FDR = 0.03) (Figure 2D).

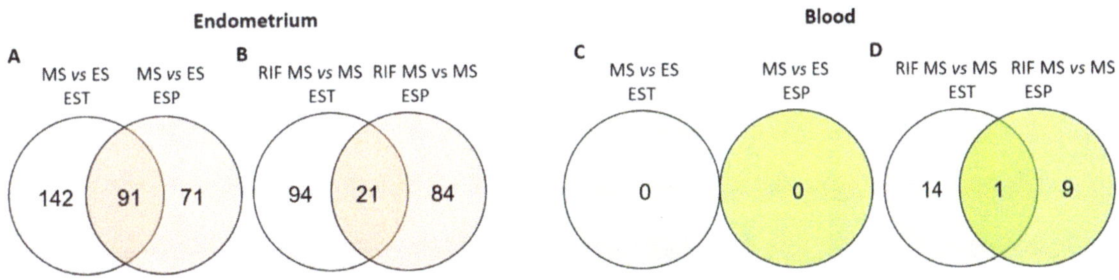

Figure 2. Number of differentially expressed miRNAs in EST and validation cohort ESP in (**A**) MS vs. ES phase endometrial samples of fertile women; (**B**) MS phase endometrial samples from infertile recurrent implantation failure (RIF) patient vs. fertile women; (**C**) MS vs. ES phase blood samples of fertile women; and (**D**) MS phase blood samples from RIF patients vs. fertile women. EST - Estonian samples; ESP—Spanish samples; RIF—recurrent implantation failure; ES—early secretory phase; MS—mid-secretory phase.

3.3. *Novel miRNAs*

In total, 18 novel miRNAs were determined from endometrial and blood samples (Tables 1 and S5). Out of these miRNAs, chr2_1900 (sequence: aucugaaauuugaaauggucc) and chr16_22077 (aggcuaggcugggccacag) were detected only from MS endometrial samples (from 7 and 4 out of 73 MS samples, respectively) and chr2_4401 (gaacacugaaguuaauggcug) was found from the majority (75.3%; 55/73) of MS endometrial samples and only from 2.7% (1/37) of ES endometrial samples. chr14_10307 (ucugagcccuguucucccuagg) was uniquely determined from blood samples of 16.7% RIF patients (4 out of 24 RIF blood samples).

We further focused on the most promising novel miRNA chr2_4401. Validation analysis by qRT-PCR confirmed the differential expression of chr2_4401 showing that the level of this novel miRNA was 37-fold higher ($p = 0.0001$) in MS compared to ES endometrium in fertile women (Figure 3A). No statistically significant differences in the expression levels of chr2_4401 between the MS endometrial samples from fertile and RIF women were detected (Figure 3A). Cell type-specific expression analysis showed very low levels of chr2_4401 in endometrial stromal cells, but 55-fold upregulation in epithelial cells was detected ($p = 0.003$, Figure 3B).

Table 1. Novel miRNAs detected from endometrial and blood samples.

Novel miRNA Provisional ID	Sample Type	Consensus Mature Sequence	Endometrium			Blood		
			Fertile Women		Recurrent Implantation Failure (RIF) Patients	Fertile Women		RIF Patients
			Detected from Early Secretory (ES) Samples (n = 37)	Detected from Mid-Secretory (MS) Samples (n = 37)	Detected from MS Samples (n = 36)	Detected from ES Samples (n = 34)	Detected from MS Samples (n = 34)	Detected from MS Samples (n = 24)
chr2_1900	Endometrium	aucugaaauuugaaauggucc	-	3	4	-	-	-
chr2_4401	Endometrium	gaacacugaaguuaauggcug	1	26	29	-	-	-
chr2_2219	Blood	cugagaagacagucgaacuugac	-	-	-	11	7	5
chr3_7058	Endometrium, blood	aaaaguaaucgcggucuuugcc	-	2	-	3	4	5
chr3_1133	Endometrium, blood	caacaggccuugcucugcucacaga	1	-	2	2	-	2
chr3_2060	Endometrium	cacaggcaggagaccccacagc	7	2	1	-	-	-
chr4_6331	Endometrium	aacugggcauguuggaacuaagc	2	1	-	-	-	-
chr4_10018	Endometrium	aucagagaaacuuucuugaac	3	1	9	-	-	-
chr8_10314	Endometrium, blood	uuauccuccaguagacuaggga	18	8	14	3	2	4
chr8_8702	Endometrium, blood	acaccggggguuagagcuucaacc	2	-	-	-	1	2
chr10_9545	Endometrium, blood	aaaaguuauugcgguuuuugc	1	-	1	1	2	2
chr12_27523	Endometrium, blood	ucuggcuccuuucuaaucac	-	2	-	3	2	6
chr13_23821	Endometrium, blood	augugccuaguggcugcuguc	5	5	4	6	6	7
chr14_10307	Blood	ucugagcccuguucucccuagg	-	-	-	-	-	4
chr14_3458	Endometrium	aaaagucaucucgguucuugcc	3	1	3	-	-	-
chr15_26374	Endometrium	aaacguaauuguggauuuugc	2	1	3	-	-	-
chr16_22077	Endometrium	cugacugcccuggccuggccag	-	2	2	-	-	-
chr17_11615	Endometrium, blood	uaacucuuagaaucccaaag	1	-	-	18	16	10

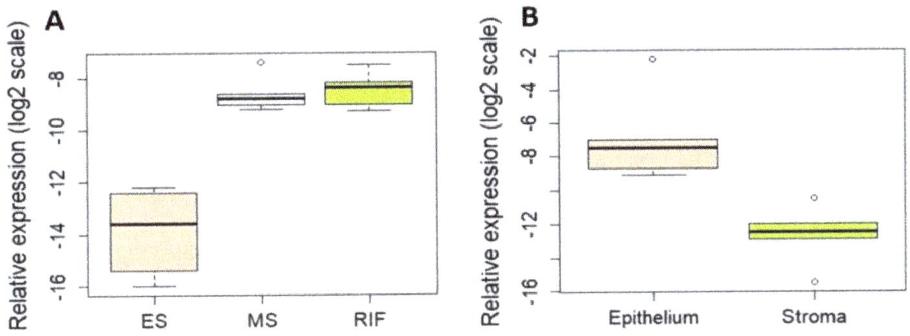

Figure 3. Validation of the novel miRNA chr2_4401 by quantitative real-time polymerase chain reaction (qRT-PCR). (**A**) miRNA expression level differences between paired ES ($n = 8$) and MS ($n = 8$) endometria of fertile women, and MS endometria of RIF ($n = 8$) patients. The relative miRNA expression levels (ΔCt) in the endometrium were 37-fold higher in MS compared to ES samples ($p = 0.0001$). No differences between MS samples of fertile and RIF patients were observed. (**B**) Relative expression level differences in epithelial ($n = 5$) and stromal ($n = 5$) fractions. The relative miRNA levels (ΔCt) were 55-fold higher in epithelial cells (CD9+) compared to stromal cells (CD13+) ($p = 0.003$). The average levels of miR-151a-5p and miR-196b-5p were used for data normalization. For illustrative purposes, relative expression levels (ΔCt) were multiplied by −1. ES—early secretory phase; MS—mid-secretory phase.

The novel miRNA chr2_4401 precursor sequence (Figure 4A) is located within the short arm of chromosome 2 (p11.2) and is transcribed from intergenic region. No human miRNAs with high similarity sequences for the predicted novel miRNA was found from miRBase database, however, chr2_4401 shares the seed region (2–7 nt) with miR-200 family miRNAs (Figure 4B) and, therefore, target genes of the predicted miRNA are expected to be common with those of the miR-200 family. miRDB revealed more than 800 potential target genes for chr2_4401 (Table S6A). For pathway analysis, we focussed only on these potential targets that were detected as down-regulated in the MS endometrium in our mRNA-seq analysis from the same women (98 genes, Table S6B). The predicted target genes of this novel miRNA identified in our dataset were involved in pathways including estrogen-mediated S-phase entry ($p = 0.006$), cell cycle regulation by B-cell translocation gene (BTG) family proteins ($p = 0.01$), role of checkpoint kinase (CHK) proteins in cell cycle checkpoint control ($p = 0.03$), and epithelial adherens junction signalling ($p = 0.03$) (Table S4B).

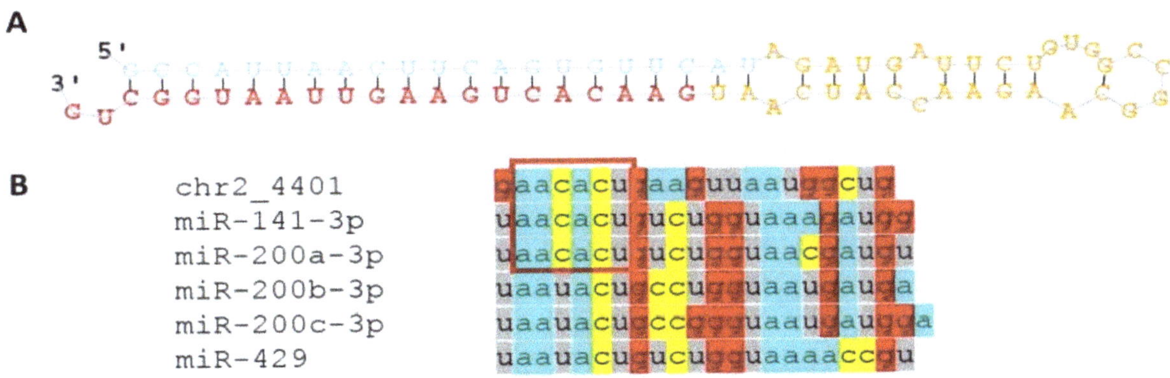

Figure 4. (**A**) Hairpin structure of the novel miRNA chr2_4401 precursor predicted by miRDeep2. (**B**) Sequence alignment between chr2_4401 and five miR-200 family miRNAs. The seed region (nucleotides 2–7, red square) of chr2_4401 is identical with miR-141-3p and miR-200a-3p.

3.4. *miRNA–mRNA Interaction Prediction*

The joint analysis of the miRNA and mRNA RNA-seq data from MS vs. ES endometrial samples by IPA software detected 81 (out of 91) DE miRNAs and 865 (out of 4240) mRNAs (Table S3A). Further analysis demonstrated that these 865 target mRNAs are involved in canonical pathways important in several MS endometrial functions (Table S4C). According to the calculated z-scores, differential miRNA expression leads to the up-regulation of gene expression in the glucocorticoid, G-protein coupled receptor, IGF-1 and JAK/STAT signalling pathways (obtaining the highest z-score), while the most significant down-regulation of gene expression was observed within Wnt/beta-catenin signalling pathway (as demonstrated by the lowest z-score) (Figure 5).

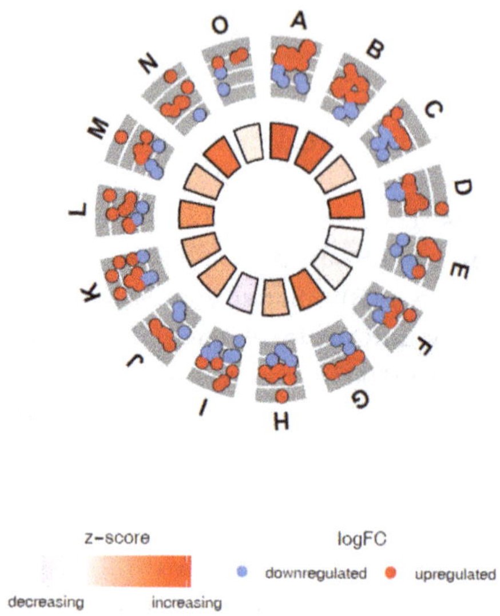

ID	Description
A	Glucocorticoid Receptor Signalling
B	G-Protein Coupled Receptor Signalling
C	Leukocyte Extravasation Signalling
D	IGF-1 Signalling
E	Calcium Signalling
F	Gap Junction Signalling
G	Integrin Signalling
H	Growth Hromone Signalling
I	Wnt/beta-catenin Signalling
J	VEGF Signalling
K	Tight Junction Signalling
L	Epithelial Adherenc Junction Signalling
M	Leptin Signalling
N	JAK/STAT Signalling
O	Estrogen Receptor Signalling

Figure 5. Circular plot of IPA canonical pathways enriched with miRNA targets differentially expressed between MS and ES endometrial samples of fertile women. The inner circle represents the overall up- or down-regulation of gene expression in each pathway according to the z-score. The outer circle depicts the fold change of expression level for each gene.

The analysis of our miRNA and mRNA sequencing data from MS endometrial samples from fertile vs. RIF women detected 21 DE miRNAs and 14 mRNAs. Out of those the IPA program identified 4 miRNAs in interaction with 5 mRNAs (Table S3B). The miRNA targets were detected to be involved in signal transducer and activator of transcription 3 (STAT3) and cyclin-dependent kinase 5 (CDK5) signalling pathways (Table S4D).

4. Discussion

Current knowledge about the involvement of miRNAs in endometrial transformation from ES (pre-receptive) to MS (receptive) stage both in fertility as well as in infertility conditions is limited. Moreover, it is not known whether the overall systemic changes in women during these processes are also reflected in blood miRNA profiles, which could serve as non-invasive biomarker candidates for evaluating fertility status. Therefore, using small RNA-seq technology, we profiled and validated the complete endometrial and blood miRNome from ES and MS phase samples from healthy fertile women and infertile RIF patients in two independent sample cohorts. This is the largest miRNA endometrial receptivity study to date with novel aspects of combining matched samples of the endometrium and blood from two different populations.

We identified 91 DE miRNAs between the ES and MS endometrium of healthy fertile individuals, showing the same directions in miRNA expression in both cohorts. Several miRNAs that have repeatedly been associated with endometrial receptivity were also confirmed by our study, including miR-30b-5p, miR-30d-3p, miR-30d-5p and miR-30a-5p that were up-regulated in the MS phase endometrium. Endometrial expression level changes of the miR-30 family members have been described in a number of previous endometrial receptivity studies [4,15–17,22,23,38]. Furthermore, it has been previously shown that endometrial miR-30d is taken up by the pre-implantation embryo, resulting in modified transcriptome and embryo adhesion [16]. Therefore, our data together with previous studies collectively indicate that the miR-30 family is crucial in the regulation of endometrial receptivity and implantation processes in the endometrium.

Another interesting group of miRNAs in endometrial receptivity is the miR-200 family. We detected several miR-200 family members, such as miR-200a-3p, miR-200c-3p, miR-200c-5p, miR-141-3p, miR-141-5p and miR-429 to be up-regulated in the MS phase endometrium, being in line with previously published studies [16,23]. miR-200 family members are shown to target several genes in the endometrium that influence cell proliferation, migration and inflammation [39], which are all important cellular processes in MS endometrial functions. Interestingly, the miR-200 family is predominantly expressed in endometrial epithelial cells [36], which is the first cellular layer to interact with the implanting embryo. Nevertheless, as whole tissue biopsies were analysed in our study, we cannot confirm that the detected findings are epithelial cell specific.

Our study results also identified miRNAs that, to the best of our knowledge, have not been associated with human endometrial receptivity before (Table S2A). For instance, miR-873-3p was one of the most up-regulated and miR-3131 one of the most down-regulated miRNAs in the MS phase endometrium from fertile women. The roles of these miRNAs in endometrial tissue remain to be elucidated, but it has been demonstrated that miR-873-3p plays a role in the selection of bovine dominant follicle [40].

Interaction analysis of the DE miRNAs and their target genes identified the involvement of several canonical pathways important in endometrial receptivity, including JAK/STAT signalling, leptin signalling and growth hormone signalling. JAK/STAT signalling pathway transmits information from extracellular signals to the nucleus influencing transcription, and its involvement in embryo implantation is widely acknowledged [41–44]. Leptin signalling, regulated by cytokines and playing a role in inflammatory response, is another classically known signalling pathway to be involved in endometrial receptivity [43,45]. Furthermore, leptin is known to mediate the effects of growth hormone [46]. The beneficial effects of growth hormone administration on endometrial receptivity have recently been published, where growth hormone increased endometrial blood perfusion and the expression of different cytokines [47]. Growth hormone administration has also improved implantation, pregnancy and live birth rates in RIF patients [48]. The Wnt/beta-catenin signalling pathway has been demonstrated to control estrogen-dependent endometrial cell proliferation, decidualisation, trophoblast attachment and invasion. Fine-tuning of this pathway is particularly important as failures in Wnt signalling are associated with infertility, endometriosis, endometrial cancer and gestational diseases such as complete mole placentae and choriocarcinomas (reviewed in [49]). miRNAs differentially expressed upon the establishment of receptivity participate in the overall down-regulation of Wnt/beta-catenin signalling in the normal endometrium according to our results.

The endometrial biopsies from healthy participants were all taken at two time-points (ES and MS) within the same menstrual cycle. A recent study by Evans et al. demonstrated that the prior collection of an LH + 2 sample does not affect the general gene expression level of a LH + 7 sample [50]. Their study data suggested that the expression of tested genes exhibited a stable pattern, which was not affected even when sampled twice in one cycle and is valid to predict endometrial receptivity in the subsequent cycles. No such studies have been performed regarding miRNA expression, but similar variability in expression levels is expected within and between cycles as for mRNAs.

Besides the considerable changes in endometrial tissue, cyclic alterations take place throughout the female body during the menstrual cycle under the governance of steroid hormones. In order to find systemic miRNA changes as indirect markers that could reflect the receptive stage of the endometrium, we analysed blood samples that were collected simultaneously to endometrial biopsies. However, no DE blood miRNAs were found between the samples corresponding to ES and MS time-points in either of the cohorts, regardless of the sample isolation protocol. Prior to our current study, miRNAs from whole blood have not been investigated in abovementioned time-points, but we have previously shown that there are no differences in circulating plasma miRNA profile throughout the menstrual cycle [51]. Therefore, it is probable that there are no blood-derived miRNAs that could reflect the changes which occur in female body during the establishment of the receptive endometrium.

Another interesting aspect of our study was the identification of dysregulated miRNAs among RIF women. Although this study group comprised of patients with various causes of infertility, they all had in common at least three unsuccessful IVF treatments with embryo transfers. We identified 21 DE miRNAs in the MS endometrium of RIF patients compared to fertile women. Notably, as has also been previously reported [21–23], miR-424-5p was up-regulated in RIF patients, while being down-regulated in the mid-secretory endometrium of healthy fertile women. Therefore, miR-424-5p may be a useful marker to determine endometrial insufficiency for embryo implantation. miR-424-5p has been previously shown to stimulate cell-to-cell adhesion during embryo implantation by targeting osteopontin (encoded by *SPP1* gene) [22,52,53]. Aberrant osteopontin levels in the endometrium have been linked to infertility [54].

Our target prediction analysis for miR-424-5p detected genes involved in numerous interleukins signalling pathways in the MS endometrium of fertile women, while the differential expression of a single miR-424-5p target gene, *SGK2*, was detected in the MS endometrium of RIF patients. Interleukins are a group of cytokines that regulate cell growth, differentiation, and they are particularly important in stimulating immune responses. The importance of immune responses in the pre- and peri-implantation periods in endometrial functions are widely acknowledged [4]. In order to provide a hospitable environment for the embryo, the balance between the maternal immune tolerance toward a semi-allogeneic implanting embryo and the protective anti-infectious mechanisms in the receptive phase uterus must be established [55]. Indeed, a recent study detected a range of interleukins to be down-regulated in infertile women with implantation failure in IVF [56]. SGK2, the potential target of miR-424-5p in RIF patients, is a protein kinase having powerful stimulating effect on K^+ channels with a possible role in the regulation of epithelial transport and cell proliferation [57]. The importance of ion channels, including K^+ in the regulation of endometrial receptivity has been summarised in a previous review [58]. However, to the best of our knowledge, *SGK2* has not been identified before in endometrial functions, and it could serve as an interesting target for future studies in women experiencing implantation failure.

Interaction analysis of our DE miRNAs and mRNAs identified the involvement of STAT3 and CDK5 signalling pathways in the development of RIF. CDK5 participates in a variety of cellular processes via the effects on angiogenesis, cell proliferation, cell adhesion, migration, and immune system (summarised in a recent review [59]).

Despite the differences in sample isolation protocols applied in the two cohorts, miR-30a-5p was elevated in the blood of RIF patients in both datasets. As miRNA levels in blood reflect its cellular composition, it is possible that the observed differences in miR-30a-5p levels between RIF patients and controls are derived from the alterations in the ratio of different cell types in blood samples. Although there are studies indicating that infertile women have altered levels of immune cells in blood, the results are highly conflicting [60], and therefore we cannot confirm nor rule out the possibility that miR-30a-5p alteration is due to variability in immune cell levels in RIF patients. Nevertheless, this miRNA could indirectly imply to the dysregulated physiological conditions resulting in implantation failure and serves as a potential minimally invasive biomarker in this regard. miR-30a-5p acts as a tumour suppressor by inhibiting cell proliferation and invasion [61–63]. In blood cells, miR-30a takes part in the regulation of erythrocyte maturation [64]. However, the precise mechanism that leads to the observed elevated levels in RIF patients' blood in our study remains obscure.

As an additional finding in our study, we identified 18 novel miRNAs. Recently, a publication involving nearly 500 small RNA libraries from different mammalian primary cells analysed whether previously annotated and unannotated short RNA sequences serve as valid miRNAs [65]. Several sequences proposed as novel unannotated miRNAs in our study matched the genomic coordinates of proposed precursor sequences (including chr2_1900 and chr2_4401) and some met the multiple high-confidence criteria set for miRNAs in the aforementioned publication [65] (Table S5). This adds further confidence that the reported novel sequences in our study are genuine miRNAs. The most interesting finding to emerge from novel miRNA analysis was sequence chr2_4401 that showed higher expression levels in the MS vs. ES phase endometrium. Sequence alignment of chr2_4401 to several miR-200 family members demonstrates that their seed regions are identical. Therefore, we conclude that chr2_4401 plays similar important roles in MS endometrial functions as described above for the miR-200 family.

5. Conclusions

Our study approach, involving 230 samples, provides additional knowledge about the complex regulation of endometrial receptivity for successful embryo implantation. Our study results highlight the involvement of miR-30 and miR-200 family members in endometrial receptivity development, and miR-424-5p as dysregulated in RIF. The current study also identified several miRNAs not previously known to be involved in endometrial receptivity. We discovered several novel miRNAs, among which validated miRNA chr2_4401 is of most interest in MS endometrial functions. The results of our study provide new aspects of miRNA functions in MS endometrial processes in health and disease and provide means for further studies for identifying molecular biomarkers of fertility/infertility from endometrial and/or blood samples.

Supplementary Materials:
Table S1: (a) miRNAs detected in EST endometrial samples. (b) miRNAs detected in ESP endometrial samples. (c) miRNAs detected in EST blood samples. (d) miRNAs detected in ESP blood samples. Table S2: (a) Differentially expressed miRNAs between MS and ES endometrial samples from fertile women. (b) Differentially expressed miRNAs from MS endometrial samples between RIF and fertile women. Table S3: (a) Predicted target genes for differentially expressed miRNAs in MS vs. ES phase endometrium from fertile women. (b) Predicted target genes for differentially expressed miRNAs in MS endometrium from fertile vs. infertile women with RIF. Table S4: (a) miR-424-5p target genes involved in different canonical pathways in MS phase endometrium. (b) Novel miRNA chr2_4401 target genes involved in different canonical pathways in MS phase endometrium. (c) Canonical pathways involved in MS endometrial functions (MS vs. ES endometrium from fertile women). (d) Canonical pathways involved in mid-secretory endometrial functions from infertile RIF women. Table S5: Novel miRNAs detected from endometrial and blood samples. Table S6: (a) All predicted targets for novel chr2_4401 miRNA (sequence: gaacacugaaguuaauggcug) according to miRDB. (b) Predicted targets for novel miRNA chr2_4401 down-regulated in MS endometrium.

Author Contributions: K.R. performed experiments and data analysis, and wrote the manuscript; S.A performed data analysis, and wrote the manuscript; M.S. participated in designing the study, and performed experiments; M.P. contributed to data analysis and interpretation, and critically revised the manuscript; J.F.M-B. performed experiments; F.M.C. performed experiments; F.V. performed experiments; C.S. participated in designing the study; A.S. participated in designing the study, and critically revised the manuscript; A.V.-M. participated in designing the study, performed experiments and data analysis, and critically revised the manuscript. All authors contributed to drafting the manuscript and have approved the final version of the article.

Acknowledgments: We are thankful to the personnel at the Nova Vita Clinic and Elite Clinic in Estonia as well as IVI Clinic Valencia for their involvement in sample collection, to Katrin Kepp and Pilar Alamá for the recruitment of healthy women to the study, and to Sergio Cabanillas for the recruitment of RIF patients. We are grateful to all study participants. We acknowledge Tatjana Jatsenko for her help with figures. We thank Parameswaran Grace Lalitkumar from Karolinska Institutet, Sweden for assisting with IPA analyses.

References

1. Simon, A.; Laufer, N. Repeated implantation failure: Clinical approach. *Fertil. Steril.* **2012**, *97*, 1039–1043. [CrossRef] [PubMed]

2. Macklon, N.S.; Stouffer, R.L.; Giudice, L.C.; Fauser, B.C. The science behind 25 years of ovarian stimulation for in vitro fertilization. *Endocr. Rev.* **2006**, *27*, 170–207. [CrossRef] [PubMed]
3. Altmäe, S.; Esteban, F.J.; Stavreus-Evers, A.; Simon, C.; Giudice, L.; Lessey, B.A.; Horcajadas, J.A.; Macklon, N.S.; D'Hooghe, T.; Campoy, C.; et al. Guidelines for the design, analysis and interpretation of 'omics' data: Focus on human endometrium. *Hum. Reprod. Update* **2014**, *20*, 12–28. [CrossRef] [PubMed]
4. Altmäe, S.; Koel, M.; Vosa, U.; Adler, P.; Suhorutsenko, M.; Laisk-Podar, T.; Kukushkina, V.; Saare, M.; Velthut-Meikas, A.; Krjutskov, K.; et al. Meta-signature of human endometrial receptivity: A meta-analysis and validation study of transcriptomic biomarkers. *Sci. Rep.* **2017**, *7*, 10077. [CrossRef] [PubMed]
5. Koler, M.; Achache, H.; Tsafrir, A.; Smith, Y.; Revel, A.; Reich, R. Disrupted gene pattern in patients with repeated in vitro fertilization (IVF) failure. *Hum. Reprod.* **2009**, *24*, 2541–2548. [CrossRef] [PubMed]
6. Altmäe, S.; Martinez-Conejero, J.A.; Salumets, A.; Simon, C.; Horcajadas, J.A.; Stavreus-Evers, A. Endometrial gene expression analysis at the time of embryo implantation in women with unexplained infertility. *Mol. Hum. Reprod.* **2010**, *16*, 178–187. [CrossRef] [PubMed]
7. Altmäe, S.; Tamm-Rosenstein, K.; Esteban, F.J.; Simm, J.; Kolberg, L.; Peterson, H.; Metsis, M.; Haldre, K.; Horcajadas, J.A.; Salumets, A.; et al. Endometrial transcriptome analysis indicates superiority of natural over artificial cycles in recurrent implantation failure patients undergoing frozen embryo transfer. *Reprod. Biomed. Online* **2016**, *32*, 597–613. [CrossRef] [PubMed]
8. Ledee, N.; Munaut, C.; Aubert, J.; Serazin, V.; Rahmati, M.; Chaouat, G.; Sandra, O.; Foidart, J.M. Specific and extensive endometrial deregulation is present before conception in IVF/ICSI repeated implantation failures (IF) or recurrent miscarriages. *J. Pathol.* **2011**, *225*, 554–564. [CrossRef] [PubMed]
9. Koot, Y.E.; van Hooff, S.R.; Boomsma, C.M.; van Leenen, D.; Koerkamp, M.J.G.; Goddijn, M.; Eijkemans, M.J.; Fauser, B.C.; Holstege, F.C.; Macklon, N.S. An endometrial gene expression signature accurately predicts recurrent implantation failure after IVF. *Sci. Rep.* **2016**, *6*, 19411. [CrossRef] [PubMed]
10. Fan, L.J.; Han, H.J.; Guan, J.; Zhang, X.W.; Cui, Q.H.; Shen, H.; Shi, C. Aberrantly expressed long noncoding RNAs in recurrent implantation failure: A microarray related study. *Syst. Biol. Reprod. Med.* **2017**, *63*, 269–278. [CrossRef] [PubMed]
11. Carthew, R.W.; Sontheimer, E.J. Origins and mechanisms of miRNas and siRNAs. *Cell* **2009**, *136*, 642–655. [CrossRef] [PubMed]
12. Lewis, B.P.; Burge, C.B.; Bartel, D.P. Conserved seed pairing, often flanked by adenosines, indicates that thousands of human genes are microRNA targets. *Cell* **2005**, *120*, 15–20. [CrossRef] [PubMed]
13. Lam, E.W.; Shah, K.; Brosens, J.J. The diversity of sex steroid action: The role of micro-RNAs and FOXO transcription factors in cycling endometrium and cancer. *J. Endocrinol.* **2012**, *212*, 13–25. [CrossRef] [PubMed]
14. Liu, W.; Niu, Z.; Li, Q.; Pang, R.T.; Chiu, P.C.; Yeung, W.S. MicroRNA and embryo implantation. *Am. J. Reprod. Immunol.* **2016**, *75*, 263–271. [CrossRef] [PubMed]
15. Altmäe, S.; Martinez-Conejero, J.A.; Esteban, F.J.; Ruiz-Alonso, M.; Stavreus-Evers, A.; Horcajadas, J.A.; Salumets, A. MicroRNAs miR-30b, miR-30d, and miR-494 regulate human endometrial receptivity. *Reprod. Sci.* **2013**, *20*, 308–317. [CrossRef] [PubMed]
16. Vilella, F.; Moreno-Moya, J.M.; Balaguer, N.; Grasso, A.; Herrero, M.; Martinez, S.; Marcilla, A.; Simon, C. Hsa-miR-30d, secreted by the human endometrium, is taken up by the pre-implantation embryo and might modify its transcriptome. *Development* **2015**, *142*, 3210–3221. [CrossRef] [PubMed]
17. Moreno-Moya, J.M.; Vilella, F.; Martinez, S.; Pellicer, A.; Simon, C. The transcriptomic and proteomic effects of ectopic overexpression of miR-30d in human endometrial epithelial cells. *Mol. Hum. Reprod.* **2014**, *20*, 550–566. [CrossRef] [PubMed]
18. Choi, Y.; Kim, H.R.; Lim, E.J.; Park, M.; Yoon, J.A.; Kim, Y.S.; Kim, E.K.; Shin, J.E.; Kim, J.H.; Kwon, H.; et al. Integrative analyses of uterine transcriptome and microRNAome reveal compromised LIF-STAT3 signaling and progesterone response in the endometrium of patients with recurrent/repeated implantation failure (RIF). *PLoS ONE* **2016**, *11*, e0157696. [CrossRef] [PubMed]
19. Li, R.; Qiao, J.; Wang, L.; Li, L.; Zhen, X.; Liu, P.; Zheng, X. microRNA Array and microarray evaluation of endometrial receptivity in patients with high serum progesterone levels on the day of hCG administration. *Reprod. Biol. Endocrinol.* **2011**, *9*, 29. [CrossRef] [PubMed]
20. Revel, A.; Achache, H.; Stevens, J.; Smith, Y.; Reich, R. microRNAs are associated with human embryo implantation defects. *Hum. Reprod.* **2011**, *26*, 2830–2840. [CrossRef] [PubMed]

21. Shi, C.; Shen, H.; Fan, L.J.; Guan, J.; Zheng, X.B.; Chen, X.; Liang, R.; Zhang, X.W.; Cui, Q.H.; Sun, K.K.; et al. Endometrial microRNA signature during the window of implantation changed in patients with repeated implantation failure. *Chin. Med. J. (Engl.)* **2017**, *130*, 566–573. [PubMed]
22. Sha, A.G.; Liu, J.L.; Jiang, X.M.; Ren, J.Z.; Ma, C.H.; Lei, W.; Su, R.W.; Yang, Z.M. Genome-wide identification of micro-ribonucleic acids associated with human endometrial receptivity in natural and stimulated cycles by deep sequencing. *Fertil. Steril.* **2011**, *96*, 150–155. [CrossRef] [PubMed]
23. Sigurgeirsson, B.; Amark, H.; Jemt, A.; Ujvari, D.; Westgren, M.; Lundeberg, J.; Gidlof, S. Comprehensive RNA sequencing of healthy human endometrium at two time points of the menstrual cycle. *Biol. Reprod.* **2017**, *96*, 24–33. [PubMed]
24. Suhorutshenko, M.; Kukushkina, V.; Velthut-Meikas, A.; Altmae, S.; Peters, M.; Magi, R.; Krjutskov, K.; Koel, M.; Codoner, F.M.; Martinez-Blanch, J.F.; et al. Endometrial receptivity revisited: Endometrial transcriptome adjusted for tissue cellular heterogeneity. *Hum. Reprod.* **2018**, *33*, 2074–2086. [CrossRef] [PubMed]
25. Bolger, A.M.; Lohse, M.; Usadel, B. Trimmomatic: A flexible trimmer for Illumina sequence data. *Bioinformatics* **2014**, *30*, 2114–2120. [CrossRef] [PubMed]
26. Friedlander, M.R.; Mackowiak, S.D.; Li, N.; Chen, W.; Rajewsky, N. miRDeep2 accurately identifies known and hundreds of novel microRNA genes in seven animal clades. *Nucl. Acids Res.* **2012**, *40*, 37–52. [CrossRef] [PubMed]
27. Griffiths-Jones, S. The microRNA registry. *Nucl. Acids Res.* **2004**, *32*, D109–D111. [CrossRef] [PubMed]
28. Robinson, M.D.; McCarthy, D.J.; Smyth, G.K. edgeR: A Bioconductor package for differential expression analysis of digital gene expression data. *Bioinformatics* **2010**, *26*, 139–140. [CrossRef] [PubMed]
29. McCarthy, D.J.; Chen, Y.; Smyth, G.K. Differential expression analysis of multifactor RNA-Seq experiments with respect to biological variation. *Nucl. Acids Res.* **2012**, *40*, 4288–4297. [CrossRef] [PubMed]
30. Robinson, M.D.; Oshlack, A. A scaling normalization method for differential expression analysis of RNA-seq data. *Genom. Biol.* **2010**, *11*, R25. [CrossRef] [PubMed]
31. Friedlander, M.R.; Chen, W.; Adamidi, C.; Maaskola, J.; Einspanier, R.; Knespel, S.; Rajewsky, N. Discovering microRNAs from deep sequencing data using miRDeep. *Nat. Biotechnol.* **2008**, *26*, 407–415. [CrossRef] [PubMed]
32. Wong, N.; Wang, X. miRDB: An online resource for microRNA target prediction and functional annotations. *Nucl. Acids Res.* **2015**, *43*, D146–D152. [CrossRef] [PubMed]
33. Kim, D.; Pertea, G.; Trapnell, C.; Pimentel, H.; Kelley, R.; Salzberg, S.L. TopHat2: Accurate alignment of transcriptomes in the presence of insertions, deletions and gene fusions. *Genom. Biol.* **2013**, *14*, R36. [CrossRef] [PubMed]
34. Anders, S.; Pyl, P.T.; Huber, W. HTSeq—A Python framework to work with high-throughput sequencing data. *Bioinformatics* **2015**, *31*, 166–169. [CrossRef] [PubMed]
35. Walter, W.; Sanchez-Cabo, F.; Ricote, M. GOplot: An R package for visually combining expression data with functional analysis. *Bioinformatics* **2015**, *31*, 2912–2914. [CrossRef] [PubMed]
36. Saare, M.; Rekker, K.; Laisk-Podar, T.; Soritsa, D.; Roost, A.M.; Simm, J.; Velthut-Meikas, A.; Samuel, K.; Metsalu, T.; Karro, H.; et al. High-throughput sequencing approach uncovers the miRNome of peritoneal endometriotic lesions and adjacent healthy tissues. *PLoS ONE* **2014**, *9*, e112630. [CrossRef] [PubMed]
37. Livak, K.J.; Schmittgen, T.D. Analysis of relative gene expression data using real-time quantitative PCR and the $2^{-\Delta\Delta Ct}$ method. *Methods* **2001**, *25*, 402–408. [CrossRef] [PubMed]
38. Kuokkanen, S.; Chen, B.; Ojalvo, L.; Benard, L.; Santoro, N.; Pollard, J.W. Genomic profiling of microRNAs and messenger RNAs reveals hormonal regulation in microRNA expression in human endometrium. *Biol. Reprod.* **2010**, *82*, 791–801. [CrossRef] [PubMed]
39. Panda, H.; Pelakh, L.; Chuang, T.D.; Luo, X.; Bukulmez, O.; Chegini, N. Endometrial miR-200c is altered during transformation into cancerous states and targets the expression of *ZEBs*, *VEGFA*, *FLT1*, *IKKβ*, *KLF9*, and *FBLN5*. *Reprod. Sci.* **2012**, *19*, 786–796. [CrossRef] [PubMed]
40. Sontakke, S.D.; Mohammed, B.T.; McNeilly, A.S.; Donadeu, F.X. Characterization of microRNAs differentially expressed during bovine follicle development. *Reproduction* **2014**, *148*, 271–283. [CrossRef] [PubMed]
41. Aghajanova, L.; Altmae, S.; Bjuresten, K.; Hovatta, O.; Landgren, B.M.; Stavreus-Evers, A. Disturbances in the LIF pathway in the endometrium among women with unexplained infertility. *Fertil. Steril.* **2009**, *91*, 2602–2610. [CrossRef] [PubMed]

42. Pawar, S.; Starosvetsky, E.; Orvis, G.D.; Behringer, R.R.; Bagchi, I.C.; Bagchi, M.K. STAT3 regulates uterine epithelial remodeling and epithelial-stromal crosstalk during implantation. *Mol. Endocrinol.* **2013**, *27*, 1996–2012. [CrossRef] [PubMed]
43. Altmäe, S.; Reimand, J.; Hovatta, O.; Zhang, P.; Kere, J.; Laisk, T.; Saare, M.; Peters, M.; Vilo, J.; Stavreus-Evers, A.; et al. Research resource: Interactome of human embryo implantation: Identification of gene expression pathways, regulation, and integrated regulatory networks. *Mol. Endocrinol.* **2012**, *26*, 203–217. [CrossRef] [PubMed]
44. Aghajanova, L. Update on the role of leukemia inhibitory factor in assisted reproduction. *Curr. Opin. Obstet. Gynecol.* **2010**, *22*, 213–219. [CrossRef] [PubMed]
45. Gonzalez, R.R.; Simon, C.; Caballero-Campo, P.; Norman, R.; Chardonnens, D.; Devoto, L.; Bischof, P. Leptin and reproduction. *Hum. Reprod. Update* **2000**, *6*, 290–300. [CrossRef] [PubMed]
46. Margetic, S.; Gazzola, C.; Pegg, G.G.; Hill, R.A. Leptin: A review of its peripheral actions and interactions. *Int. J. Obes. Relat. Metab. Disord.* **2002**, *26*, 1407–1433. [CrossRef] [PubMed]
47. Wang, X.-M.; Hong, J.; Zhang, W.-X.; Li, Y. The effects of growth hormone on clinical outcomes after frozen-thawed embryo transfer. *Int. J. Gynaecol. Obstet.* **2016**, *133*, 347–350.
48. Altmäe, S.; Mendoza-Tesarik, R.; Mendoza, C.; Mendoza, N.; Cucinelli, F.; Tesarik, J. Effect of growth hormone on uterine receptivity in women with repeated implantation failure in an oocyte donation program: A randomized controlled trial. *J. Endocr. Soc.* **2017**, *2*, 96–105. [CrossRef] [PubMed]
49. Sonderegger, S.; Pollheimer, J.; Knofler, M. Wnt signalling in implantation, decidualisation and placental differentiation—Review. *Placenta* **2010**, *31*, 839–847. [CrossRef] [PubMed]
50. Evans, G.E.; Phillipson, G.T.M.; Sykes, P.H.; McNoe, L.A.; Print, C.G.; Evans, J.J. Does the endometrial gene expression of fertile women vary within and between cycles? *Hum. Reprod.* **2018**, *33*, 452–463. [CrossRef] [PubMed]
51. Rekker, K.; Saare, M.; Roost, A.M.; Salumets, A.; Peters, M. Circulating microRNA profile throughout the menstrual cycle. *PLoS ONE* **2013**, *8*, e81166. [CrossRef] [PubMed]
52. Johnson, G.A.; Burghardt, R.C.; Bazer, F.W. Osteopontin: A leading candidate adhesion molecule for implantation in pigs and sheep. *J. Anim. Sci. Biotechnol.* **2014**, *5*, 56. [CrossRef] [PubMed]
53. Kang, Y.J.; Forbes, K.; Carver, J.; Aplin, J.D. The role of the osteopontin-integrin $\alpha v\beta 3$ interaction at implantation: Functional analysis using three different in vitro models. *Hum. Reprod.* **2014**, *29*, 739–749. [CrossRef] [PubMed]
54. Casals, G.; Ordi, J.; Creus, M.; Fabregues, F.; Carmona, F.; Casamitjana, R.; Balasch, J. Osteopontin and $\alpha v\beta 3$ integrin as markers of endometrial receptivity: The effect of different hormone therapies. *Reprod. Biomed. Online* **2010**, *21*, 349–359. [CrossRef] [PubMed]
55. Haller-Kikkatalo, K.; Altmae, S.; Tagoma, A.; Uibo, R.; Salumets, A. Autoimmune activation toward embryo implantation is rare in immune-privileged human endometrium. *Semin. Reprod. Med.* **2014**, *32*, 376–384. [CrossRef] [PubMed]
56. Pathare, A.D.S.; Zaveri, K.; Hinduja, I. Downregulation of genes related to immune and inflammatory response in IVF implantation failure cases under controlled ovarian stimulation. *Am. J. Reprod. Immunol.* **2017**, *78*. [CrossRef] [PubMed]
57. Gamper, N.; Fillon, S.; Feng, Y.; Friedrich, B.; Lang, P.A.; Henke, G.; Huber, S.M.; Kobayashi, T.; Cohen, P.; Lang, F. K^+ channel activation by all three isoforms of serum- and glucocorticoid-dependent protein kinase SGK. *Pflugers Arch.* **2002**, *445*, 60–66. [PubMed]
58. Ruan, Y.C.; Chen, H.; Chan, H.C. Ion channels in the endometrium: Regulation of endometrial receptivity and embryo implantation. *Hum. Reprod. Update* **2014**, *20*, 517–529. [CrossRef] [PubMed]
59. Shupp, A.; Casimiro, M.C.; Pestell, R.G. Biological functions of CDK5 and potential CDK5 targeted clinical treatments. *Oncotarget* **2017**, *8*, 17373–17382. [CrossRef] [PubMed]
60. Seshadri, S.; Sunkara, S.K. Natural killer cells in female infertility and recurrent miscarriage: A systematic review and meta-analysis. *Hum. Reprod. Update* **2014**, *20*, 429–438. [CrossRef] [PubMed]
61. Ortega, M.; Bhatnagar, H.; Lin, A.P.; Wang, L.; Aster, J.C.; Sill, H.; Aguiar, R.C. A microRNA-mediated regulatory loop modulates NOTCH and MYC oncogenic signals in B- and T-cell malignancies. *Leukemia* **2015**, *29*, 968–976. [CrossRef] [PubMed]
62. Zhang, S.; Liu, Q.; Zhang, Q.; Liu, L. microRNA-30a-5p suppresses proliferation, invasion and tumor growth of hepatocellular cancer cells via targeting FOXA1. *Oncol. Lett.* **2017**, *14*, 5018–5026. [CrossRef] [PubMed]

63. Liu, X.; Ji, Q.; Zhang, C.; Liu, Y.; Liu, N.; Sui, H.; Zhou, L.; Wang, S.; Li, Q. miR-30a acts as a tumor suppressor by double-targeting COX-2 and BCL9 in *H. pylori* gastric cancer models. *Sci. Rep.* **2017**, *7*, 7113. [CrossRef] [PubMed]
64. Rouzbeh, S.; Kobari, L.; Cambot, M.; Mazurier, C.; Hebert, N.; Faussat, A.M.; Durand, C.; Douay, L.; Lapillonne, H. Molecular signature of erythroblast enucleation in human embryonic stem cells. *Stem Cells* **2015**, *33*, 2431–2441. [CrossRef] [PubMed]
65. de Rie, D.; Abugessaisa, I.; Alam, T.; Arner, E.; Arner, P.; Ashoor, H.; Astrom, G.; Babina, M.; Bertin, N.; Burroughs, A.M.; et al. An integrated expression atlas of miRNAs and their promoters in human and mouse. *Nat. Biotechnol.* **2017**, *35*, 872–878. [CrossRef] [PubMed]

11

Single Nucleotide Polymorphisms in *MIR143* Contribute to Protection against Non-Hodgkin Lymphoma (NHL) in Caucasian Populations

Gabrielle Bradshaw, Larisa M. Haupt, Eunise M. Aquino, Rodney A. Lea, Heidi G. Sutherland and Lyn R. Griffiths *

Genomics Research Centre, School of Biomedical Sciences, Institute of Health and Biomedical Innovation, Queensland University of Technology, Brisbane, QLD 4001, Australia; gabrielle.bradshaw@hdr.qut.edu.au (G.B.); larisa.haupt@qut.edu.au (L.M.H.); eunise.aquino@connect.qut.edu.au (E.M.A.); rodney.a.lea@gmail.com (R.A.L.); heidi.sutherland@qut.edu.au (H.G.S.)

* Correspondence: lyn.griffiths@qut.edu.au

Abstract: Recent studies show an association of microRNA (miRNA) polymorphisms (miRSNPs) in different cancer types, including non-Hodgkin lymphoma (NHL). The identification of miRSNPs that are associated with NHL susceptibility may provide biomarkers for early diagnosis and prognosis for patients who may not respond well to current treatment options, including the immunochemotherapy drug combination that includes rituximab, cyclophosphamide, doxorubicin, vincristine and prednisome (R-CHOP). We developed a panel of miRSNPs for genotyping while using multiplex PCR and chip-based mass spectrometry analysis in an Australian NHL case-control population (300 cases, 140 controls). Statistical association with NHL susceptibility was performed while using Chi-square (χ^2) and logistic regression analysis. We identified three SNPs in *MIR143* that are to be significantly associated with reduced risk of NHL: rs3733846 (odds ratio (OR) [95% confidence interval (CI)] = 0.54 [0.33–0.86], p = 0.010), rs41291957 (OR [95% CI] = 0.61 [0.39–0.94], p = 0.024), and rs17723799 (OR [95% CI] = 0.43 [0.26–0.71], p = 0.0009). One SNP, rs17723799, remained significant after correction for multiple testing (p = 0.015). Subsequently, we investigated an association between the rs17723799 genotype and phenotype by measuring target gene Hexokinase 2 (*HKII*) expression in cancer cell lines and controls. Our study is the first to report a correlation between miRSNPs in *MIR143* and a reduced risk of NHL in Caucasians, and it is supported by significant SNPs in high linkage disequilibrium (LD) in a large European NHL genome wide association study (GWAS) meta-analysis.

Keywords: biomarker; cancer; *MIR143*; miRSNP; non-Hodgkin lymphoma; rs17723799; Hexokinase 2; HKII

---

1. Introduction

The global incidence of non-Hodgkin lymphoma (NHL) varies with the geographical region, with the highest incidence rates occurring in North America, Europe, and Australasia [1]. According to the American Cancer Society, non-Hodgkin lymphoma (NHL) is one of the most common cancers in the United States, with around 75,000 people being predicted to be diagnosed in 2018, resulting in around 20,000 deaths [2]. The European Cancer Information System predicts approximately 115,000 new NHL cases in 2018, resulting in 48,000 deaths [3]. Currently, in Australia, lymphoma is the sixth most commonly diagnosed cancer (most common is breast cancer, followed by prostate, colorectal, melanoma, and lung cancer) [4]. NHL is comprised of a large group of diverse lymphoid malignancies involving either the lymph nodes or extranodal sites and it is more prevalent in males living in more developed countries over the age of 60 years [5]. The two most common NHL subtypes are diffuse large B-cell lymphoma (DLBCL), which is highly aggressive with a poor prognosis [6] and follicular lymphoma (FL), which is more indolent, but it can undergo a high-grade transformation into DLBCL [7]. Molecular mechanisms that are involved in NHL pathogenicity have been widely investigated; however, there is still a need to identify improved biomarkers for clinical diagnostic and therapeutic use.

Mature microRNAs (miRNAs) are short sequences of non-coding RNA that function to regulate the translational expression of their target genes through complementary binding to the 3′-untranslated regions (3′-UTRs) of the target mRNA, in what is referred to as the miRNA recognition element (MRE) [8,9]. The processing pathway and biogenesis of miRNAs has been well characterised [10–12], with several miRNAs being involved in targeting and regulating the expression of oncogenes and/or tumour suppressor genes [13]. Single nucleotide polymorphisms (SNPs) in miRNA biogenesis genes and miRNA genes themselves influence miRNA biogenesis, the processing of precursors to mature miRNAs, regulatory function, and stability of the miRNAs, and, if located in the mRNA 3′-UTR binding sites, can also disrupt miRNA binding and affect the expression of these target genes [14]. According to the miRNA SNP Disease Database (MSDD) [15] SNPs that occur in miRNA-related functional regions, such as mature miRNAs, promoter regions, pri- and pre-miRNAs, and target gene 3′UTR binding sites are collectively referred to as "miRSNPs" (miRNA polymorphisms). These polymorphisms represent a novel category of functional molecules that regulate gene expression [16]. SNPs in miRNA genes have been shown to affect miRNA biogenesis, with these SNPs potentially resulting in reduced and increased mature miRNA levels. It has been predicted that, if a SNP occurs in the miRNA stem region and decreases the stability of the hairpin structure, then it will reduce the amount of mature miRNA produced [16]; however, if it fails to decrease stability, it will tend to increase the amount of mature miRNA produced and lead to disease pathogenesis [16]. Furthermore, SNPs occurring in the non-coding regulatory regions of genes, such as the miRNA-binding sites in the 3′-UTRs of target genes, can cause mRNA instability changes, leading to disease pathogenesis. A number of studies have identified roles for miRNAs [17], as well as miRSNPs [18,19] in NHL. These studies have examined the suitability of miRSNPs as diagnostic and prognostic biomarkers for improved clinical treatment and survival, particularly as NHL is a heterogeneous disease with outcomes that vary from patient to patient. However, few studies have demonstrated the association of SNPs in miRNAs with NHL risk.

Serum lactate dehydrogenase A (LDH-A), which is an enzyme converting pyruvate to lactate, is commonly elevated in aggressive NHL cases, such as DLBCL and Burkitt's lymphoma, and it has been used as a prognostic indicator for overall survival [5]. Cancer cells are able to produce lactate in a high oxygen environment (aerobic glycolysis), causing a phenomenon of cancer metabolism, known as the Warburg effect. The unusual utilisation of glucose by cancer cells results in the production of lactate, which is normally only produced during anaerobic respiration [20]. Hexokinase 2 (HKII) is a tissue-specific isoenzyme that phosphorylates glucose to glucose-6-phosphate (G-6-P) at the start of the glycolysis pathway, and its upregulation contributes to cancer cell glycolysis and the Warburg effect [21–23]. Hexokinase 2 maintains the high rate of glucose catabolism that is required for the survival of tumour cells, allowing them to sustain a higher rate of proliferation and resistance to cell death signals [21,23]. A variety of tumours are characterised by upregulated HKII expression, making it an attractive therapeutic target [23–25]. Oncogenes, such as *MYC* and *RAS*, and tumour suppressor genes, such as *p53*, are known to regulate metabolic enzyme expression and they can cause cancer if they happen to undergo mutations [20]. In addition, hypoxia-inducible factor 1-alpha (HIF1A) and MYC proteins cooperate to regulate the expression of *HKII* and *PDK1* genes [26]. An investigation of the Warburg effect in Burkitt's lymphoma (BL) cells, where the HIF1A protein was highly expressed in EBV-positive BL cell lines, showed that the inhibition of MYC activity led to decreased expression of MYC-dependent genes and LDH-A activity, implicating MYC as the master regulator of aerobic glycolysis in these cells [26].

Most recently, Bhalla et al. investigated the role of hypoxia in DLBCL [27]. They demonstrated that the up-regulation of HIF1A resulted in repressed protein translation, however HKII was selectively translated by eIF4E1 to promote DLBCL growth in vitro and in vivo under hypoxic stress. Their findings suggest HKII as a key metabolic driver of the DLBCL phenotype. It has also been shown that acquired resistance in rituximab-resistant lymphoma cell lines (RRCL) was associated with the deregulation of glucose metabolism and an increase in the apoptotic threshold, leading to chemotherapy resistance, where RRCL expressed higher levels of HKII. Targeting HKII in these cells led to decreased resistance, implying that increased HKII levels in aggressive lymphoma causes chemotherapy resistance, while also identifying this as a potential therapeutic target [22]. Many HKII inhibitors have been effective in anti-cancer therapies, such as 3-bromopyruvate (3-BP), which was found to inhibit HKII, activate the mitochondrial cell death pathway, and deplete levels of ATP [22], and it was also shown to induce apoptosis in a breast cancer cell line (MDA-MB-231) [28].

The main aim of this study was to investigate the genetic association between miRSNPs that were previously implicated in tumourigenesis and/or NHL susceptibility and prognosis, based on a comprehensive review of recent literature [19,29,30]. We genotyped 39 miRSNPs using a multiplex PCR and matrix assisted laser desorption time-of-flight (MALDI-TOF) mass spectrometry (MS) MassARRAY® system in our Genomics Research Centre Genomics Lymphoma Population (GRC GLP-non-Hodgkin lymphoma) cohort. After basic association testing, three SNPs in *MIR143* were identified to be significantly associated with NHL, with one SNP, rs17723799, remaining significant after Bonferroni correction for multiple testing (p-value = 0.015). After logistic regression testing the same three SNPs in *MIR143* were significantly associated with reduced risk of NHL in the Additive model: rs3733846 (Odds ratio (OR) [95% confidence interval (CI)] = 0.54 [0.33–0.86], $p = 0.010$), rs41291957 (OR [95% CI] = 0.61 [0.39–0.94], $p = 0.024$), and rs17723799 (OR [95% CI] = 0.43 [0.26–0.71], $p = 0.0009$). As *HKII* is a known target gene for mature hsa-miR-143 (miR-143), our secondary aim examined HKII expression in four patient-derived NHL cell lines, as compared to a metastatic breast cancer (MDA-MB-231) and melanoma (MDA-MB-435) cell line, as well as two healthy control subjects to assess the potential functional link between miR-143 regulation and HKII levels in NHL.

2. Materials and Methods

2.1. Study Population

The GRC-GLP retrospective cohort consists of 300 NHL cases and 140 healthy controls. All of the samples are of Caucasian origin with Australian/British/European grandparents with no family history of a haematological malignancy. The cases were matched according to age- (within five years), sex-, and ethnicity with healthy cancer-free controls. Cases were collected between 2010 and 2014 from the Princess Alexandra Hospital in Brisbane, and the GRC clinic in Mermaid Waters on the Gold Coast. The case cohort mainly consists of FL ($n = 95$) and DLBCL ($n = 88$), with 79 cases being unclassified as NHL or "Other B-cell". B-cell chronic lymphocytic lymphoma (CLL), cutaneous T-cell lymphoma, Mantle cell lymphoma (MCL), Splenic marginal zone lymphoma (SMZL), Mucosa-associated lymphoid tissue lymphoma (MALT), and Burkitt's lymphoma (BL) make up the remaining subtypes, in the order of frequency from highest to lowest (Table 1). Patients and healthy volunteers were required to complete a personal questionnaire and provide written consent to participate in research. The cohort is comprised of 48% male and 52% female participants, with the mean age of cases 63.72 years (standard deviation (SD) = 12.95 years) and the mean age of controls 63.14 years (SD = 13.03 years). In addition, 35 new NHL cases were received in 2016 (collected in 2014), comprised of 71% male and 29% female, with an average age of 59.6 years (SD = 13.6 years). Ethics for the collection and the use of participant samples was approved by the Queensland University of Technology Research Ethics Committee (approval number 1400000125). All of the subjects gave written informed consent, in accordance with the Declaration of Helsinki.

Table 1. Retrospective Genomics Lymphoma Project-non-Hodgkin lymphoma (GLP-NHL) cohort comprising of different NHL subtypes and the number of cases for each subtype.

NHL Subtype	No. of Cases in the Cohort
FL	95
DLBCL	88
Other B-cell/NHL/unclassified	79
B-CLL	16
T-cell lymphoma	7
Mantle cell lymphoma (MCL)	6
Splenic marginal zone lymphoma (SMZL)	4
Mucosa-associated lymphoid tumour (MALT)	3
Burkitt's lymphoma (BL)	2
Total	300

GLP: Genomics lymphoma population; NHL: Non-Hodgkin lymphoma; FL: Follicular lymphoma; DLBCL: Diffuse large B cell lymphoma; CLL: Chronic lymphocytic lymphoma.

2.2. *Genomic DNA Extraction*

Genomic DNA (gDNA) was extracted from whole blood collected into EDTA tubes using an in-house salting-out method, as evaluated by Chacon et al. [31]. DNA concentration and purity was measured using the NanoDrop™ ND-1000 spectrophotometer (ThermoFisher Scientific Inc., Waltham, MA, USA) before dilution to 15–20 ng/µL and storage as stock gDNA at 4 °C.

2.3. *miRSNP Selection and iPlex Primer Design*

39 cancer-related microRNA-related SNPs (miRSNPs) were selected for genotyping by chip-based multiplex PCR and MALDI-TOF MS (Agena MassARRAY®, San Diego, CA, USA), following a comprehensive review of the recent literature related to miRNA biomarkers in tumourigenesis, including breast cancer and/or non-Hodgkin lymphoma [18]. Forward, reverse, and extension primers were designed for each SNP using the MassARRAY® AgenaCx design software. Primers were manufactured by Integrated DNA Technologies (IDT, Singapore) and primer sequences are available on request.

2.4. *Primary Multiplex PCR*

Genotyping was performed using the iPlex™ GOLD Reagent Kit (Agena), according to the manufacturer's instructions. Forward and reverse primers were pooled, with extension primers being pooled according to mass using the linear primer adjustment method. All of the PCR reactions were performed in 96-well reaction plates in a Veriti™ 96-well Thermal Cycler (ThermoFisher Scientific Inc., Waltham, MA, USA).

2.5. *MALDI-TOF MS and Data Analysis*

Prior to each run, the extension products were spotted on to the SpectroCHIP® (Agena) using the Agena MassARRAY® Nanodispenser (Agena), which was immediately loaded into the Agena MassARRAY® Analyser 4 for genotype detection of each SNP in the assay by the SpectroAcquire v4.0 software (Agena). Final data analysis was performed using the MassARRAY® Typer software v4.0 (Agena).

2.6. *In-Vitro Culture of Cell Lines and Primary Lymphocytes*

All of the commercial patient-derived NHL B-lymphoid (SU-DHL-4, Raji, Mino, Toledo), breast cancer (MDA-MB-231), and melanoma (MDA-MB-435) cell lines were previously acquired from the American Type Culture Collection (ATCC®, Manassas, VA, USA). The cell lines were authenticated by the GenePrint® 10 System, according to the manufacturer's instructions as a service provided by the School of Biomedical Sciences, QUT. For experiments, the cells were seeded into T-75 flasks with 20 mL RPMI-1640 or DMEM growth media (Invitrogen, ThermoFisher), supplemented with 10% foetal bovine serum and 1% Penicillin-Streptomycin antibiotic. Cultures were incubated at 37 °C under 5% CO_2 conditions. The cells were harvested into TRIzol® and Runx protein-lysis buffer for RNA and protein extraction (see below) and then pelleted for DNA extraction at 90% confluence and 95% viability. Cell counts and viability were assessed via the Trypan Blue exclusion method on a TC10™ automated cell counter (Bio-Rad, Hercules, CA, USA). Primary lymphocytes from healthy subjects were isolated from peripheral blood mononuclear cells (PBMCs) using the Ficoll-Histopaque centrifugation method (Sigma-Aldrich, St. Louis, MI, USA). Cells were washed with 1x PBS and then plated into T-75 culture flasks in RPMI-1640 with Phytohaemagglutinin (PHA) (Sigma-Aldrich) mitogen for up to 24 h to allow for monocytes to adhere to the flask. After a minimum of three hours, the lymphocytes in suspension were removed and plated into new flasks with PHA and interleukin 2 (IL-2) (Sigma-Aldrich). The lymphocytes were left to proliferate at 37 °C under 5% CO_2 conditions for five days. Live lymphocyte counts were performed on a haemocytometer and viability assessed by Trypan blue dye.

2.7. Validation of Genotyping by MassARRAY® and Genotyping of Cell Lines and Healthy Controls by Sanger Sequencing

The genotype plots on the MassARRAY® system were assessed and manually genotyped for scattered data not assigned by the analysis software. For statistically significant SNPs ($p < 0.05$), a subset of samples was validated by Sanger sequencing (SS) using primers (Table 2) to ensure the accuracy of genotype calls by the Agena MassARRAY® system. Briefly, amplified PCR products were treated with Exo-SAP for use in Big Dye Terminator v3.1 (ThermoFisher Scientific, Life Technologies) sequencing reactions, and analysed on the 3500 Genetic Analyser (ThermoFisher Scientific, Life Technologies). Sequencing data for each sample chromatogram was assessed using Chromas Lite 2.1.1 software). The genotypes obtained were found to be 100% concordant between the MassArray® and SS. gDNA was extracted from NHL cell line pellets using the GenePrint® 10 System, according to the manufacturer's instructions.

Table 2. Sanger sequencing primers for three analysed *MIR143* single nucleotide polymorphisms.

SNP	Forward Primer (5′–3′)	Reverse Primer (5′–3′)	Accession ID
rs3733846	TGTTTGCCTCCATCTCCTCT	CCTTCCCATGGAGCTTTGT	NC_000005.1
rs41291957	CAGGAAACACAGTTGTGAGG [1]	AGGAGAAGGGGTGTTAGAGG [1]	NC_000005.1
rs17723799	TGGTCATCCAATCAGCCACC	GGAAGGGACCCTGTCAACTG	NC_000005.1

[1] Same sequence as MassARRAY® primer design (AgenaCx); SNP: Single Nucleotide Polymorphism.

2.8. RNA Extraction, cDNA Synthesis and q-PCR

Total RNA was extracted from cell lines and control primary lymphocyte homogenates using the Direct-zol™ RNA MiniPrep extraction kit (Zymo Research, Irvine, CA, USA), according to the manufacturer's instructions. Total RNA was converted to complementary DNA (cDNA) while using the iScript™ cDNA Synthesis Kit (Bio-Rad Laboratories), according to the manufacturer's instructions. Each q-PCR was performed in triplicate, with 120 ng cDNA being amplified with SYBR® Green PCR master mix (Promega, Madison, Wisconsin, USA) and *HKII* forward (5′-CCCGGGAAAGCAACTGTTTG-3′) and reverse (5′-ACCGGTGTTGAGAAGCTCTG-3′) primers in a 10ul final reaction volume on the QuantStudio 7 instrument under the following reaction conditions: 50 °C for 2 min (×1 cycle), 95 °C for 3 min (×1 cycle), 95 °C for 3 s, and 60 °C for 30 s (×50 cycles). PCR efficiency for the *HKII* primer pair was 103%, with a single peak melt curve. *18S* was used as an endogenous control and gene expression was calculated using the relative quantification method ($^{\Delta\Delta}$Ct). A student's *t*-test was used to calculate the significant differences between *HKII* expression profiles in cell lines and healthy controls.

2.9. Protein Detection by Western Blot

Total protein was isolated from cell lysates using the RUNX protein-lysis buffer (20 mM HEPES, 25% glycerol, 1.5 mM $MgCl_2$, 420 mM NaCl, 0.5 mM DTT, 0.2 mM EDTA, 0.5% Igepal CA-630, 0.2 mM Na_3VO_4, 1 mM PMSF, and dH_20 containing a protease and phosphatase inhibitor cocktail). Protein concentration was measured using the Qubit™ Protein Assay Kit (Invitrogen). For SDS-PAGE (Sodium Dodecyl Sulphate-Polyacrylamide Gel Electrophoresis), 30 µg total protein was separated for 1.5 h at a constant 120 V in a mini-PROTEAN Tetra unit. The HKII protein (109 kDa) was detected using a mouse anti-HKII primary antibody (Abcam, ab104836) and an HRP-conjugated anti-mouse IgG secondary antibody (Cell Signaling, #7076). Sample loading was normalised using secondary HRP-conjugated anti-Beta-actin (Cell Signaling, #5125S). Enhanced chemiluminescence detection (Clarity™, ECL, Bio-Rad) of HKII and Beta-actin was performed using the Fusion Spectra chemiluminescent system (Vilber Lourmat, Fisher Biotec) and optical density quantitation was assessed using Image J software [32].

2.10. Statistical Analysis

Chi-square (χ^2) analysis was used to determine the differences in genotype and allele frequencies between case and control samples, where a p-value < 0.05 was considered to be statistically significant. Genetic association and haplotype analysis of genotyping data was performed using the latest Plink v1.09 [33], RStudio [34], and HaploView [35] software. Quality control (QC) filters were applied to the full dataset consisting of 39 variants, 300 cases, and 140 controls. Following basic association testing, where correction for multiple testing was applied in the filtered dataset, a mixed model logistic regression test adjusting for the covariate 'sex' was performed in association with NHL with risk determined by OR and CI that were set at 95%.

2.11. NHL-GWAS Replication Dataset

Cerhan et al. [36] conducted a meta-analysis of three new NHL genome-wide association studies (GWAS): NCI, GELA, and MAYO_DLBCL, and one previous scan (SF), including around 3857 cases and 7665 healthy controls of European ancestry (InterLymph Consortium), to identify the genetic susceptibility loci for DLBCL, with further genotyping of nine SNPs in 1359 cases and 4557 healthy controls. For the first three groups, NCI, GELA, and MAYO_DLBCL, data was genotyped and for the scan, SF, data was imputed. We requested genotype summary statistics from the authors, for SNPs to be included in the GWAS that were in high linkage disequilibrium (LD) (D′ = 1.00), with significantly associated SNPs in our SNP MassARRAY® panel, i.e., rs12659504 and rs878008, in order to investigate the replication of our findings.

3. Results

3.1. Genetic Association of MIR143 SNPs and NHL Risk

In this study, 39 cancer-related microRNA-related SNPs (miRSNPs) were selected for genotyping by chip-based multiplex PCR and MALDI-TOF MS in the GLP-NHL population of 300 cases and 140 healthy controls. Allele frequencies for each variant in the GLP-NHL case-control cohort were established and the minor allele frequencies (MAFs) were compared to those that were observed in European and British populations in the Hapmap 1000G database (Table 3). After the QC filters were applied, basic allelic association testing revealed two SNPs in *MIR143/145* (rs3733846 [OR = 0.56, $p = 0.012$], rs41291957 [OR = 0.56, $p = 0.008$]), and one SNP in the promoter region of *MIR143* (rs17723799 [OR = 0.42, $p = 0.0004$]) on chromosome 5 to be significantly associated with a reduced risk of NHL (Table 3). After correction for multiple testing, the rs17723799 SNP remained statistically significant using the Bonferroni ($p = 0.015$) adjustment method (Table 3). A logistic regression test adjusted for the covariate 'sex' revealed the same three SNPs in *MIR143* to be significantly associated with the reduced risk of NHL in the Additive model: rs3733846 (OR [95% CI] = 0.54 [0.33–0.86],

$p = 0.010$), rs41291957 (OR [95% CI] = 0.61 [0.39–0.94], $p = 0.024$), and rs17723799 (OR [95% CI] = 0.43 [0.26–0.71], $p = 0.0009$) (Table 4). Genotype and allele frequencies were determined for the three *MIR143* SNPs and were compared with 1000G and gnomAD MAF percentages (Table 5). The rs17723799 SNP was identified to have a significant difference in allele ($p = 0.013$) and genotype ($p = 0.039$) frequencies between the case and control sample cohorts (Table 5). A mixed model logistic regression test adjusted for the covariate 'sex' revealed the rs17723799 SNP to be significantly associated with a reduced risk of NHL in the Additive (OR [95% CI] = 0.43 [0.26–0.71], $p = 0.0009$), Dominant (OR [95% CI] = 0.54 [0.33–0.88], $p = 0.015$), Over-dominant (OR [95% CI] = 0.56 [0.33–0.92], $p = 0.024$), and Log-additive (OR [95% CI] = 0.58 [0.38–0.90], $p = 0.017$) models of inheritance (Table 6). The two other *MIR143* SNPs in LD with rs17723799, i.e., rs3733846, and rs41291957 (Table 7) were also shown to be associated with a significantly reduced risk of NHL after logistic regression testing in the Additive model only: rs3733846 (OR [95% CI] = 0.54 [0.33–0.86], $p = 0.010$) and rs41291957 (OR [95% CI] = 0.61 [0.39–0.94], $p = 0.024$). Analysis of the three SNPs was also performed in association with the two most common NHL subtypes that are present in the cohort, i.e., DLBCL and FL, however the results were non-significant for all three SNPs ($p > 0.05$). This could be due to the small number of cases representing these subtypes and the reduced statistical power. All three SNPs were validated by Sanger sequencing in 5–10 randomly selected cases and controls, with the genotype of all samples being confirmed.

Table 3. Basic association testing for 39 genotyped SNPs showing minor allele frequencies (MAFs), Hardy-Weinberg Equilibrium (HWE) score for controls, unadjusted odds ratios, and *p*-values and adjusted *p*-values after correction for multiple testing. A1, minor allele; A2, major allele; MAF, minor allele frequency; NCHROBS, number of chromosomes observed; 1000G, 1000 Genomes Database; HWE, Hardy-Weinberg Equilibrium; OR, odds ratio.

Chr	miRNA/Target Gene	SNP	A1	A2	MAF/ NCHROBS	MAF 1000G	HWE UNAFF	OR	*p*-Value	Adjusted *p*-Value
1	*E2F2*	rs2075993	G	A	0.5/862	G = 0.3488/1747	0.2368	0.885	0.4535	1
1	*GEMIN3* 3′-UTR	rs197412	C	T	0.4092/870	C = 0.4744/2376	0.6013	0.8779	0.4301	1
2	hsa-miR-155-3p	rs4672612	A	G	0.338/858	A = 0.3878/1942	0.5476	1.095	0.6006	1
4	*TET2*	rs7670522	A*	C	0.4701/870	C = 0.3600/1803	1.000	1.089	0.6041	1
4	hsa-miR-4330/ 5100	rs2647257	T	A	0.408/848	T = 0.2386/1195	0.3732	0.9403	0.7094	1
5	hsa-miR-224-5p	rs12719481	G	A	0.2719/868	G = 0.3670/1838	1.000	1.072	0.7052	1
5	hsa-miR-143	rs3733846	G	A	0.1367/878	G = 0.2063/1033	1.000	0.5646	0.012	0.467
5	hsa-miR-143	rs17723799	T	C	0.1129/868	T = 0.1118/560	1.000	0.423	0.0004	0.015
5	hsa-miR-143	rs41291957	A	G	0.1412/878	A = 0.1214/608	1.000	0.5624	0.008	0.326
5	hsa-miR-145	rs353291	C	T	0.4237/826	C = 0.3608/1807	0.1473	1.204	0.2648	1
5	hsa-miR-146-a	rs2910164	C	G	0.2204/862	C = 0.2797/2881	0.5978	1.19	0.3908	1
5	hsa-miR-218-2	rs11134527	A	G	0.2189/868	A = 0.3462/1734	0.7982	1.061	0.7635	1
6	*XPO5*	rs11077	G	T	0.4255/872	G = 0.4036/2021	0.3862	0.9566	0.7869	1
6	*TAB2*	rs9485372	A	G	0.1965/850	A = 0.2408/1206	1.000	0.844	0.4084	1
6	*ESR1, C6orf97*	rs2046210	A	G	0.3353/850	A = 0.4121/2064	1.000	1.277	0.1659	1
8	*TP53*	rs896849	G	A	0.1501/866	G = 0.2183/1093	0.4915	0.9369	0.7695	1
8	*CASC21*	rs13281615	G	A	0.4417/840	G = 0.4912/2460	0.3853	0.7541	0.08388	1
8	*AGO2*	rs3864659	C	A	0.1023/860	C = 0.1436/719	0.6026	1.147	0.6275	1
8	*AGO2*	rs4961280	A	C	0.1835/872	A = 0.1490/746	0.7390	1.416	0.1075	1
9	hsa-miR-101-2	rs462480	G	T	0.3984/876	G = 0.4451/2229	0.7246	1.001	0.9948	1
10	hsa-miR-608	rs4919510	G	C	0.1979/874	G = 0.3638/1822	1.000	1.115	0.6034	1
10	hsa-miR-202	rs12355840	C	T	0.1368/848	C = 0.3189/1597	0.6598	0.494	0.1049	1
11	hsa-miR-210	rs1062099	C	G	0.1701/876	C = 0.1649/826	0.3067	1.29	0.2536	1
11	*LSP1*	rs3817198	C	T	0.3289/836	C = 0.2155/1079	0.7152	0.7085	0.04181	1
11	*TMEM45, BARX2*	rs7107217	A	C	0.4883/858	A = 0.4876/2442	0.5987	0.8992	0.514	1
12	*KRAS 3′-UTR*	rs61764370	C	A	0.0962/852	C = 0.0347/174	1.000	0.6795	0.1473	1
12	hsa-miR-196-a2	rs11614913	T	C	0.4205/880	T = 0.333/1666	0.226	0.928	0.6501	1
12	pre-miR-618	rs2682818	A	C	0.1465/874	A = 0.2424/1214	0.3061	1.115	0.6448	1
14	*HIF1A 3′-UTR*	rs2057482	T	C	0.1250/856	T = 0.2424/1214	0.6797	1.106	0.6831	1
14	*DICER1*	rs3742330	G	A	0.0878/854	G = 0.1382/692	0.2499	1.261	0.4423	1
14	*DICER1*	rs1057035	C	T	0.3709/852	C = 0.1723/863	0.854	0.8726	0.4174	1
16	*TOX3*	rs8051542	T	C	0.4322/856	T = 0.3133/1569	1.000	0.7698	0.1095	1

Table 3. *Cont.*

Chr	miRNA/ Target Gene	SNP	A1	A2	MAF/ NCHROBS	MAF 1000G	HWE UNAFF	OR	*p*-Value	Adjusted *p*-Value
16	*TOX3*	rs3803662	A	G	0.2207/852	A = 0.4403/2205	1.000	0.9065	0.6119	1
18	hsa-miR-143-5p	rs4987859	T	C	0.0631/856	T = 0.0477/239	0.597	0.744	0.3466	1
18	hsa-miR-27-a-5p	rs4987852	C	T	0.0667/854	C = 0.0190/95	0.4495	1.081	0.811	1
18	hsa-miR-27-a-5p	rs1016860	T	C	0.1175/868	T = 0.1166/584	1.000	0.8423	0.4808	1
21	hsa-miR-155 HG	rs987195	G	C	0.0917/840	G = 0.1472/737	0.361	0.6702	0.1265	1
21	hsa-miR-155	rs12482371	C	T	0.1632/858	C = 0.4151/2079	0.5662	0.7506	0.1748	1
X	hsa-miR-221/222	rs34678647	T	G	0.0375/667	T = 0.1423/537	0.1782	0.433	0.05456	1

Table 4. Logistic regression analysis of *MIR143* SNPs in association with NHL showing *p*-values after adjusting for covariate 'sex'. A1, minor allele.

Chr	Gene	SNP	A1	Model	OR (CI 95%)	*p*-Value
5	*MIR143*	rs3733846	G	Additive	0.54 (0.34–0.87)	**0.010**
5	*MIR143*	rs41291957	A	Additive	0.61 (0.39–0.94)	**0.024**
5	*MIR143*	rs17723799	T	Additive	0.61 (0.26–0.71)	**0.0009**

Table 5. Allele and genotype frequencies of SNPs located in *MIR143* in the GLP-NHL case and control populations in comparison to allele frequencies obtained from HapMap 1000G and gnomAD databases. HWE, Hardy-Weinberg Equilibrium (HWE) test score.

	rs17723799							
	Allele			*Genotype*				
	C (%)	T (%)	*p*-Value	C/C (%)	C/T (%)	T/T (%)	*p*-Value	HWE
Controls	234 (84.8)	42 (15.2)	**0.013**	99 (71.7)	36 (26.1)	3 (2.2)	**0.039**	1.000
Cases	536 (90.5)	56 (9.5)		244 (82.4)	48 (16.2)	4 (1.4)		0.311
MAF	770 (88.7)	98 (11.3)						
1000G (%)	88.8	11.2						
gnomAD (%)	86.9	13.1						
	rs3733846							
	Allele			*Genotype*				
	A (%)	G (%)	*p*-Value	A/A (%)	A/G (%)	G/G (%)	*p*-Value	HWE
Controls	231 (83)	47 (17)	0.057	96 (69)	39 (28)	4 (3)	0.167	1.000
Cases	527 (88)	73 (12)		232 (77)	63 (21)	5 (2)		0.785
MAF (%)	758 (86.3)	120 (13.7)						
1000G (%)	79.4	20.6						
gnomAD (%)	84.4	15.6						
	rs41291957							
	Allele			*Genotype*				
	G (%)	A (%)	*p*-Value	G/G (%)	G/A (%)	A/A (%)	*p*-Value	HWE
Controls	229 (82.4)	49 (17.6)	**0.043**	94 (67.6)	41 (29.5)	4 (2.9)	0.070	1.000
Cases	525 (87.5)	75 (12.5)		233 (77.7)	59 (19.7)	8 (2.6)		0.106
MAF (%)	754 (85.9)	124 (14.1)						
1000G (%)	87.9	12.1						
gnomAD (%)	84	16						

Table 6. Mixed model logistic regression analysis adjusted for the covariate 'sex' for the rs17723799 SNP in *MIR143*.

Model	Genotype	Controls (%)	Cases (%)	χ^2	OR (95% CI)	*p*-Value
Allelic	T vs. C	39/459	36/178	12.84	-	**0.0003**
Additive	-	-	-		0.43 (0.26–0.71)	**0.0009**
Co-dominant	CC	97 (71.9)	243 (82.4)		1.00 (reference)	
	CT	35 (25.9)	48 (16.3)		0.55 (0.33–0.91)	
	TT	3 (2.2)	4 (1.4)		0.47 (0.10–2.23)	0.051
Dominant	CC	97 (71.9)	243 (82.4)		1.00 (reference)	
	CT–TT	38 (28.1)	52 (17.6)		0.54 (0.33–0.88)	**0.015**
Recessive	CC–CT	132 (97.8)	291 (98.6)		1.00 (reference)	
	TT	3 (2.2)	4 (1.4)		0.53 (0.11–2.52)	0.438
Over-dominant	CC–TT	100 (74.1)	247 (83.7)		1.00 (reference)	
	CT	35 (25.9)	48 (16.3)		0.56 (0.33–0.92)	**0.024**
Log-additive	-	135 (31.4)	295 (68.6)		0.58 (0.38–0.90)	**0.017**

Table 7. Linkage disequilibrium (D') values for SNPs on chromosome 5 in HapMap 1000G European (EUR) Central European (CEU)/Great Britain (GBR) populations.

SNP	Location (GRChg38)	rs3733846	rs12659504	rs878008	rs17723799	rs41291957
rs3733846	149,425,059	-	1.000/1.000	1.000/1.000	1.000/1.000	0.721/0.941
rs12659504	149,425,442	-	-	1.000/1.000	1.000/1.000	0.721/0.942
rs878008	149,425,488	-	-	-	1.000/1.000	0.721/0.942
rs17723799	149,427,514	-	-	-	-	0.999/0.999
rs41291957	149,428,827	-	-	-	-	-

3.2. Replication of Summary Statistics in a Large EUROPEAN NHL GWAS Meta-Analysis

As we did not have access to an in-house replication NHL population, we interrogated the complete list of SNPs (700,000+ loci) that are available on the HumanOmniExpress-12-v1-1-C chip (Illumina) studied in three NHL GWAS for the presence of SNPs found to be significantly associated with NHL in our study. Unfortunately, our three SNPs in *MIR143* were not included in the GWAS, however we were able to find two SNPs (rs12659504 and rs878008) on chromosome 5 in the GWAS in high LD (r^2 = 1.00, D' = 1.00) with rs3733846 and rs17723799 (Table 7). A high linkage disequilibrium (LD) score identified strong non-random association of nearby variants on the same chromosome. Additionally, a meta-analysis of three individual GWAS and one imputation scan performed by Cerhan et al. [37], which included 3855 cases and 7664 controls, showed a significantly reduced risk of NHL for both of these SNPs: rs12659504 (OR [95% CI] = 0.91 [0.84–0.99], p = 0.033) and rs878008 (OR [95% CI] = 0.92 [0.84–1.00], and p = 0.041 (Table 8) (InterLymph Consortium).

3.3. Haplotype Analysis of MIR143 SNPs in LD on Chromosome 5

We performed haplotype analysis for the three SNPs (rs3733846, rs41291957, and rs17723799) that were located in *MIR143* on chromosome 5 found to be significantly associated with a reduced risk of NHL in our case-control cohort. The most commonly inherited haplotype was haplotype 3 (A-G-C), with a frequency of 0.82 (82%) in case samples and 0.79 (79%) in control samples (Table 9). A significantly reduced risk of NHL was observed with the inheritance of haplotype 4 (G-A-T) with all three minor alleles, which occurred at a frequency of 0.09 (9%): OR [95% CI] = 0.42 [0.18–1.00], p = 0.049) (Table 10). A linkage disequilibrium plot showed moderate linkage between the three SNPs when being assessed using Haploview 4.2 (Figure 1).

Table 8. Summary statistics for three European NHL-GWAS with an imputation scan (InterLymph Consortium) and meta-analysis (Cerhan et al. 2014) for two SNPs in high LD with rs17723799 on chromosome 5. EAF, effect allele frequency; OR, odds ratio, CI, confidence interval.

SNP	Chr	Location (GRChg38)	Group	Controls	Cases	Effect Allele	EAF Controls	EAF Cases	OR	CI (95%)	*p*-Value
rs12659504	5	149,425,442	NCI	6221	2661	G	0.1502	0.1359	0.93	0.84–1.02	0.120
rs12659504	5	149,425,442	GELA	525	548	G	0.1418	0.125	0.90	0.70–1.15	0.392
rs12659504	5	149,425,442	MAYO_DLBCL	171	392	G	0.173	0.1569	0.79	0.54–1.14	0.211
rs12659504	5	149,425,442	SF	747	254	G	0.133	0.1205	0.89	0.66–1.21	0.456
rs12659504	5	149,425,442	Meta-analysis	7664	3855				**0.91**	**0.84–0.99**	**0.033**
rs878008	5	149,425,488	NCI	6221	2661	C	0.1501	0.1362	0.93	0.84–1.02	0.132
rs878008	5	149,425,488	GELA	524	548	C	0.1396	0.1242	0.91	0.71–1.16	0.445
rs878008	5	149,425,488	MAYO_DLBCL	172	392	C	0.172	0.1548	0.78	0.53–1.13	0.188
rs878008	5	149,425,488	SF	747	253	C	0.1307	0.1199	0.90	0.66–1.23	0.516
rs878008	5	149,425,488	Meta-analysis	7664	3854				**0.92**	**0.84–1.00**	**0.041**

Table 9. Haplotype frequencies for analysed *MIR143* SNPs in GLP-NHL case and control populations.

Haplotype	rs3733846	rs41291957	rs17723799	Frequencies Cases	Frequencies Controls
1	A	A	C	0.03608	0.03255
2	A	A	T	0.00000	0.00000
3	A	G	C	0.81643	0.79232
4	A	G	T	0.02582	0.00517
5	G	A	C	0.02159	-
6	G	A	T	0.06733	0.14260
7	G	G	C	0.03092	0.02067
8	G	G	T	0.00182	0.00670

Table 10. Association of haplotypes for analysed *MIR143* SNPs in GLP-NHL case and control populations.

Haplotype	rs3733846	rs41291957	rs17723799	Haplotype Frequencies	OR (CI 95%)	*p*-Value
1	A	G	C	0.80883	0.90 (0.43–1.90)	0.7799
2	A	G	T	0.01898	6.34 (0.60–67.07)	0.1246
3	G	A	C	0.01461	Inf (Inf-Inf)	0.0000
4	**G**	**A**	**T**	**0.09137**	**0.42 (0.18–1.00)**	**0.0495**
5	G	G	C	0.02762	1.22 (0.38–3.88)	0.7370
rare	*	*	*	0.00359	0.12 (0.00–2.95)	0.1920

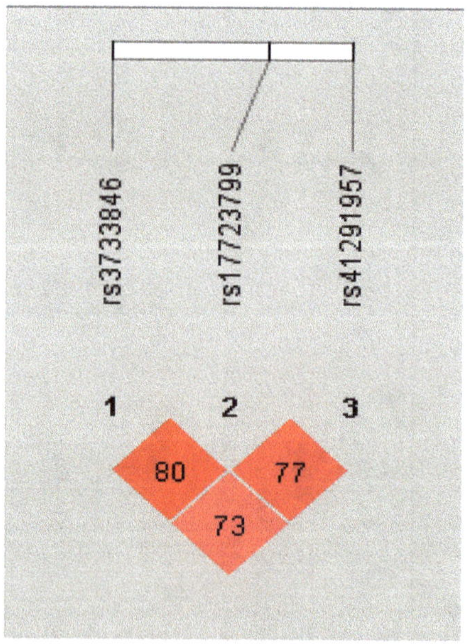

Figure 1. Linkage disequilibrium (LD) plot for analysed polymorphisms in the *MIR143* host gene using Haploview software version 4.2 (Daly Lab, Broad Institute, Cambridge, MA, USA).

3.4. Genotyping of rs17723799 (C>T) in Cell Lines and Healthy Control Samples

We genotyped four commercially sourced, immortalised NHL cell lines (SU-DHL-4, Raji, Mino and Toledo), a metastatic breast cancer cell line (MDA-MB-231), a melanoma cell line (MDA-MB-435), and two healthy cancer-free control subjects for the *MIR143* rs17723799 SNP. All of the NHL cell lines were homozygous for the wild type risk allele C (CC). The breast cancer cell line was homozygous wild type CC. The melanoma cell line was homozygous for the minor allele T (TT). Both the healthy control subjects were homozygous for the wild type C allele (CC). As the minor T allele is reported to be present at a frequency of 0.11 (11%) in the healthy European population (Table 5), these results were not surprising; however, it was interesting that all NHL cell lines were identified to be homozygous for the wild type allele C, which was observed as the risk allele for NHL in our genetic association analysis.

3.5. *miR-143 Target Gene HKII Expression in Cancer Cell Lines and Healthy Control Lymphocytes*

miR-143 inhibits the expression of its target gene *HKII* by binding to a conserved recognition motif that is located in the 3′-UTR of mRNA. We examined *HKII* expression at the mRNA and protein level in four NHL cell lines (SU-DHL-4, Raji, Mino, Toledo), a metastatic breast cancer cell line (MDA-MB-231), and a melanoma cell line (MDA-MB-435). In addition, two healthy control donors were included as a reference for basal *HKII* expression levels for statistical analysis.

3.5.1. HKII Gene Expression Is Increased in NHL Compared to Healthy Controls

q-PCR analysis of *HKII* mRNA transcript expression showed increased levels of *HKII* mRNA transcript in the four patient-derived NHL cell lines (SU-DHL-4, Raji, Mino, and Toledo) when compared to lymphocytes that are derived from two healthy control subjects, however the differences in expression levels were not statistically significant (Figure 2). Western blot and band quantitation analysis confirmed these findings, showing increased HKII protein levels in the NHL cell lines examined when compared to the controls (Figure 3).

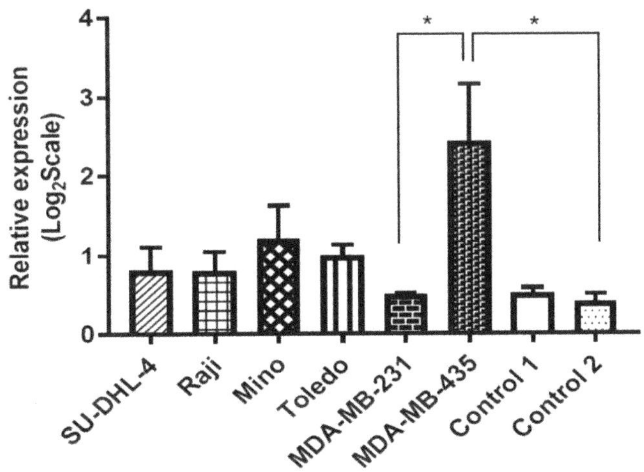

Figure 2. Hexokinase 2 (*HKII*) gene expression. q-PCR analysis of *HKII* mRNA transcript expression in NHL (SU-DHL-4, Raji, Mino, Toledo), breast cancer (MDA-MB-231) and melanoma (MDA-MB-435) cell lines as compared to healthy controls 1 and 2. *HKII* mRNA levels were increased in NHL cell lines compared to breast cancer cells and healthy controls, however expression levels were not significantly different. *HKII* mRNA levels were significantly increased in the melanoma (MDA-MB-435) cells when compared to breast cancer cells (MDA-MB-231) ($p = 0.044$), healthy control 1 ($p = 0.045$) and 2 ($p = 0.037$), but not significantly increased compared to the NHL cells. Relative expression normalized to *18S*, error bars = SEM and statistical significance calculated using paired Student's t-test and defined as: * $p < 0.05$.

3.5.2. Increased HKII Gene Expression May Be Associated with the MIR143 rs17723799 TT Genotype

q-PCR analysis demonstrated *HKII* mRNA transcript levels that were were higher in the melanoma (MDA-MB-435) cells, correlating with the observed homozygous mutant rs17723799 'TT' genotype in these cells. *HKII* mRNA transcript levels were observed to be lower in all four NHL cell lines, and even lower in the breast cancer (MDA-MB-231) and healthy control cells, which correlated with the homozygous wild type rs17723799 CC genotype of these cells (Figure 2). A Student's t-test showed a significant difference in the relative *HKII* mRNA expression between melanoma cells and the breast cancer ($p = 0.044$), control subject 1 ($p = 0.045$), and control subject 2 ($p = 0.037$) cells. Western blot and band quantitation analysis demonstrated upregulated HKII protein expression in the melanoma cells, with lower expression being observed in the breast cancer cells and even lower in the four NHL cell lines examined (Figure 3). Surprisingly, the breast cancer cells expressed low *HKII* mRNA transcript, but high protein levels.

a)

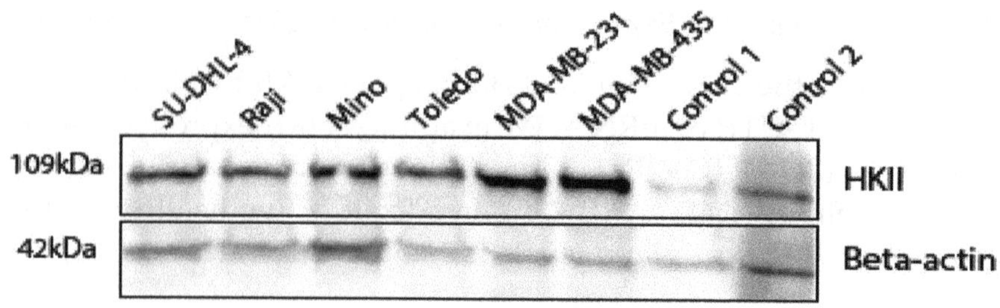

b)

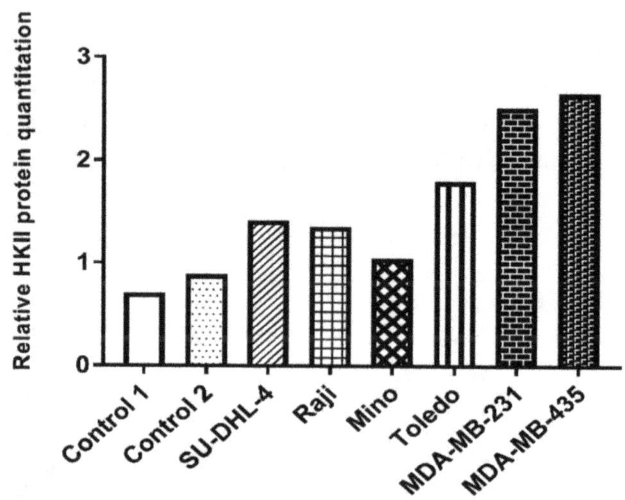

Figure 3. HKII protein detection. Western blot analysis of HKII protein (109kDa) expression in NHL (SU-DHL-4, Raji, Mino, Toledo), breast cancer (MDA-MB-231) and melanoma (MDA-MB-435) cell lines compared to healthy control subjects 1 and 2. 30 µg total protein was loaded and samples were normalised against the Beta-actin (42kDa) loading control. (**a**) The HKII protein levels were increased in all four NHL cell lines compared to healthy control subjects 1 and 2, and decreased when compared to the breast cancer (MDA-MB-231) and melanoma (MDA-MB-435) cells. (**b**) HKII protein quantitation analysis compared the normalised intensity ratios between samples and showed the breast cancer (MDA-MB-231) and melanoma (MDA-MB-435) cells to have increased HKII levels as compared to NHL cells and healthy control subjects 1 and 2.

4. Discussion

Our study investigated 39 miRSNPs previously implicated in human tumourigenesis that were identified from the literature, including those not previously reported to have an association with NHL susceptibility, but rather with disease prognosis, in Caucasians. Following genotyping by multiplex PCR and MALDI-TOF MS, subsequent genotypic association and haplotype analysis identified three SNPs on chromosome 5 (rs3433846, rs17723799, and rs41291957) in *MIR143* in moderately high LD, which together appear to confer protection against NHL in our GRC-GLP cohort. In particular, the rs17723799 SNP in the promoter region of *MIR143* remained significant at the $p < 0.05$ threshold after Bonferroni correction for multiple testing. We also provided evidence for replication in a large European NHL GWAS meta-analysis study.

One study identified the rs11614913 SNP (T > C) in hsa-miR-169-a2 (miR-196-a2) to be associated with NHL in a Chinese case-control cohort, where the presence of the risk C allele and CC genotype was shown to be more frequent in the case samples, conferring an increased risk (OR [95% CI] = 1.82 [1.16–2.85], $p = 0.009$) [36]. The authors confirmed that the variant genotype affected the expression of miR-196-a2, where the carriers of the CC genotype had significantly higher levels than those with only one C allele or the TT genotype. Interestingly, this SNP was also shown to increase risk of hepatocellular carcinoma in a Chinese cohort of 109 cases and 105 controls [38]; however, no functional analyses were performed. The rs11614913 SNP was shown to be associated with decreased central nervous system (CNS) Acquired Immunodeficiency Syndrome (AIDS)-NHL in an American case-control cohort of 180 cases and 529 controls, where the CT genotype was more frequently observed in control samples and conferred protection (OR [95% CI] = 0.52 [0.27–0.99]. This study also showed an increased risk of systemic AIDS-NHL with the presence of the T allele in rs2057482, which creates a binding site for miR-196-a2 in the *HIF1A* 3′-UTR (OR [95% CI] = 1.73 [1.12–2.67]) [39]. We were not able to show a positive significant association for miR-196-a2 rs11614913 in our Australian/European cohort (OR [95% CI] = 0.93 [0.67–1.29], $p = 0.668$) (Table 3). This is most likely due to ethnicity differences in our cohort and the Chinese cohort of Li et al. [36]. Previous studies on this SNP have shown a significant association with cancer in predominantly Asian populations, but not in Caucasians, which may explain the lack of association in our cohort [40].

Other studies have shown the association of *MIR143/145* SNPs with reduced cancer risk in Chinese cohorts. Li et al. [41] conducted a case-control analysis of 12 SNPs in the promoter region of *MIR143/145* in 242 cases with colorectal cancer (CRC) and 283 healthy controls. In support of our findings in NHL, the rs3733846 (A > G) mutant genotype A/G was associated with a significantly reduced risk of CRC (OR [95% CI] = 0.57 [0.44–0.73], $p < 0.001$). A more recent study by Wu et al. [42] showed a functional association between the rs353293 SNP (G>A) AG/AA genotypes and a reduced risk of bladder cancer in 333 Chinese cases when compared to 536 healthy controls (OR [95% CI] = 0.64 [0.46–0.90], $p = 0.008$). In their study, in vitro luciferase reporter analysis was used to show a significantly reduced effect of the protective rs353293A allele as compared with the rs353293G allele on transcriptional activity ($p < 0.001$). The promoter transcriptional activity was identified to be reduced due to the polymorphism, which could also be the case for the rs17723799 SNP that we identified, however this would imply lower miR-143/145 levels and lower tumour suppressor activity, which appears to be in contradiction with their genotypic analysis data as well as ours, showing the protective effect of the polymorphism. The authors were not able to show whether serum miR-143/145 or the target gene expression levels were altered.

Epigenetic mechanisms, such as miRNAs, have been shown to regulate aerobic glycolysis in cancer cells through the targeting and down-regulation of glycolytic enzymes [14]. miR-143, an essential regulator of glycolysis [43], inhibits *HKII* expression by binding to a conserved recognition motif that is located in the 3′-UTR of *HKII* mRNA and has been observed to be inversely correlated with HKII levels in primary keratinocytes and in head and neck squamous cell carcinoma (HNSCC)-derived cell lines [44]. miR-143 and miR-155 have both been shown to regulate glycolysis by targeting *HKII* in breast cancer, with miR-155 being able to suppress miR-143 production through targeting the *C/EBPβ* transcription activator for miR-143 [45]. In addition, the transcriptional regulation of miR-145 was recently reviewed by Zeinali et al. [46]. The authors discuss miR-145 regulation by DNA-binding factors, including c-MYC, p53, forkhead transcription factors of the O class 1 and 3 (Fox01 and Fox03), and miR-143/145 level regulation by RREB1 in other cancers but not in lymphoma.

The significant *MIR143* SNP that was identified in this study (rs17723799) has been previously reported via in-silico analysis to affect aerobic glycolysis. The authors showed that this SNP overlaps with the transcription factor binding site in the *MIR143* promoter region (148783674-148788779, UCSC Genome Browser) and concluded that this SNP could potentially affect the regulation of miRNA biogenesis and alter the hsa-miR-143-3p and hsa-miR-143-5p levels, thereby affecting its target genes that directly or indirectly control glycolysis [14].

Following our genotypic analysis, *HKII* expression at the mRNA and protein level in NHL (SU-DHL-4, Raji, Mino, Toledo), breast cancer (MDA-MB-231), and melanoma (MDA-MB-435) cell lines was assessed to determine the potential functional link between *MIR143* regulation by a promoter polymorphism and *HKII* target gene expression in these cells, along with lymphocytes from two healthy control subjects. Our results showed higher *HKII* mRNA and HKII protein levels in the NHL cell lines when compared to healthy controls, however the levels were lower in the NHL cell lines compared to the breast cancer (protein only) and melanoma cell lines. Melanoma cells carrying the

MIR143 rs17723799 homozygous recessive TT genotype expressed a significantly increased level of *HKII* mRNA transcript and HKII protein in comparison to other cancer cells that were carrying the rs17723799 homozygous wild type CC genotype, indicating that this homozygous polymorphism may have an effect on the transcriptional regulation of *MIR143*. Interrogation of the GeneCards Database identifed 38 promoters and enhancers for the *MIR143* gene with numerous transcription factor binding sites for each, implicating multiple mechanisms of regulation of this gene by multiple transcription factors in different tissue types. As we do not have access to an appropriate wild type melanoma cell line control, further studies on *MIR143* SNPs in melanoma may be beneficial. Our results also show an up-regulation of the HKII protein in the metastatic breast cancer (MDA-MB-231) cells, as reported by Palmieri et al. [24]. Interestingly, Western blot showed HKII levels were higher in the Toledo (DLBCL) cell line when compared to the other less aggressive NHL cell line subtypes.

Although not in NHL, but as a marker of non-small cell lung cancer (NSCLC), miR-143 expression in peripheral blood mononuclear cells (lymphocytes and monocytes) was significantly lower in NSCLC patients than in healthy individuals ($p < 0.0001$) [47]. Similarly, a study in renal cell carcinoma (RCC) identified the miR-143/145 cluster to be downregulated in RCC tissues when compared to adjacent non-cancerous tissues, with significantly higher HKII levels in RCC tissues as compared to non-cancerous tissues, confirming the tumour suppressive effect of the miR-143/145 cluster through targeting HKII [48]. One study in aggressive Burkitt's lymphoma (BL) showed that upregulated miR-143 by the regulation of PI3K/Akt prevented cell growth, implicating miR-143 as a tumour suppressor in NHL [49]. miR-143 has also been demonstrated to have an anti-tumour effect in leukaemia patients, where significantly lower miR-143 levels were observed in patients when compared to healthy controls, and the overexpression of miR-143 decreased DNA methyltransferase 3A (*DNMT3A)* mRNA and protein expression, thereby reducing cell proliferation, colony formation, and cell cycle progression, as well as increased apoptosis [50]. As there is a paucity of literature on *MIR143* and miR-143 expression and polymorphisms in NHL and its subtypes, we searched for significant eQTLs (expression quantitative trait loci) for the three analysed *MIR143* SNPs in the GTEx Portal [51]. GTEx analysis indicated increased *MIR143* host gene expression in skeletal muscle for heterozygous and homozygous mutant alleles in all three significant *MIR143* SNPs analysed (Figure S1). Although the relationship between miR-143/145 and its *MIR143* host gene is an open issue, the potentially increased *MIR143* host gene expression, as seen due to the polymorphisms in skeletal muscle, may enhance hsa-miR-143-3p and -5p processing, which poses a potential protective function for these SNPs in lymphocytes. The presence of the rs17723799 CT or TT genotype may increase *MIR143* expression and miR-143 transcription with the subsequent downregulation of *HKII* causing reduced glycolysis and lymphomagenesis. Although we did not measure mature miR-143 levels in the NHL cell lines or patient samples to correlate this with HKII expression, our genotypic analysis data indicate a possible protective functional effect of *MIR143* SNPs on miR-143 regulation, with a possible inhibitory effect on *HKII* expression in controls as compared to NHL cases. Further studies investigating *MIR143* and hsa-miR-143-3p/5p expression levels in different NHL subtypes and their association with HKII levels are therefore necessary. We have summarised the role of HKII and miR-143 in cancer in Figure 4a and a potential functional effect of the rs17723799 polymorphism in control cells when compared to NHL cells in Figure 4b.

A limitation of this study is the lack of expression data for mature hsa-miR-143-3p and hsa-miR-143-5p in the NHL cell lines or patient samples. We only had access to four NHL cell lines and we did not have access to RNA or protein from our retrospective patient cohort. Additionally, we were not able to compare HKII expression in NHL cell lines polymorphic for the rs17723799 SNP, as they were all wild type CC, and therefore further studies investigating miR-143 and HKII expression in NHL cell lines and/or patient-derived tumour samples with the CT or TT genotypes would be beneficial.

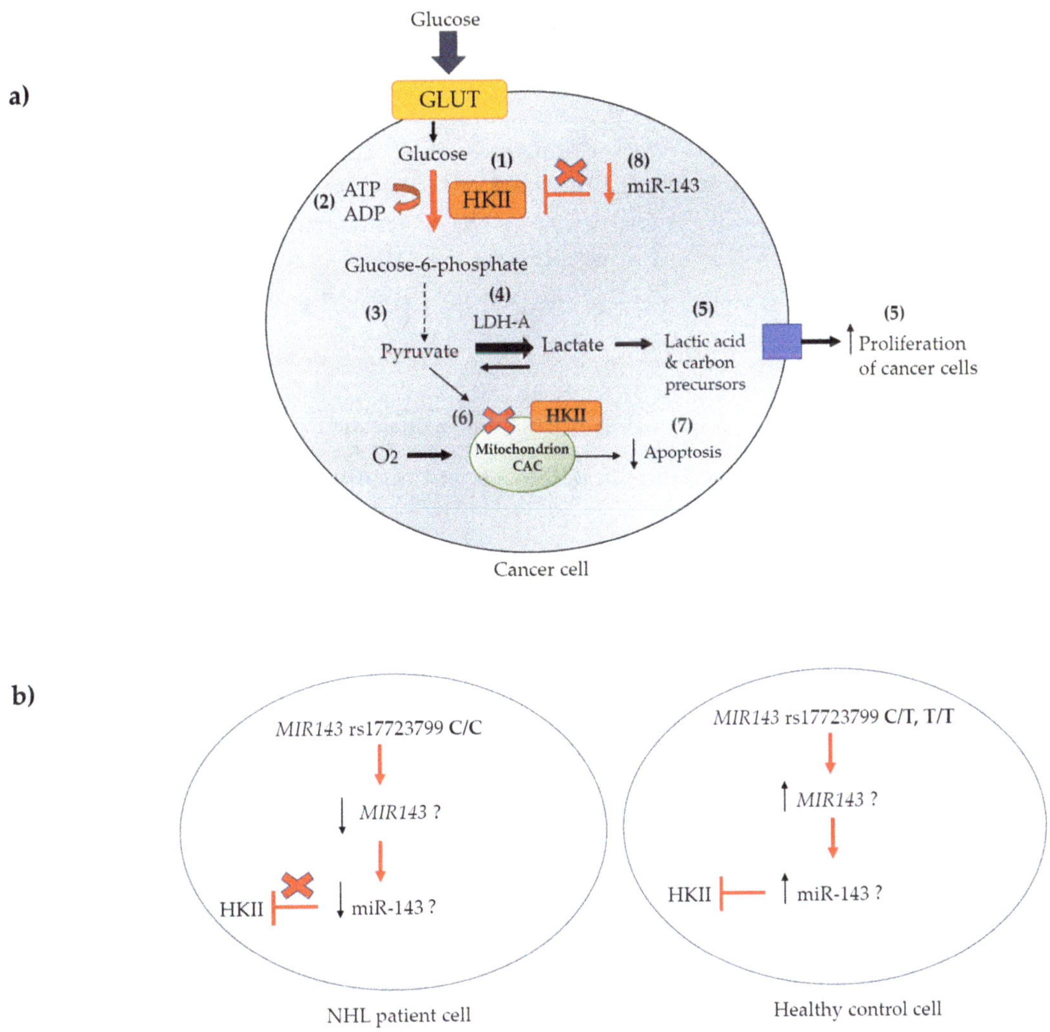

Figure 4. The role of HKII and miR-143 in cancer. (**a**) In a cancer cell, aerobic glycolysis occurs with an increased uptake of glucose and production of lactate even in the presence of oxygen (in normal cells glycolysis is anaerobic) and is known as the "Warburg effect". Upregulation of Hexokinase 2 (*HKII*) in malignant tumours catalyses the conversion of glucose to glucose-6-phosphate (G-6-P) in the first step of the glycolysis pathway (**1**) with high consumption of ATP to ADP (**2**) to maintain the high rate of glycolysis required. G-6-P undergoes multiple conversions to pyruvate (**3**), which together with high lactate dehydrogenase A (LDH-A) activity in one direction (**4**) results in lactic acid production in the cytosol, incorporation of carbon precursors and increased cancer cell proliferation (**5**). Oxidation of pyruvate in the mitochondrion is reduced resulting in reduced citrate production for the citric acid cycle (CAC) (**6**). HKII bound to the mitochondrial membrane can also enhance cancer by increasing mitochondrial resistance to cell death signals helping to reduce apoptosis (**7**). Downregulation of miR-143 has been shown to directly reduce inhibition of *HKII* (**8**) causing increased glycolysis and tumour proliferation. (**b**) In a lymphoid cell, we propose that in the presence of the protective *MIR143* SNP rs17723799 genotype CT or TT, there may be increased *MIR143* expression and miR-143 transcription with increased targeting of *HKII* and reduced glycolysis via the Warburg effect causing decreased lymphomagenesis. This hypothesis is speculative, and further functional studies are required. ATP, adenosine triphosphate; ADP, adenosine diphosphate; CAC, citric acid cycle; GLUT, glucose transporter; LDH-A, lactate dehydrogenase A. (Adapted from Wikimedia [52] and Akins et al. (2018) [23]) [43].

5. Conclusions

This study is the first to report significant statistical correlation between SNPs in *MIR143* and reduced risk of NHL in Caucasians, and it is supported by the identification of significant SNPs in high LD in a large European NHL GWAS meta-analysis study. Our findings suggest that the three SNPs in LD in *MIR143* may be novel useful biomarkers to assess the risk of NHL in a clinical setting. Furthermore, this study gives some insight into *MIR143* transcriptional regulation by a protective promoter polymorphism in NHL. We do not suggest that there is a definitive functional association between *MIR143* SNPs and NHL onset, however further functional analyses are necessary to confirm this potential mechanism.

Author Contributions: The following authors contributed to this study: Conceptualisation, G.B., L.M.H. and L.R.G.; methodology, G.B., H.G.S. and L.M.H.; analysis, G.B. and E.M.A.; validation, R.A.L.; writing–original draft preparation, G.B.; writing–review and editing, G.B, L.M.H., H.G.S. and L.R.G.; supervision, H.G.S., L.M.H. and L.R.G.; funding acquisition, L.R.G.

Acknowledgments: We would like to acknowledge Miles C. Benton for his technical support in the PLINK and Haploview analysis. We would also like to acknowledge the healthy control subjects for agreeing to donate their blood samples for the purposes of our research. We would like to thank the InterLymph Consortium for providing access to their GWAS summary statistics data for replication.

References

1. Skrabek, P.; Turner, D.; Seftel, M. Epidemiology of Non-Hodgkin Lymphoma. *Transfus. Apheresis Sci.* **2013**, *49*, 133–138. [CrossRef] [PubMed]
2. Cancer Facts & Figures 2018. Available online: https://www.cancer.org/ (accessed on 15 November 2018).
3. Lymphoma Coalition Europe. Available online: https://www.lymphomacoalition.org/europe (accessed on 15 November 2018).
4. Australian Institute of Health and Welfare 2017. Cancer in Australia 2017 Cancer Series No. 101. Cat. No. CAN 100. Canberra: AIHW. Available online: www.aihw.gov.au (accessed on 23 September 2018).
5. Yadav, C.; Ahmad, A.; D'Souza, B.; Agarwal, A.; Nandini, M.; Ashok Prabhu, K.; D'Souza, V. Serum Lactate Dehydrogenase in Non-Hodgkin's Lymphoma: A Prognostic Indicator. *Indian J. Clin. Biochem.* **2016**, *31*, 240–242. [CrossRef] [PubMed]
6. Roehle, A.; Hoefig, K.P.; Repsilber, D.; Thorns, C.; Ziepert, M.; Wesche, K.O.; Thiere, M.; Loeffler, M.; Klapper, W.; Pfreundschuh, M.; et al. MicroRNA signatures characterize diffuse large B-cell lymphomas and follicular lymphomas. *Br. J. Haematol.* **2008**, *142*, 732–744. [CrossRef] [PubMed]
7. Lawrie, C.H.; Chi, J.; Taylor, S.; Tramonti, D.; Ballabio, E.; Palazzo, S.; Saunders, N.J.; Pezzella, F.; Boultwood, J.; Wainscoat, J.S.; et al. Expression of microRNAs in diffuse large B cell lymphoma is associated with immunophenotype, survival and transformation from follicular lymphoma. *J. Cell. Mol. Med.* **2003**, *13*, 1248–1260. [CrossRef] [PubMed]
8. Bartel, D.P. MicroRNAs: Genomics, biogenesis, mechanism, and function. *Cell* **2004**, *116*, 281–297. [CrossRef]
9. Shukla, G.C.; Singh, J.; Barik, S. MicroRNAs: Processing, Maturation, Target Recognition and Regulatory Functions. *Mol. Cell. Pharmacol.* **2011**, *3*, 83–92. [PubMed]
10. Salzman, D.W.; Weidhaas, J.B. SNPing cancer in the bud: MicroRNA and microRNA-target site polymorphisms as diagnostic and prognostic biomarkers in cancer. *Pharmacol. Ther.* **2013**, *137*, 55–63. [CrossRef] [PubMed]
11. Cullen, B.R. Transcription and Processing of Human microRNA Precursors. *Mol. Cell* **2004**, *16*, 861–865. [CrossRef] [PubMed]
12. Treiber, T.; Treiber, N.; Meister, G. Regulation of microRNA biogenesis and function. *Thromb. Haemost.* **2012**, *107*, 605. [CrossRef] [PubMed]
13. Vannini, I.; Fanini, F.; Fabbri, M. Emerging roles of microRNAs in cancer. *Curr. Opin. Genet. Dev.* **2018**, *48*, 128–133. [CrossRef] [PubMed]

14. Suresh, P.S.; Venkatesh, T.; Tsutsumi, R. In silico analysis of polymorphisms in microRNAs that target genes affecting aerobic glycolysis. *Ann. Transl. Med.* **2016**, *4*, 69. [CrossRef] [PubMed]
15. Yue, M.; Zhou, D.; Zhi, H.; Wang, P.; Zhang, Y.; Gao, Y.; Guo, M.; Li, X.; Wang, Y.; Zhang, Y.; et al. MSDD: A manually curated database of experimentally supported associations among miRNAs, SNPs and human diseases. *Nucleic Acids Res.* **2018**, *46*, D181–D185. [CrossRef] [PubMed]
16. Gong, J.; Tong, Y.; Zhang, H.-M.; Wang, K.; Hu, T.; Shan, G.; Sun, J.; Guo, A.-Y. Genome-wide identification of SNPs in microRNA genes and the SNP effects on microRNA target binding and biogenesis. *Hum. Mutat.* **2012**, *33*, 254–263. [CrossRef] [PubMed]
17. Zheng, B.; Xi, Z.; Liu, R.; Yin, W.; Sui, Z.; Ren, B.; Miller, H.; Gong, Q.; Liu, C. The Function of MicroRNAs in B-Cell Development, Lymphoma, and Their Potential in Clinical Practice. *Front. Immunol.* **2018**, 9. [CrossRef] [PubMed]
18. Dzikiewicz-Krawczyk, A. MicroRNA polymorphisms as markers of risk, prognosis and treatment response in hematological malignancies. *Crit. Rev. Oncol. Hematol.* **2015**, *93*, 1–17. [CrossRef] [PubMed]
19. Bradshaw, G.; Sutherland, H.; Haupt, L.; Griffiths, L. Dysregulated MicroRNA Expression Profiles and Potential Cellular, Circulating and Polymorphic Biomarkers in Non-Hodgkin Lymphoma. *Genes* **2016**, *7*, 130. [CrossRef] [PubMed]
20. Vaitheesvaran, B.; Xu, J.; Yee, J.; Q.-Y., L.; Go, V.L.; Xiao, G.G.; Lee, W.N. The Warburg effect: A balance of flux analysis. *Metab. Off. J. Metab. Soc.* **2015**, *11*, 787–796. [CrossRef] [PubMed]
21. Wolf, A.; Agnihotri, S.; Micallef, J.; Mukherjee, J.; Sabha, N.; Cairns, R.; Hawkins, C.; Guha, A. Hexokinase 2 is a key mediator of aerobic glycolysis and promotes tumor growth in human glioblastoma multiforme. *J. Exp. Med.* **2011**, *208*, 313–326. [CrossRef] [PubMed]
22. Gu, J.J.; Singh, A.; Xue, K.; Mavis, C.; Barth, M.; Yanamadala, V.; Lenz, P.; Grau, M.; Lenz, G.; Czucman, M.S.; et al. Up-regulation of hexokinase II contributes to rituximab-chemotherapy resistance and is a clinically relevant target for therapeutic development. *Oncotarget* **2017**, 4020–4033. [CrossRef]
23. Akins, N.S.; Nielson, T.C.; Le, H.V. Inhibition of Glycolysis and Glutaminolysis: An Emerging Drug Discovery Approach to Combat Cancer. *Curr. Top. Med. Chem.* **2018**, *18*, 494–504. [CrossRef] [PubMed]
24. Palmieri, D.; Fitzgerald, D.; Shreeve, S.M.; Hua, E.; Bronder, J.L.; Weil, R.J.; Davis, S.; Stark, A.M.; Merino, M.J.; Kurek, R.; et al. Analyses of resected human brain metastases of breast cancer reveal the association between up-regulation of hexokinase 2 and poor prognosis. *Mol. Cancer Res.* **2009**, *7*, 1438–1445. [CrossRef] [PubMed]
25. Ros, S.; Schulze, A. Glycolysis Back in the Limelight: Systemic Targeting of HK2 Blocks Tumor Growth. *Cancer Discov.* **2013**, *3*, 1105. [CrossRef] [PubMed]
26. Mushtaq, M.; Darekar, S.; Klein, G.; Kashuba, E. Different Mechanisms of Regulation of the Warburg Effect in Lymphoblastoid and Burkitt Lymphoma Cells. *PLoS ONE* **2015**, *10*, e0136142. [CrossRef] [PubMed]
27. Bhalla, K.; Jaber, S.; Nahid, M.; Underwood, K.; Beheshti, A.; Landon, A.; Bhandary, B.; Bastian, P.; Evens, A.M.; Haley, J.; et al. Role of hypoxia in Diffuse Large B-cell Lymphoma: Metabolic repression and selective translation of HK2 facilitates development of DLBCL. *Sci. Rep.* **2018**, *8*, 744. [CrossRef] [PubMed]
28. Kwiatkowska, E.; Wojtala, M.; Gajewska, A.; Soszyński, M.; Bartosz, G.; Sadowska-Bartosz, I. Effect of 3-bromopyruvate acid on the redox equilibrium in non-invasive MCF-7 and invasive MDA-MB-231 breast cancer cells. *J. Bioenergy Biomembr.* **2016**, *48*, 23–32. [CrossRef] [PubMed]
29. Chacon-Cortes, D.; Smith, R.A.; Haupt, L.M.; Lea, R.A.; Youl, P.H.; Griffiths, L.R. Genetic association analysis of miRNA SNPs implicates MIR145 in breast cancer susceptibility. *BMC Med. Genet.* **2015**, *16*, 107. [CrossRef] [PubMed]
30. Upadhyaya, A.; Smith, R.A.; Chacon-Cortes, D.; Revêchon, G.; Bellis, C.; Lea, R.A.; Haupt, L.M.; Chambers, S.K.; Youl, P.H.; Griffiths, L.R. Association of the microRNA-Single Nucleotide Polymorphism rs2910164 in miR146a with sporadic breast cancer susceptibility: A case control study. *Gene* **2016**, *576*, 256–260. [CrossRef] [PubMed]
31. Chacon-Cortes, D.; Haupt, L.M.; Lea, R.A.; Griffiths, L.R. Comparison of genomic DNA extraction techniques from whole blood samples: A time, cost and quality evaluation study. *Mol. Biol. Rep.* **2012**, *39*, 5961–5966. [CrossRef] [PubMed]
32. Schneider, C.A.; Rasband, W.S.; Eliceiri, K.W. NIH Image to ImageJ: 25 years of image analysis. *Nat. Methods* **2012**, *9*, 671–675. [CrossRef] [PubMed]

33. Purcell, S.; Neale, B.; Todd-Brown, K.; Thomas, L.; Ferreira, M.A.R.; Bender, D.; Maller, J.; Sklar, P.; de Bakker, P.I.W.; Daly, M.J.; et al. PLINK: A Tool Set for Whole-Genome Association and Population-Based Linkage Analyses. *Am. J. Hum. Genet.* **2007**, *81*, 559–575. [CrossRef] [PubMed]
34. Team, R. *RStudio: Integrated Development for R*; RStudio, Inc.: Boston, MA, USA, 2015. Available online: http://www.rstudio.com.
35. Fry, B.; Maller, J.; Barrett, J.C.; Daly, M.J. Haploview: Analysis and visualization of LD and haplotype maps. *Bioinformatics* **2004**, *21*, 263–265. [CrossRef]
36. Li, T.; Niu, L.; Wu, L.; Gao, X.; Li, M.; Liu, W.; Yang, L.; Liu, D. A functional polymorphism in microRNA-196a2 is associated with increased susceptibility to non-Hodgkin lymphoma. *Tumor Biol.* **2015**, *36*, 3279–3284. [CrossRef] [PubMed]
37. Cerhan, J.R.; Berndt, S.I.; Vijai, J.; Ghesquières, H.; McKay, J.; Wang, S.S.; Wang, Z.; Yeager, M.; Conde, L.; de Bakker, P.I.W.; et al. Genome-wide association study identifies multiple susceptibility loci for diffuse large B-cell lymphoma. *Nat. Genet.* **2014**, *46*, 1233–1238. [CrossRef] [PubMed]
38. Li, J.; Cheng, G.; Wang, S. A Single-Nucleotide Polymorphism of miR-196a2T>C rs11614913 Is Associated with Hepatocellular Carcinoma in the Chinese Population. *Genet. Test Mol. Biomark.* **2016**, *20*, 213–215. [CrossRef] [PubMed]
39. Peckham-Gregory, E.C.; Thapa, D.R.; Martinson, J.; Duggal, P.; Penugonda, S.; Bream, J.H.; Chang, P.-Y.; Dandekar, S.; Chang, S.-C.; Detels, R.; et al. MicroRNA-related polymorphisms and non-Hodgkin lymphoma susceptibility in the Multicenter AIDS Cohort Study. *Cancer Epidemiol.* **2016**, *45*, 47–57. [CrossRef] [PubMed]
40. Liu, Y.; He, A.; Liu, B.; Zhong, Y.; Liao, X.; Yang, J.; Chen, J.; Wu, J.; Mei, H. rs11614913 polymorphism in miRNA-196a2 and cancer risk: An updated meta-analysis. *OncoTargets Ther.* **2018**, *11*, 1121–1139. [CrossRef] [PubMed]
41. Li, L.; Pan, X.; Li, Z.; Bai, P.; Jin, H.; Wang, T.; Song, C.; Zhang, L.; Gao, L. Association between polymorphisms in the promoter region of miR-143/145 and risk of colorectal cancer. *Hum. Immunol.* **2013**, *74*, 993–997. [CrossRef] [PubMed]
42. Wu, J.; Huang, Q.; Meng, D.; Huang, M.; Li, C.; Qin, T. A Functional rs353293 Polymorphism in the Promoter of miR-143/145 Is Associated with a Reduced Risk of Bladder Cancer. *PLoS ONE* **2016**, *11*, e0159115. [CrossRef] [PubMed]
43. Fang, R.; Xiao, T.; Fang, Z.; Sun, Y.; Li, F.; Gao, Y.; Feng, Y.; Li, L.; Wang, Y.; Liu, X.; et al. MicroRNA-143 (miR-143) Regulates Cancer Glycolysis via Targeting Hexokinase 2 Gene. *J. Biol. Chem.* **2012**, *287*, 23227–23235. [CrossRef] [PubMed]
44. Peschiaroli, A.; Giacobbe, A.; Formosa, A.; Markert, E.K.; Bongiorno-Borbone, L.; Levine, A.J.; Candi, E.; D'Alessandro, A.; Zolla, L.; Finazzi Agrò, A.; et al. miR-143 regulates hexokinase 2 expression in cancer cells. *Oncogene* **2012**, *32*, 797. [CrossRef] [PubMed]
45. Jiang, S.; Zhang, L.F.; Zhang, H.W.; Hu, S.; Lu, M.H.; Liang, S.; Li, B.; Li, Y.; Li, D.; Wang, E.D.; et al. A novel miR-155/miR-143 cascade controls glycolysis by regulating hexokinase 2 in breast cancer cells. *Embo J.* **2012**, *31*, 1985–1998. [CrossRef] [PubMed]
46. Zeinali, T.; Mansoori, B.; Mohammadi, A.; Baradaran, B. Regulatory mechanisms of miR-145 expression and the importance of its function in cancer metastasis. *Biomed. Pharmacother.* **2019**, *109*, 195–207. [CrossRef] [PubMed]
47. Zeng, X.-L.; Zhang, S.-Y.; Zheng, J.-F.; Yuan, H.; Wang, Y. Altered miR-143 and miR-150 expressions in peripheral blood mononuclear cells for diagnosis of non-small cell lung cancer. *Chin. Med. J.* **2013**, *126*, 4510–4516. [CrossRef] [PubMed]
48. Yoshino, H.; Enokida, H.; Itesako, T.; Kojima, S.; Kinoshita, T.; Tatarano, S.; Chiyomaru, T.; Nakagawa, M.; Seki, N. Tumor-suppressive microRNA-143/145 cluster targets hexokinase-2 in renal cell carcinoma. *Cancer Sci.* **2013**, *104*, 1567–1574. [CrossRef] [PubMed]
49. Dos Santos Ferreira, A.C.; Robaina, M.C.; de Rezende, L.M.M.; Severino, P.; Klumb, C.E. Histone deacetylase inhibitor prevents cell growth in Burkitt's lymphoma by regulating PI3K/Akt pathways and leads to upregulation of miR-143, miR-145, and miR-101. *Ann. Hematol.* **2014**, *93*, 983–993. [CrossRef] [PubMed]
50. Shen, J.; Zhang, Y.; Fu, H.; Wu, D.; Zhou, H. Overexpression of microRNA-143 inhibits growth and induces apoptosis in human leukemia cells. *Oncol. Rep.* **2014**, *31*, 2035–2042. [CrossRef] [PubMed]
51. Carithers, L.J.; Ardlie, K.; Barcus, M.; Branton, P.A.; Britton, A.; Buia, S.A.; Compton, C.C.; DeLuca, D.S.;

Peter-Demchok, J.; Gelfand, E.T.; et al. A Novel Approach to High-Quality Postmortem Tissue Procurement: The GTEx Project. *Biopreserv. Biobank.* **2015**, *13*, 311–319. [CrossRef] [PubMed]

52. Bcndoye. Comparison of LDH Activity in Normal and Cancerous Cell Metabolism. Available online: https://commons.wikimedia.org/wiki/File:LDH_activity-_normal_vs_canceous_cells.png (accessed on 9 November 2018).

12

Systems Analysis of Transcriptomic and Proteomic Profiles Identifies Novel Regulation of Fibrotic Programs by miRNAs in Pulmonary Fibrosis Fibroblasts

Steven Mullenbrock [1,†], Fei Liu [1,‡], Suzanne Szak [1], Xiaoping Hronowski [1], Benbo Gao [1], Peter Juhasz [1], Chao Sun [1], Mei Liu [1], Helen McLaughlin [1], Qiurong Xiao [1], Carol Feghali-Bostwick [2] and Timothy S. Zheng [1,*,§]

1 Biogen, Cambridge, MA 02142, USA; steven.mullenbrock@cellsignal.com (S.M.); fliu@admirxtx.com (F.L.); suzanne.szak@biogen.com (S.S.); xiaoping.hronowski@biogen.com (X.H.); benbo.gao@biogen.com (B.G.); peter.juhasz@biogen.com (P.J.); chao.sun@biogen.com (C.S.); mei.liu@biogen.com (M.L.); helen.mclaughlin@biogen.com (H.M.); qiurong.xiao@biogen.com (Q.X.)

2 Division of Rheumatology and Immunology, Department of Medicine, The Medical University of South Carolina, Charleston, SC 29425, USA; feghalib@musc.edu

* Correspondence: timothy.zheng@boehringer-ingelheim.com

† Current address: Cell Signaling Technology, Danvers, MA 01923, USA.

‡ Current address: Admirx Inc., Cambridge, MA 02139, USA.

Abstract: Fibroblasts/myofibroblasts are the key effector cells responsible for excessive extracellular matrix (ECM) deposition and fibrosis progression in both idiopathic pulmonary fibrosis (IPF) and systemic sclerosis (SSc) patient lungs, thus it is critical to understand the transcriptomic and proteomic programs underlying their fibrogenic activity. We conducted the first integrative analysis of the fibrotic programming in these cells at the levels of gene and microRNA (miRNA) expression, as well as deposited ECM protein to gain insights into how fibrotic transcriptional programs culminate in aberrant ECM protein production/deposition. We identified messenger RNA (mRNA), miRNA, and deposited matrisome protein signatures for IPF and SSc fibroblasts obtained from lung transplants using next-generation sequencing and mass spectrometry. SSc and IPF fibroblast transcriptional signatures were remarkably similar, with enrichment of WNT, TGF-β, and ECM genes. miRNA-seq identified differentially regulated miRNAs, including downregulation of miR-29b-3p, miR-138-5p and miR-146b-5p in disease fibroblasts and transfection of their mimics decreased expression of distinct sets of fibrotic signature genes as assessed using a Nanostring fibrosis panel. Finally, proteomic analyses uncovered a distinct "fibrotic" matrisome profile deposited by IPF and SSc fibroblasts compared to controls that highlights the dysregulated ECM production underlying their fibrogenic activities. Our comprehensive analyses of mRNA, miRNA, and matrisome proteomic profiles in IPF and SSc lung fibroblasts revealed robust fibrotic signatures at both the gene and protein expression levels and identified novel fibrogenesis-associated miRNAs whose aberrant downregulation in disease fibroblasts likely contributes to their fibrotic and ECM gene expression.

Keywords: interstitial lung disease; idiopathic pulmonary fibrosis; systemic sclerosis; myofibroblast; gene expression; proteomics

---

1. Introduction

Interstitial lung disease (ILD) associated with idiopathic pulmonary fibrosis (IPF) and systemic sclerosis (SSc) has a critical impact on a patient's quality of life and is the predominant cause of mortality in these diseases. Although the pathogenesis of pulmonary fibrosis in IPF and SSc remains incompletely understood, it is generally accepted that they stem from different root causes, with clinical and genetic evidence supporting epithelial injury/dysfunction and vasculopathy/inflammation as the underlying pathogenic triggers for IPF and SSc, respectively [1]. Despite their distinct origins, the resulting persistent tissue injury events converge on pathological fibroblast/myofibroblast activation, culminating in excessive extracellular matrix (ECM) deposition and ultimately progressive loss of lung function in both IPF and SSc [2].

To understand mechanisms underlying pulmonary fibrosis, several groups have undertaken transcriptomic analyses of both tissue and fibroblasts/myofibroblasts derived from fibrotic lung of IPF and SSc patients [3–9]. These studies found that a limited number of genes, pathways, and functions are altered in pulmonary fibrosis, including TGF-β and WNT, as well as altered expression of ECM genes such as collagens, crosslinking enzymes, TIMPs, and MMPs, many of which have been shown to be functionally relevant in fibrogenesis by subsequent in vitro and in vivo studies. Importantly, key fibrosis-associated pathways (e.g., ECM, WNT, and TGF-β) identified from transcriptomic analyses of fibrotic lung tissue were also captured in the gene signatures of isolated fibroblasts, suggesting that the molecular programming in pulmonary fibrosis is driven in large part by fibroblasts/myofibroblasts.

While these studies have shed insights into mechanisms underlying fibrogenic activation of fibroblasts/myofibroblasts, their fibrotic programs remain incompletely understood. For example, most early studies were performed using microarrays, which are less able to accurately detect low abundance transcripts and are limited to interrogating the expression of the transcripts present on each array platform, which precludes measuring microRNAs (miRNAs) in many cases [3–9]. To our knowledge, no global miRNA studies have been reported for either IPF or SSc lung fibroblasts and how miRNAs regulate their fibrotic program remains to be characterized. In addition, excessive ECM protein deposition by fibroblasts/myofibroblasts is directly responsible for IPF and SSc disease pathology, yet surprisingly little has been done to characterize the aberrant matrisome protein profile of these disease fibroblasts, which is not only affected by transcriptional changes, but is also subjected to post-transcriptional regulation [10].

To thoroughly interrogate mechanisms by which pathological activation of fibroblasts/ myofibroblasts is regulated in IPF and SSc, we performed genome-wide analyses of both messenger RNA (mRNA) and miRNA in these cells, and characterized their ECM deposition properties by proteomic analysis. Altogether, this current study is the first integrative analysis of fibrotic gene and protein signatures, providing novel insights into the multitude of regulatory mechanisms governing the fibrogenic potential of fibroblasts/myofibroblasts in pulmonary fibrosis.

2. Materials and Methods

2.1. Cell Culture

Primary fibroblasts were isolated from lung tissues of normal donors whose lungs were not used for transplantation and SSc or IPF patients who underwent lung transplantation at the University of Pittsburgh Medical Center under a protocol approved by the institution's Institutional Review Board. Isolation and subsequent culture of lung fibroblasts was previously described [4].

2.2. RNA-seq and miRNA-seq Analysis

RNA-seq libraries were prepared using the Illumina (San Diego, CA, USA) TruSeq RNA Sample kit with poly-T selection and sequenced using a HiSeq (75-bp paired-end reads). Reads were mapped using Tophat (version 2.0.8), transcripts were assembled using Cufflinks (version 2.2.1), and differential expression was calculated using CuffDiff.

miRNA-seq libraries were prepared using the Illumina TruSeq Small RNA Sample Kit and sequenced using an Illumina miSeq (51-bp single-end reads). Bowtie (version 0.12.5) was used to perform a stepwise alignment of fastq files to Illumina databases. Differentially expressed miRNAs were identified using Bioconductor's limma package.

2.3. Real-Time Reverse Transcription-Polymerase Chain Reaction

RNA was reverse transcribed using the High Capacity cDNA Reverse Transcription Kit with RNase inhibitor (Life Technologies/Thermo Fisher Scientific (Waltham, MA, USA) #4374966) or TaqMan MicroRNA Reverse Transcription Kit (Life Technologies/Thermo Fisher Scientific (Waltham, MA, USA #4366596) in conjunction with appropriate miRNA reverse transcription primers (Table S1). Real-time quantitative PCR (see Table S1 for TaqMan probes) was run on a Life Technologies QuantStudio 12K Flex.

2.4. Gene Ontology and Signature Analysis

Gene ontology (GO) analysis was conducted using DAVID Bioinformatics Database [11] and gene signature analysis was conducted using NextBio curated studies [12] and pre-ranked gene set enrichment analysis (GSEA) [13]. Ingenuity Pathway Analysis (IPA) was utilized to interrogate pathways upstream of IPF and SSc differentially expressed gene sets.

2.5. miRNA Mimic Transfection and Nanostring Gene Expression Analysis

Primary lung fibroblasts were reverse transfected with a miRNA mimic or negative control mimic (2.5 nM final concentration) using Lipofectamine RNAiMAX (Invitrogen/Thermo Fisher Scientific, Waltham, MA, USA) for 48 h, after which gene expression was assayed with the Nanostring (Seattle, WA, USA) platform using a custom codeset of common "fibrosis" genes.

2.6. Extracellular matrix Proteomic Profiling

Extracellular matrix proteins deposited by patient-derived fibroblasts after 4 weeks of culture were enriched by a sequential extraction of cellular and extracellular proteins as described previously [14,15]. The extracted soluble and insoluble ECM samples were further processed for proteomic analysis using Thermo Fisher Scientific's (Waltham, MA, USA) TMT10plex label reagent. The labeled samples were pooled together and fractionated into three fractions. LC-MS of the fractions were acquired using a Thermo Fisher Scientific QE HF mass spectrometer. Peptide identification and quantification were performed using Maxquant [16] searched against the human Swiss-Prot reference database (https://www.uniprot.org/). Protein levels among fibroblast groups were compared by ANOVA test followed by Tukey's HSD test.

Additional details for Materials and Methods are provided in Supplementary Materials.

3. Results

3.1. Transcriptional Profiling Identified Similar Dysregulated Gene Expression Programs in Idiopathic Pulmonary Fibrosis and Systemic Sclerosis Lung Fibroblasts

Lung fibroblasts isolated from IPF and SSc patients undergoing lung transplant and unused healthy donor lung were grown under similar culture conditions at low passage number (passage 2-3) prior to performing RNA-seq and miRNA-seq. As reflected by the low forced vital capacity % (FVC) (FVC < 60%) and the patients' requirement for a transplant, this study provides a snapshot of fibroblasts from patients with severe, end-stage lung disease (Tables S2 and S3).

RNA-seq analysis of the 30 primary lung fibroblasts ($n = 10$ for each group) identified 297 differentially expressed genes (DEGs) across all cohort comparisons ($\geq \pm 1.5$-fold, false discovery rate (FDR) $q < 0.05$) (Figure 1A and Table S4). We confirmed differential expression of several genes by qPCR (Figure 1B) and expression measured by qPCR and RNA-seq were highly correlated (Figure S1).

The majority of DEGs were observed between disease and control fibroblasts, with 168 DEGs for the IPF vs. Normal and 176 DEGs for the SSc vs. Normal comparison. The IPF and SSc fibroblast gene signatures are very similar, as over 40% of the DEGs overlap between them ($p < 0.001$), and for those that do not overlap their expression changes generally trend in a similar direction (Figure 1A,C). This is further supported by the principle component analysis (PCA) of the patient samples using the

297 DEGs across all comparisons, which illustrates that the IPF and SSc patients cluster close to one another along the first principle component (Figure 1D).

Interestingly we did identify expression differences between IPF and SSc fibroblasts (68 DEGs IPF vs. SSc, $\geq \pm 1.5$-fold, FDR $q < 0.05$, Figure S2), although this seemed to be driven by a small subset of SSc patients that were among those with notes of pulmonary hypertension, as shown by their clear separation along the second principle component of the PCA plot (Figure 1D), although it should be noted that other SSc patients with pulmonary hypertension did not separate out similarly.

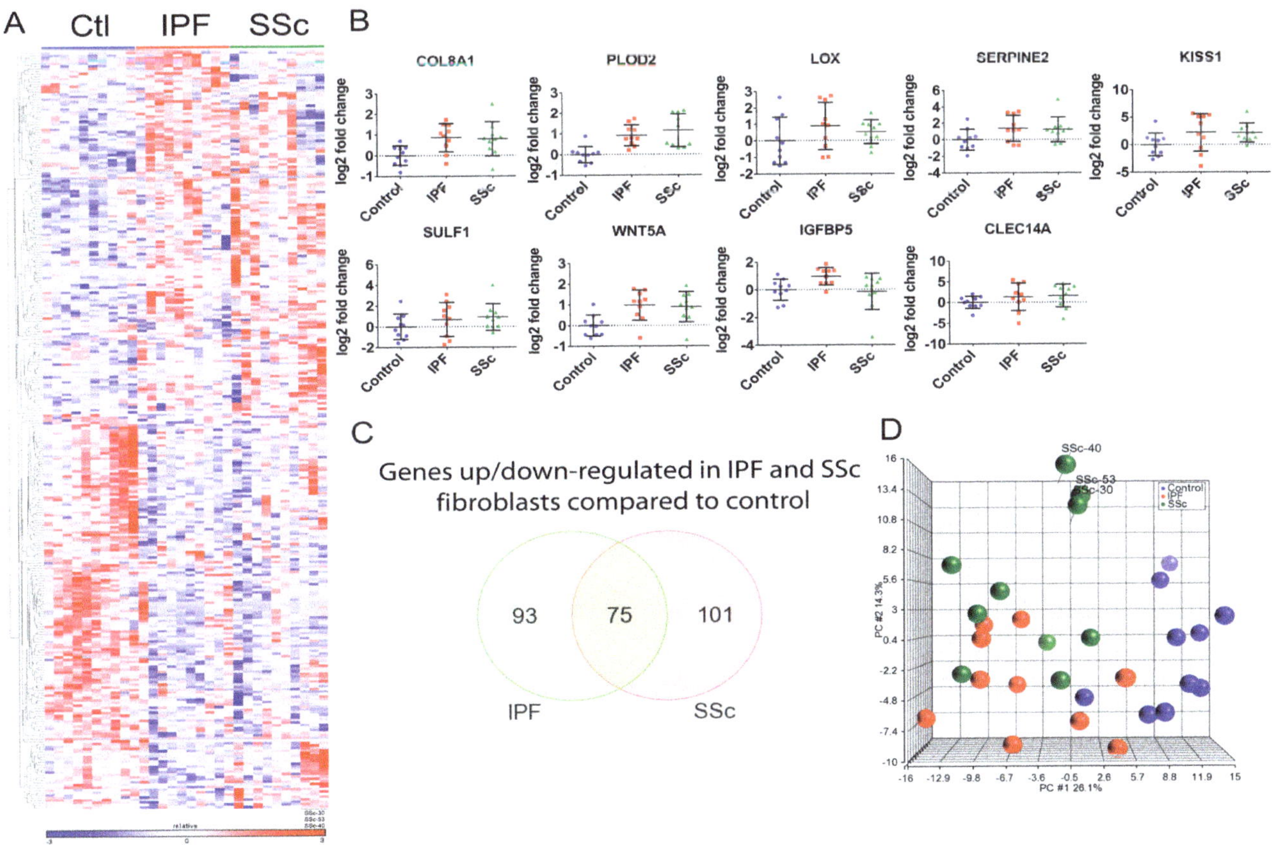

Figure 1. RNA-seq identifies 297 differentially expressed genes (DEGs) across idiopathic pulmonary fibrosis (IPF), systemic sclerosis (SSc), and healthy control primary lung fibroblast comparisons and demonstrates similar disease signatures between IPF and SSc lung fibroblasts. (**A**) Heatmap depicting 297 differentially expressed genes as determined from RNA-seq analysis using Cuffdiff (fold-change $\geq$+1.5-fold or $\leq$−1.5-fold, $q < 0.05$) across all comparisons (IPF vs. Control, SSc vs. Control, SSc vs. IPF). (**B**) Quantitative polymerase chain reaction (qPCR) analysis validated differential expression of several genes involved in profibrotic pathways. Gene expression analysis using real-time qPCR on the 30 patient fibroblast samples was conducted as described in Materials and Methods and normalized to *GAPDH*. Data is plotted as a log2 fold-change relative to the mean of the healthy control samples. (**C**) Venn diagram demonstrating overlap of statistically significant differentially expressed genes between IPF and SSc fibroblasts. (**D**) Principle component analysis on the 30 patient fibroblasts was run using the 297 genes that were significantly different across all patient group comparisons. This allows for the visualization of how similar/different the patient groups are from one another based on disease genes as well as IPF or SSc-specific genes. Healthy control fibroblasts= Ctl.

3.2. Idiopathic Pulmonary Fibrosis and Systemic Sclerosis Lung Fibroblast Disease Signatures are Associated with Pro-Fibrotic Pathways and Extracellular Matrix

The highly similar gene expression profiles of IPF and SSc lung fibroblasts likely represent a fibrotic disease signature that reflects aberrant activation of upstream signaling that sustains these fibrotic programs and that is directly involved with the pathological function of fibroblasts/myofibroblasts in ILD. Utilizing computational analyses such as GO enrichment analysis, IPA, and gene signature analysis using rank-based directional enrichment tools such as NextBio [12] and GSEA [13], we characterized the IPF and SSc fibroblast signatures to determine how they may relate to both upstream signaling pathways and potential downstream functions.

These tools revealed that IPF and SSc fibroblast signatures are associated with the activation of several profibrotic signaling pathways such as WNT (Figure 2B), TGF-β (Figure 2C, Figure S3), NOTCH1 (Figure 2D), and HIF1A (Figure 2D), as well as inhibition of the anti-fibrotic PPARG pathway (Figure 2D).

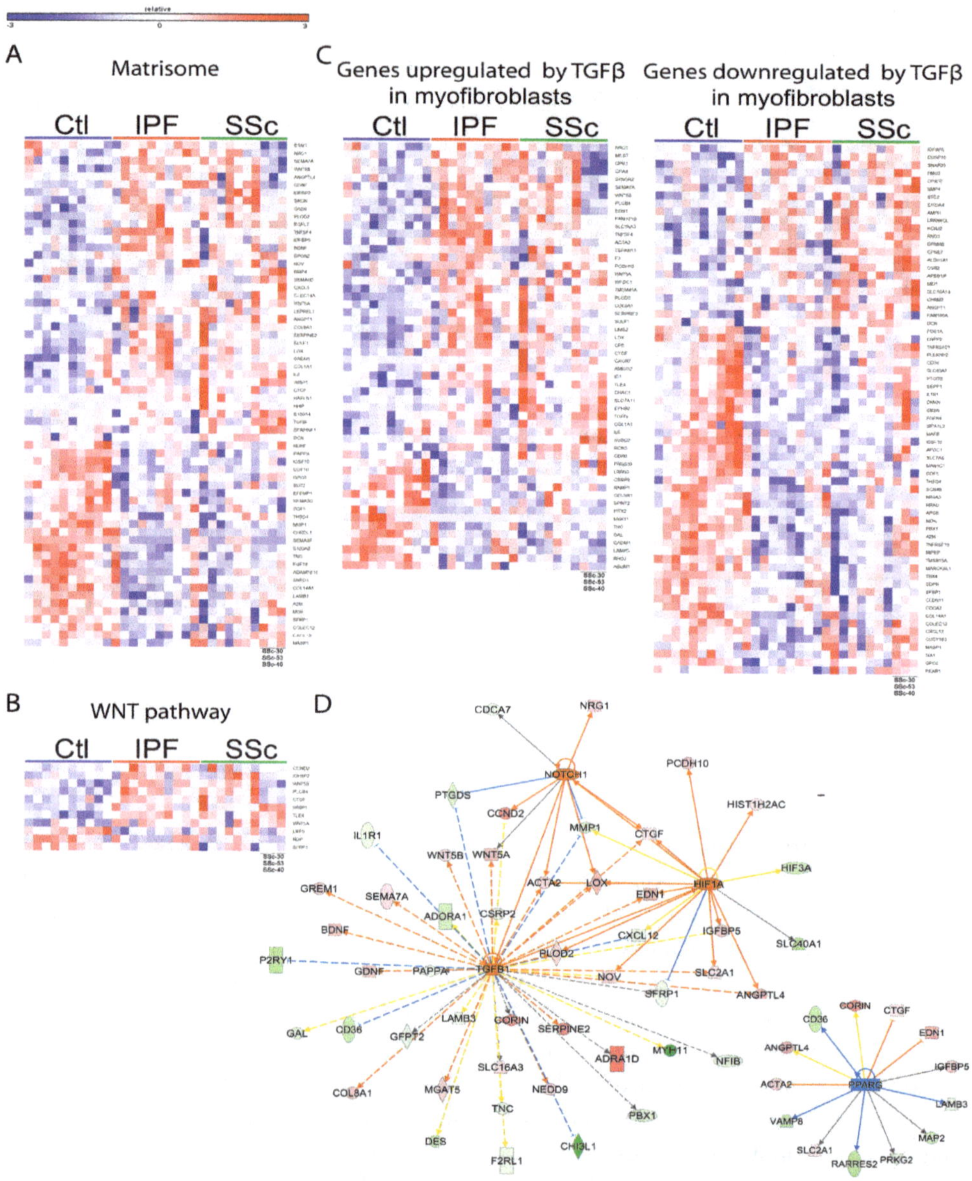

Figure 2. The dysregulated gene expression program in disease fibroblasts is composed of altered matrisome genes and associated with signaling pathways reflective of pathological fibroblast activation. Shown are heatmaps depicting expression of genes differentially expressed in IPF or SSc fibroblasts that overlapped significantly with various gene/pathway signatures such as matrisome (**A**), WNT (**B**), or TGF-β (**C**). (**D**) IPA upstream regulator analysis was used to predict pathway/transcription factor activation state upstream of the IPF or SSc DEGs. The pathways shown here are based on data from IPF DEGs (SSc DEGs had similar results). Shown is the predicted activation/repression of TGF-β, NOTCH1, HIF1A, and PPARG with lines connecting to genes differentially expressed in IPF fibroblasts that are downstream of these pathways. Healthy control fibroblasts= Ctl.

To further probe potential downstream functions of genes altered in IPF and SSc fibroblasts, we used GO analysis and found that the disease fibroblast gene signatures are enriched for genes associated with "cell proliferation" (IPF and SSc upregulated genes), "muscle contraction" (IPF upregulated genes), "response to wounding" (IPF downregulated genes), and "metallopeptidase activity" (IPF downregulated genes) (Figure 3). In addition, GO terms associated with ECM were the most significant and frequently observed terms for the upregulated and downregulated gene sets for both IPF and SSc. Consistent with this, we also found that the IPF and SSc fibroblast signatures are significantly enriched for components of the in silico matrisome derived by Naba et al. [17], with 64 of the 269 DEGs being associated with the matrisome (Figure 2A, $p < 0.001$).

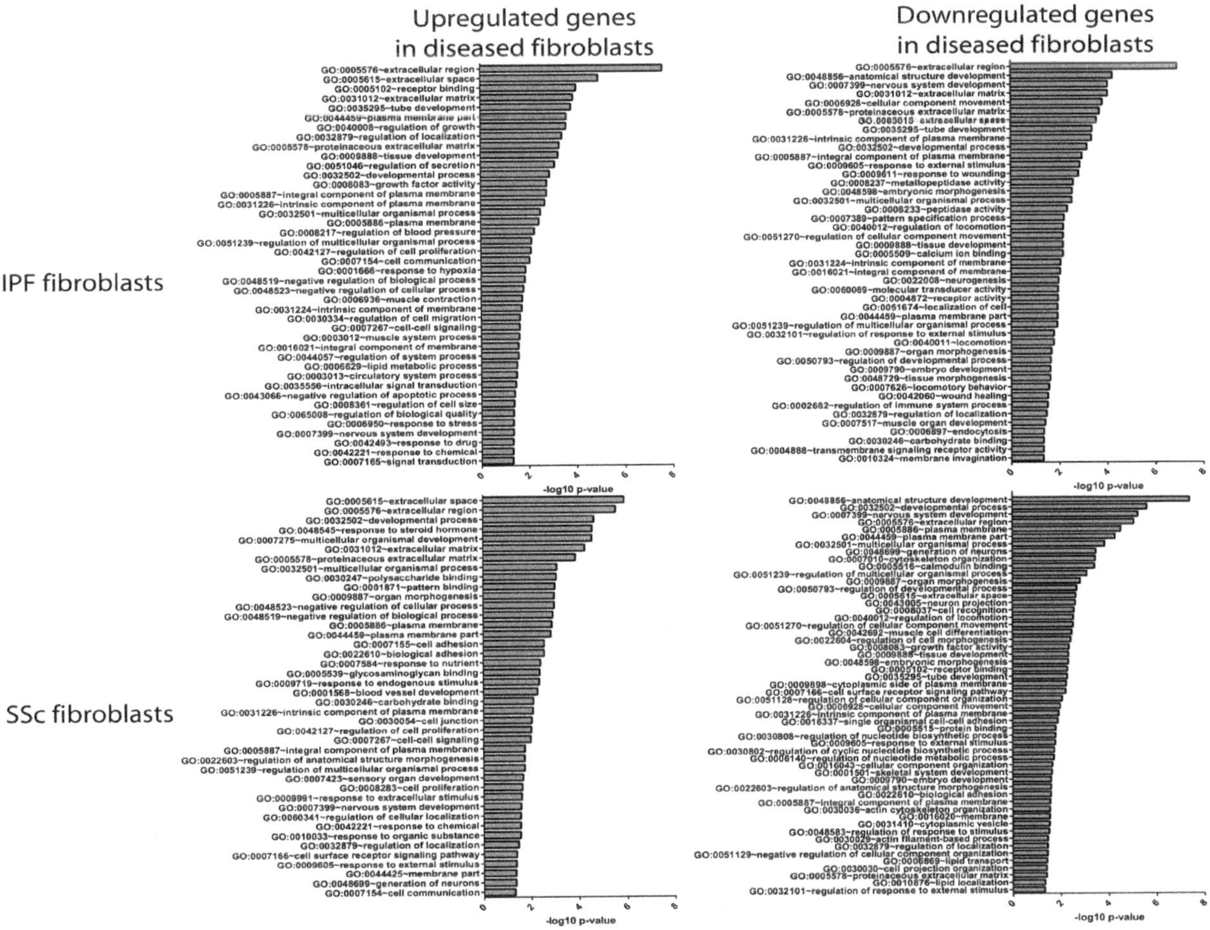

Figure 3. IPF and SSc fibroblast signatures are primarily enriched for genes associated with the extracellular matrix. Gene ontology (GO) analysis was conducted to identify over-represented GO terms for the genes upregulated in IPF or SSc fibroblasts (left panel) and downregulated in IPF or SSc fibroblasts (right panel). Shown are the significantly enriched GO terms ($p \leq 0.05$, $\geq$5% of genes had to be classified by a GO term), with highly similar GO terms being collapsed using REVIGO [18].

3.3. Global microRNA Profiling Identified Similar Fibrotic microRNA Signatures in Idiopathic Pulmonary Fibrosis and Systemic Sclerosis Lung Fibroblasts

In order to get a more complete picture of the transcriptional profiles of IPF and SSc lung fibroblasts and to identify potentially novel mechanisms by which their expression programs are regulated, we analyzed their miRNA expression using miRNA-seq.

miRNA-seq revealed miRNA expression differences between normal and disease fibroblasts that were relatively moderate (generally <2-fold), which is typical for miRNA expression data. Because of these moderate changes, we utilized a less stringent expression cutoff to minimize the possibility of false negatives. Compared to control fibroblasts, IPF fibroblasts exhibited 3 upregulated and

16 downregulated miRNAs, whereas there were 12 upregulated and 12 downregulated miRNAs for SSc fibroblasts (± 1.35-fold, $p < 0.1$, Figure 4A, Table S5). Analogous to the mRNA expression data, the IPF and SSc miRNA signatures were very similar to one another, with ~half of the differentially expressed miRNAs overlapping between them (Figure 4C). Interestingly, as was observed for mRNA expression, there were also minor differences in miRNA expression between SSc and IPF fibroblasts, with 6 miRNAs having higher and 1 miRNA having lower expression in SSc fibroblasts. To increase our confidence in these modest differences observed, we performed a second miRNA-seq and the technical replicate demonstrated high reproducibility of the data, and qPCR analysis further verified differential expression of several miRNAs (Figure 4B).

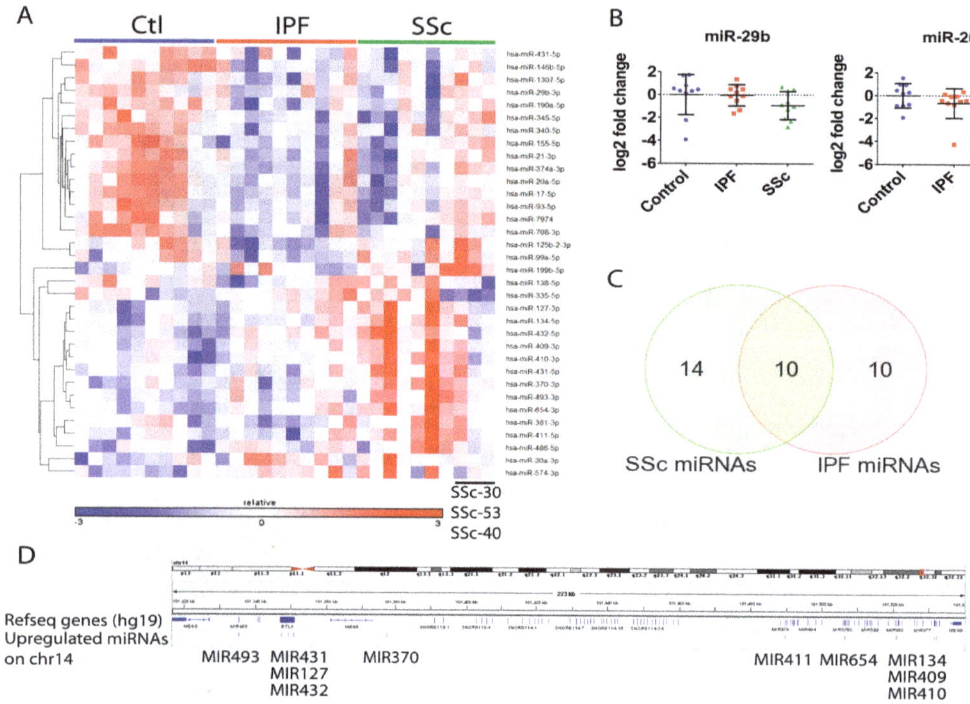

Figure 4. miRNA-seq identified similar miRNA signatures for IPF and SSc lung fibroblasts. (**A**) Heatmap depicting differentially expressed miRNAs across all patient group comparisons (fold-change $\geq +1.35$-fold or ≤ -1.35-fold, $p < 0.1$). (**B**) miRNA expression analysis using real-time qPCR on the 30 patient fibroblast samples was conducted as described in Materials and Methods for miR-20a, miR-155 and miR-29b in order to demonstrate similar expression changes as observed by miRNA-seq. (**C**) Venn diagram depicting overlap between miRNAs that were differentially expressed in SSc and IPF fibroblasts. (**D**) Genome view (hg19) of the Chr14q32 region where a cluster of 10 miRNAs were observed to be upregulated in SSc fibroblasts. Annotated Refseq genes are shown with the location of the indicated upregulated miRNAs marked directly below.

The resulting miRNA signature contained many dysregulated miRNAs with either established links to fibrosis (e.g., miR-29b and the miR-17~92 cluster), or affecting pathways and processes relevant to IPF and SSc fibroblasts (e.g., TGF-β) (Table S5). Additionally, a number of the miRNAs upregulated in SSc fibroblasts have previously been reported to be upregulated in IPF lung tissue (several miRNAs in the Chr14q32 miRNA cluster, Figure 4D).

3.4. Differentially Expressed microRNAs in Idiopathic Pulmonary Fibrosis and Systemic Sclerosis Fibroblasts Regulate Expression of Fibrosis-Associated Genes

To characterize the functional relevance of these dysregulated miRNAs to fibroblast/myofibroblast pathology, we examined whether modulating several miRNAs (miR-29b-3p, miR-138-5p, and miR-146b-5p) in disease fibroblasts would affect expression of ECM and other profibrotic genes. These miRNAs were chosen either because of their extensive associations with fibrotic disease/pathways/ECM (miR-29b-3p) or because their role in fibrosis is relatively uncharacterized, yet they were among the most highly downregulated miRNAs in disease fibroblasts (miR-138-5p, miR-146b-5p).

Transfection of miR-29b-3p, miR-146b-5p, or miR-138-5p mimics into IPF and SSc lung fibroblasts all had significant effects on ECM and fibrosis gene expression using a "fibrosis" nanostring panel, with 175 genes being modulated by at least one of the miRNAs (Figure 5). As the genes in this panel are generally pro-fibrotic, these miRNA mimics resulted mostly in their downregulation in both IPF and SSc fibroblasts.

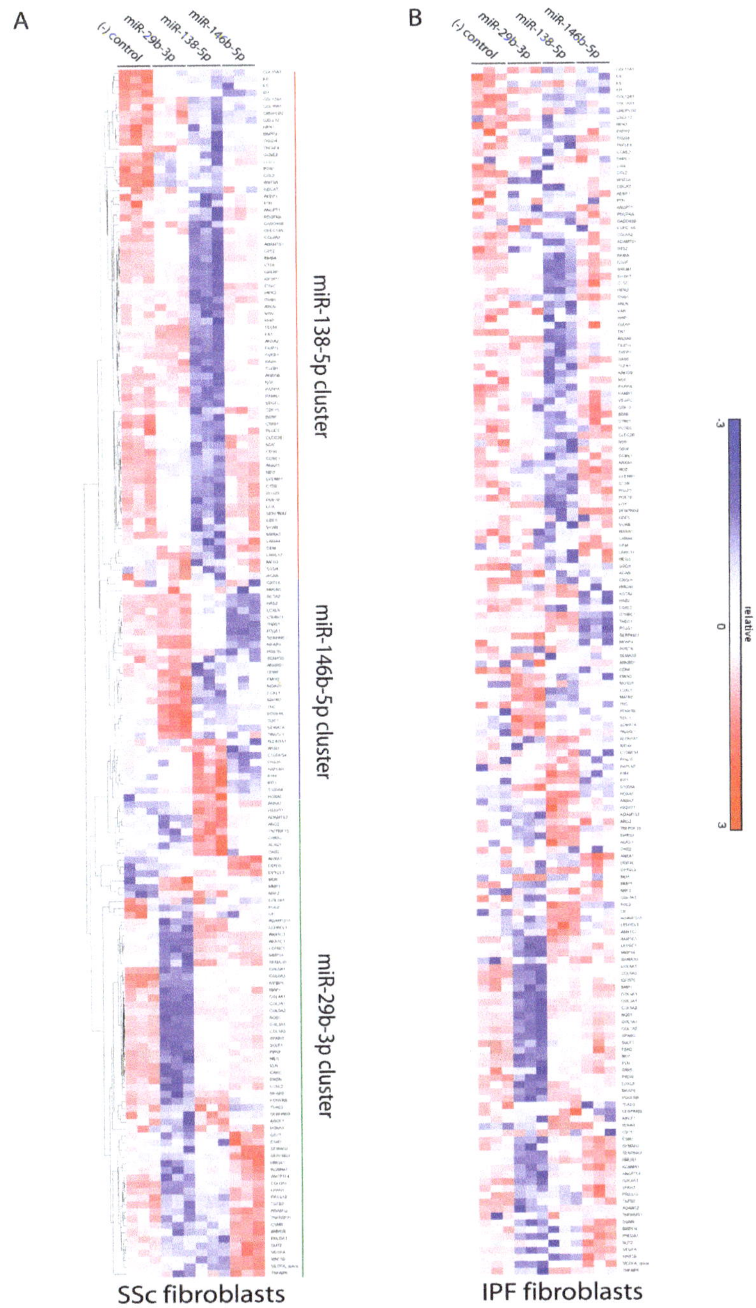

Figure 5. miR-29b-3p, miR-138-5p, and miR-146b-5p mimics downregulate expression of pro-fibrotic genes. miRNA mimics for miR-29b-3p, miR-138-5p, or miR-146b-5p were transfected into primary SSc (**A**) or IPF (**B**) lung fibroblasts as described in Materials and Methods for 48 hours prior to collecting RNA. Gene expression was measured using Nanostring analysis of a panel of 500+ "fibrosis" genes. Shown is a heatmap depicting expression of genes whose expression was significantly affected ($\geq +1.5$-fold or ≤ -1.5-fold, $p \leq 0.05$) by at least one of the miRNA mimics in either the SSc or IPF fibroblasts.

As we have gene and miRNA expression from parallel samples, we next aimed to determine if these miRNAs could contribute to the dysregulated fibrotic signature we observed in IPF and SSc fibroblasts. The fibrosis nanostring panel includes 46 genes that were also upregulated in the IPF and SSc fibroblast gene signatures. Of these, 30/46 (65%) were affected by at least one of the miRNA mimics in either SSc or IPF fibroblasts (Figure S4), and in nearly all cases the mimic resulted in downregulation.

Interestingly, each miRNA preferentially affected distinct clusters of genes (Figure 5 and Figure S4). As expected for miR-29b-3p, a number of pro-fibrotic and ECM genes including *COL1A1*, a well-studied target of miR-29, were downregulated. miR-138-5p also reversed expression of a large subset of the disease fibroblast signature, including key pro-fibrotic players such as *LOX*, *CTGF*, and *GREM1*. miR-146b-5p had a more subtle effect on downstream gene expression, however it was the only mimic that significantly reduced *ACTA2* levels in SSc fibroblasts.

3.5. Proteomic Profiling Characterized a Distinct Fibrotic Matrisome Deposited by Idiopathic Pulmonary Fibrosis and Systemic Sclerosis Lung Fibroblasts

Aberrant secretion of matrisome and matrisome-associated proteins is both the primary means and end outcome by which fibroblasts drive disease pathology in IPF and SSc ILD. This was reflected in our mRNA profiling data, where ECM/matrisome components were the most prevalent within the fibrotic gene signatures, and in our miRNA-seq data, in which several miRNAs affected matrisome gene expression. As there are additional mechanisms beyond transcriptional regulation that affect ECM production/deposition, we characterized directly the matrisome deposited by IPF, SSc, and normal fibroblasts at the protein level.

Extracellular matrix deposited by the 30 patient fibroblasts (n = 10 for each group) was collected and guanidine-soluble and insoluble fractions were subjected to mass spectrometric proteomic profiling separately, as the insoluble fraction is thought to contain more highly cross-linked ECM that could be more relevant in the context of fibrotic disease. In total, 277 matrisome proteins were detected in at least one of the samples. Importantly, using the detected matrisome proteins in the soluble fraction, hierarchical clustering of the fibroblasts resulted in a distinct clustering of IPF and SSc fibroblasts away from controls, strongly suggesting that disease fibroblasts secrete a distinct fibrotic matrisome (Figure S5A). While similar separation was not observed for the insoluble fraction (Figure S5B), the protein recovery for the insoluble fraction was highly variable between different samples, which may have confounded our analysis.

Across all fibroblast group comparisons, the majority of differences were observed between disease vs. normal fibroblasts, with SSc vs. Normal yielding 26 and 18 differentially expressed proteins (DEPs) for the soluble and insoluble fractions respectively and IPF vs. Normal yielding 28 and 6 DEPs for the soluble and insoluble fractions respectively (Figure S5C). Analogous to what we observed at the transcriptional level, the IPF and SSc matrisomes tended to be more similar than different as demonstrated by the observed DEP overlap (Figure S5D) and by the observation that IPF and SSc fibroblasts generally clustered together (Figure S5A and Figure 6). Interestingly we did identify minor differences between SSc and IPF matrisome protein expression, with 8 and 2 DEPs being observed for the soluble and insoluble fractions respectively. Hierarchical clustering using DEPs across all cohort comparisons resulted in a cluster of ~half of the disease fibroblasts when using either the soluble (Figure 6A) or insoluble (Figure 6B) fraction data, demonstrating that disease fibroblasts secrete a distinct matrisome compared to normal fibroblasts.

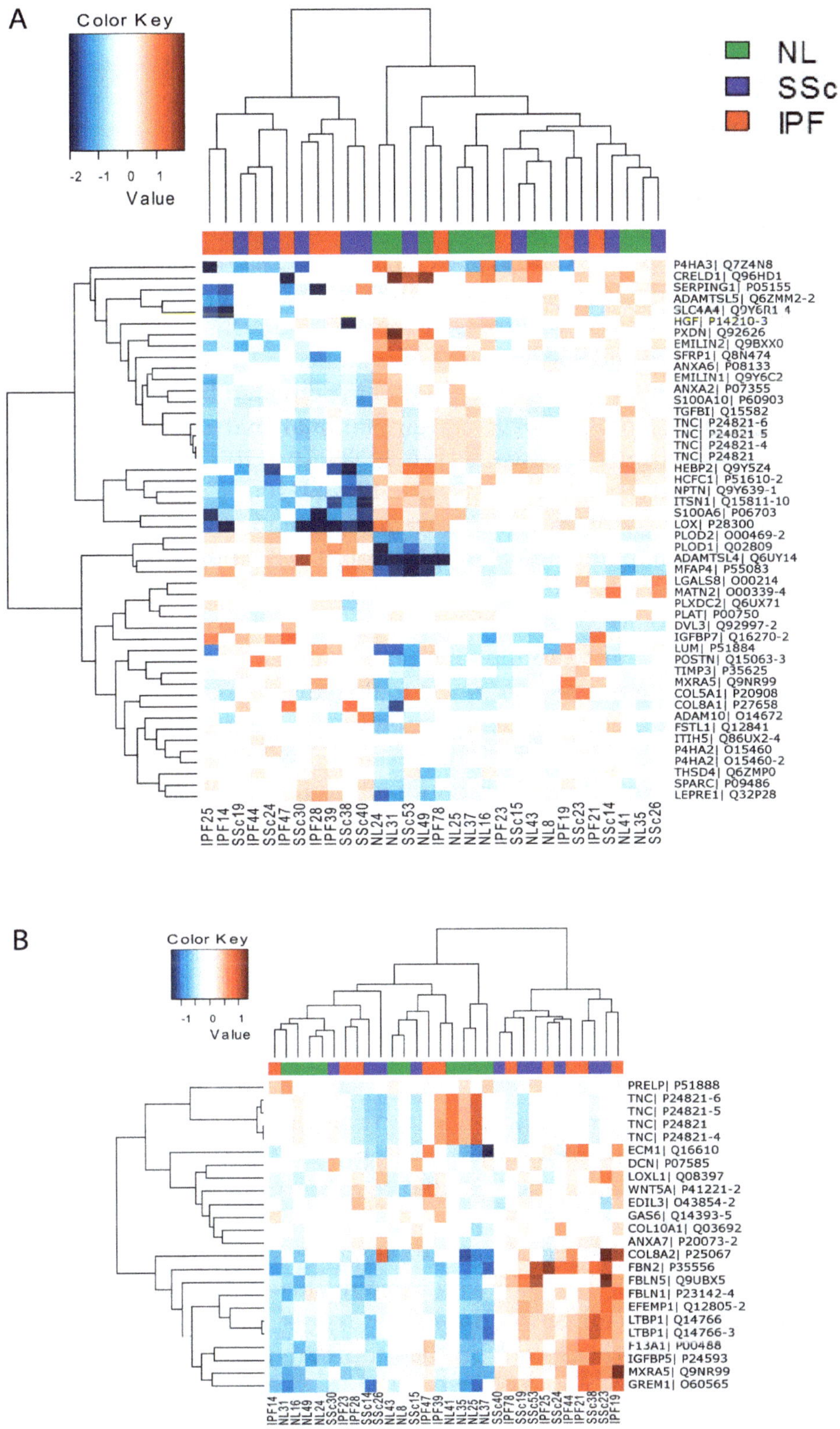

Figure 6. Hierarchical clustering using differentially expressed matrisome proteins is able to delineate normal-like and disease-like fibroblast groups. Shown are heatmaps of z-scored normalized mass spectrometry protein intensity values of matrisome proteins that were differentially expressed across all patient comparisons (fold-change $\geq$+1.2-fold or $\leq$−1.2-fold, $p \leq 0.05$) and then clustered. Shown is the data from the soluble fraction (**A**) and from the insoluble fraction (**B**).

4. Discussion

Fibroblasts/myofibroblasts are key effector cells that directly contribute to the pro-fibrotic milieu, aberrant ECM deposition, increased tissue stiffening, disruption of tissue architecture, and ultimately impaired organ function in pulmonary fibrosis [2]. Understanding the transcriptomic and proteomic profiles of these disease fibroblasts provides insights into their dysregulated fibrotic program at multiple levels and sheds light on potential therapeutic strategies. To our knowledge, the current study is the first attempt at integrating data from mRNA, miRNA, and secreted matrisome of IPF and SSc derived lung fibroblasts and our results revealed a number of novel mechanistic insights into the fibrotic programming of these effector cells underlying pulmonary fibrosis and its complex regulation.

Consistent with previous microarray-based studies, our RNA-seq clearly identified similar "fibrotic gene signatures" for IPF and SSc fibroblasts that reflect a signature of activated myofibroblasts, as illustrated by the correlation for genes upregulated in activated hepatic stellate cells (HSCs) with both the IPF and SSc DEGs (Figure S3). Further computational analysis revealed that these signatures are associated with known fibrotic pathways (WNT, TGF-β, HIF1A, NOTCH1, and PPARG) and effector functions (ECM) of activated fibroblasts/myofibroblasts.

Although the fibrotic transcriptomes from our and previous transcriptomic analyses of fibrotic lung fibroblasts are indicative of pathological myofibroblast activation, our current study also yielded interesting novel findings. First, the exact gene makeup of the reported signatures varied considerably among studies [3–7], which could result from study-specific differences related to patient profiles, source of tissue, culture methods, and profiling methods. However, disease stage is likely a significant driver as we observed stark differences compared to the study reported by Lindahl et al. [6], which used fibroblasts from earlier stage patients, as opposed to end-stage patients in our study. While their study identified a broader set of pathway signatures that were both inflammatory and fibrotic in nature, ECM changes were the most predominant feature in our disease signatures. This suggests that inflammatory changes, such as downregulation of the interferon signature, may be important for progression during early but not at later stages of disease when ECM changes predominate. Second, while previous studies have observed only minimal expression differences between IPF and SSc fibroblasts, our study successfully identified 68 genes that were differentially expressed between them, including inflammation genes (*TNFRSF21*, *CXCL5*, *IL8*) and genes associated with the GO function "oxidoreductase activity" (Figure S2). Interestingly, these differences appear to be driven by a subset of SSc patients (SSc-53, SSc-40, SSc-30). While the precise reason for this is not understood, it could be reflective of concomitant pathological changes frequently associated with SSc, such as pulmonary hypertension and inflammation.

The general similarity between IPF and SSc transcriptomes with subtle differences was also mirrored in the miRNA and matrisome data. Interestingly, for the three SSc patients noted above, we also observed that their miRNA expression tended to differ from the other SSc patients (Figure 4), raising the intriguing possibility that SSc-associated expression changes, such as those that might reflect pathological changes other than fibrosis (e.g., hypertension), could be governed in part at the miRNA level. At the protein level, the secreted matrisome also exhibited a few differences between IPF and SSc. However, unlike the mRNA and miRNA expression data, those three SSc patients did not appear to exhibit distinct matrix protein profiles compared to the other SSc fibroblasts, possibly because the matrisome signature is more reflective of fibrosis and less likely to be indicative of other SSc-associated pathologies.

As the most salient effector mechanism underlying pulmonary fibrosis, ECM deposition and remodeling by IPF and SSc fibroblasts remain poorly understood. Although the ECM/matrisome transcriptomic signature was the most predominant signal altered in both IPF and SSc fibroblasts, matrix genes were not universally upregulated and in fact many were downregulated (Figure 2A). Thus our data indicate that the fibrotic matrisome does not just result from increased expression of ECM genes but it involves complex dysregulated expression patterns, including both increased and decreased mRNA expression of different collagen types (*COL1A1* upregulation vs. *COL14A1* downregulation), downregulation of ECM-degradation enzymes (*MMP1* and *ADAMTS15*), and upregulation of ECM crosslinking/assembly enzymes (*LOX*, *LEPREL1* and *PLOD2*). Importantly, we confirmed aberrant dysregulation of ECM at the protein level as well by conducting the first proteomic analysis of IPF and SSc fibroblast-deposited matrix. We successfully identified an ECM protein signature shared between IPF and SSc that differentiates disease from healthy fibroblasts, which included several proteins implicated in fibrogenesis such as PLOD2, LUM, POSTN, IGFBP5, GREM1, and SPARC, as well as less characterized ECM proteins such as MXRA5, LEPRE1, MFAP4, and FSTL1. Upon comparing our RNA-seq and mass spectrometry results, we observed similar dysregulation at the protein and mRNA levels for such matrisome components as WNT5A, GREM1, DCN, IGFBP5, COL8A1, PLOD2, SFRP1 and TNC. Surprisingly, these were the only shared dysregulated genes/proteins between the RNA-seq and mass spectrometry data suggesting that differences between protein and mRNA levels of matrisome components are likely due to the presence of a myriad of post-transcriptional regulatory mechanisms at the levels of protein translation, secretion, intracellular and extracellular assembly, as well as enzymatic crosslinking and degradation. However we cannot rule out that the observed differences between the RNA-seq and mass spectrometry results are due to differences in the length of time the fibroblasts were cultured for each experiment, (short-term for RNA-seq, long-term to accumulate sufficient deposited matrix for mass-spectrometry experiments) or that exposure of these cells to tissue culture plastic influenced their protein expression profile.

miRNAs can affect both mRNA levels and protein translation, and a handful of miRNAs have been identified as key players in lung fibrosis. Our current study is the first comprehensive characterization of the global miRNA profiles in fibrotic lung fibroblasts, which included the identification of miRNAs capable of regulating ECM gene expression. Our miRNA-seq analysis identified a number of aberrantly expressed miRNAs in IPF and SSc fibroblasts and several lines of evidence support the relevance of these miRNAs in their fibrotic programming. In particular, the miRNA signature includes many miRNAs previously linked to fibrosis through in vitro and in vivo studies (Table S5). Of particular interest, miR-29b-3p has reduced expression in multiple fibrosis models and human fibrotic disease and can inhibit fibrosis in several mouse models [19–32]. miR-138-5p and miR-146b-5p, which demonstrated some of the highest levels of downregulation in the disease fibroblasts, have been implicated in processes relevant to fibrosis. For example, miR-138-5p has been suggested to play a role in hypertrophic scar fibroblasts during abnormal wound healing [33], osteogenic differentiation and bone formation [34,35], and epithelial-mesenchymal transition [36]. In addition to miR-146b-5p's effect on TGF-β-signaling [37], its closely-related family member miR-146a inhibits TGF-β-mediated activation of dermal fibroblasts and HSCs, as well as renal fibrosis in the unilateral ureteral obstruction model [38–40]. Importantly, we also demonstrated experimentally that several "fibrotic miRNAs" identified in our study (miR-29b-3p, miR-138-5P and miR-146b-5p) regulate the expression of fibrotic/ECM genes. Their identification as important miRNAs modulating distinct aspects of the fibrotic transcriptomes in IPF and SSc lung fibroblasts is highly significant in our view, as this further supports a role for miR-29 as a master ECM regulator, and reveals novel roles for miR-138-5p and miR-146b-5p in pulmonary fibrosis. Since we only measured their effects on transcript levels, our data understates the potential contribution these miRNAs have in the pathology of disease fibroblasts as they could have additional effects on the deposited fibrotic matrisome signature via translation inhibition. It is important to note that these dysregulated miRNAs could be affecting fibrotic/ECM genes either directly or indirectly through other transcriptional regulators.

Equally intriguing is our finding that 10 miRNAs in the Chr14q32 locus appear to be a fibrotic "miRNA module" in SSc fibroblasts, similar to the previously described Chr14q32 region in IPF lung tissue [41]. Additional miRNA clusters, whereby miRNAs/genes exhibited similar differential gene expression and genomic localization, were identified for miR-17-5p and miR-20a-5p, miR-125b-2-3p and miR-99a-5p, and for miR-335-5p and *MEST*. This implies that miRNA modules are dysregulated in IPF and SSc fibroblasts and indeed we see significant expression level correlations for the genes within these clusters (data not shown).

Altogether, through an integrative approach, we successfully characterized distinct mRNA, miRNA, and deposited matrix protein signatures for IPF and SSc fibroblasts, and in doing so identified novel-regulation of fibrotic gene expression by aberrantly expressed miRNAs.

Supplementary Materials:
Supplementary Text: Supplementary Materials and Methods, Figure S1: Gene expression values are highly correlative between RNA-seq and qPCR, Figure S2: RNA-seq identifies 68 DEGs between IPF and SSc lung fibroblasts, Figure S3: IPF and SSc fibroblast gene expression profiles reflect activated hepatic stellate cell (HSC) and TGF-β gene signatures, Figure S4: miR-29b-3p, miR-138-5p, and miR-146b-5p mimics downregulate expression of pro-fibrotic genes that are dysregulated in disease fibroblasts, Figure S5: Proteomic profiling of ECM deposited by IPF and SSc lung fibroblasts reveals a fibrotic matrisome signature that delineates them from normal fibroblasts, Table S1: List of Taqman probes and miRNA mimics used, Table S2: Patient information summary, Table S3: Individual patient information, Table S4: RNAseq expression data for DEGs across all cohort comparisons, Table S5: Differentially expressed miRNAs across all patient lung fibroblast comparisons.

Author Contributions: Conceptualization, S.M., T.S.Z., F.L., and C.-F.B. Investigation, S.M., F.L., X.H., M.L., H.M., and Q.X. Formal Analysis, S.M., F.L., S.S., B.G., C.S., and P.J. Resources, C.-F.B. Writing—Original Draft Preparation, S.M. and T.S.Z. Writing—Review & Editing, all authors.

References

1. Herzog, E.L.; Mathur, A.; Tager, A.M.; Feghali-Bostwick, C.; Schneider, F.; Varga, J. Review: interstitial lung disease associated with systemic sclerosis and idiopathic pulmonary fibrosis: how similar and distinct? *Arthritis Rheumatol.* **2014**, *66*, 1967–1978. [CrossRef] [PubMed]
2. Kendall, R.T.; Feghali-Bostwick, C.A. Fibroblasts in fibrosis: novel roles and mediators. *Front. Pharmacol.* **2014**, *5*, 123. [CrossRef] [PubMed]
3. Emblom-Callahan, M.C.; Chhina, M.K.; Shlobin, O.A.; Ahmad, S.; Reese, E.S.; Iyer, E.P.; Cox, D.N.; Brenner, R.; Burton, N.A.; Grant, G.M.; et al. Genomic phenotype of non-cultured pulmonary fibroblasts in idiopathic pulmonary fibrosis. *Genomics* **2010**, *96*, 134–145. [CrossRef] [PubMed]
4. Hsu, E.; Shi, H.; Jordan, R.M.; Lyons-Weiler, J.; Pilewski, J.M.; Feghali-Bostwick, C.A. Lung tissues in patients with systemic sclerosis have gene expression patterns unique to pulmonary fibrosis and pulmonary hypertension. *Arthritis Rheum.* **2011**, *63*, 783–794. [CrossRef] [PubMed]
5. Kabuyama, Y.; Oshima, K.; Kitamura, T.; Homma, M.; Yamaki, J.; Munakata, M.; Homma, Y. Involvement of selenoprotein P in the regulation of redox balance and myofibroblast viability in idiopathic pulmonary fibrosis. *Genes Cells* **2007**, *12*, 1235–1244. [CrossRef] [PubMed]
6. Lindahl, G.E.; Stock, C.J.; Shi-Wen, X.; Leoni, P.; Sestini, P.; Howat, S.L.; Bou-Gharios, G.; Nicholson, A.G.; Denton, C.P.; Grutters, J.C.; et al. Microarray profiling reveals suppressed interferon stimulated gene program in fibroblasts from scleroderma-associated interstitial lung disease. *Respir. Res.* **2013**, *14*, 80. [CrossRef] [PubMed]
7. Peng, R.; Sridhar, S.; Tyagi, G.; Phillips, J.E.; Garrido, R.; Harris, P.; Burns, L.; Renteria, L.; Woods, J.; Chen, L.; et al. Bleomycin induces molecular changes directly relevant to idiopathic pulmonary fibrosis: a model for "active" disease. *PLoS ONE* **2013**, *8*, e59348. [CrossRef] [PubMed]
8. Kass, D.J.; Kaminski, N. Evolving genomic approaches to idiopathic pulmonary fibrosis: moving beyond genes. *Clin. Transl. Sci.* **2011**, *4*, 372–379. [CrossRef] [PubMed]
9. Christmann, R.B.; Sampaio-Barros, P.; Stifano, G.; Borges, C.L.; de Carvalho, C.R.; Kairalla, R.; Parra, E.R.; Spira, A.; Simms, R.; Capellozzi, V.L.; et al. Association of Interferon- and transforming growth factor beta-regulated genes and macrophage activation with systemic sclerosis-related progressive lung fibrosis. *Arthritis Rheumatol.* **2014**, *66*, 714–725. [CrossRef] [PubMed]
10. Parker, M.W.; Rossi, D.; Peterson, M.; Smith, K.; Sikström, K.; White, E.S.; Connett, J.E.; Henke, C.A.; Larsson, O.; Bitterman, P.B. Fibrotic extracellular matrix activates a profibrotic positive feedback loop. *J. Clin. Invest.* **2014**, *124*, 1622–1635. [CrossRef] [PubMed]
11. Huang da, W.; Sherman, B.T.; Lempicki, R.A. Systematic and integrative analysis of large gene lists using DAVID bioinformatics resources. *Nat. Protoc.* **2009**, *4*, 44–57. [CrossRef] [PubMed]

12. Kupershmidt, I.; Su, Q.J.; Grewal, A.; Sundaresh, S.; Halperin, I.; Flynn, J.; Shekar, M.; Wang, H.; Park, J.; Cui, W.; et al. Ontology-based meta-analysis of global collections of high-throughput public data. *PLoS ONE* **2010**, *5*. [CrossRef] [PubMed]
13. Subramanian, A.; Tamayo, P.; Mootha, V.K.; Mukherjee, S.; Ebert, B.L.; Gillette, M.A.; Paulovich, A.; Pomeroy, S.L.; Golub, T.R.; Lander, E.S.; et al. Gene set enrichment analysis: a knowledge-based approach for interpreting genome-wide expression profiles. *Proc. Natl. Acad. Sci. USA* **2005**, *102*, 15545–15550. [CrossRef] [PubMed]
14. Didangelos, A.; Yin, X.; Mandal, K.; Baumert, M.; Jahangiri, M.; Mayr, M. Proteomics characterization of extracellular space components in the human aorta. *Mol. Cell Proteomics* **2010**, *9*, 2048–2062. [CrossRef] [PubMed]
15. Decaris, M.L.; Gatmaitan, M.; FlorCruz, S.; Luo, F.; Li, K.; Holmes, W.E.; Hellerstein, M.K.; Turner, S.M.; Emson, C.L. Proteomic analysis of altered extracellular matrix turnover in bleomycin-induced pulmonary fibrosis. *Mol. Cell Proteomics* **2014**, *13*, 1741–1752. [CrossRef] [PubMed]
16. Cox, J.; Mann, M. MaxQuant enables high peptide identification rates, individualized p.p.b.-range mass accuracies and proteome-wide protein quantification. *Nat. Biotechnol.* **2008**, *26*, 1367–1372. [CrossRef] [PubMed]
17. Naba, A.; Clauser, K.R.; Hoersch, S.; Liu, H.; Carr, S.A.; Hynes, R.O. The matrisome: in silico definition and in vivo characterization by proteomics of normal and tumor extracellular matrices. *Mol. Cell Proteomics* **2012**, *11*, M111 014647. [CrossRef] [PubMed]
18. Supek, F.; Bosnjak, M.; Skunca, N.; Smuc, T. REVIGO summarizes and visualizes long lists of gene ontology terms. *PLoS ONE* **2011**, *6*, e21800. [CrossRef] [PubMed]
19. He, Y.; Huang, C.; Lin, X.; Li, J. MicroRNA-29 family, a crucial therapeutic target for fibrosis diseases. *Biochimie* **2013**, *95*, 1355–1359. [CrossRef] [PubMed]
20. Montgomery, R.L.; Yu, G.; Latimer, P.A.; Stack, C.; Robinson, K.; Dalby, C.M.; Kaminski, N.; van Rooij, E. MicroRNA mimicry blocks pulmonary fibrosis. *EMBO Mol. Med.* **2014**, *6*, 1347–1356. [CrossRef] [PubMed]
21. Xiao, J.; Meng, X.M.; Huang, X.R.; Chung, A.C.; Feng, Y.L.; Hui, D.S.; Yu, C.M.; Sung, J.J.; Lan, H.Y. miR-29 inhibits bleomycin-induced pulmonary fibrosis in mice. *Mol. Ther.* **2012**, *20*, 1251–1260. [CrossRef] [PubMed]
22. Cushing, L.; Kuang, P.P.; Qian, J.; Shao, F.; Wu, J.; Little, F.; Thannickal, V.J.; Cardoso, W.V.; Lu, J. miR-29 is a major regulator of genes associated with pulmonary fibrosis. *Am. J. Respir. Cell Mol. Biol.* **2011**, *45*, 287–294. [CrossRef] [PubMed]
23. Qin, W.; Chung, A.C.; Huang, X.R.; Meng, X.M.; Hui, D.S.; Yu, C.M.; Sung, J.J.; Lan, H.Y. TGF-beta/Smad3 signaling promotes renal fibrosis by inhibiting miR-29. *J. Am. Soc. Nephrol.* **2011**, *22*, 1462–1474. [CrossRef] [PubMed]
24. Zhang, Y.; Huang, X.R.; Wei, L.H.; Chung, A.C.; Yu, C.M.; Lan, H.Y. miR-29b as a therapeutic agent for angiotensin II-induced cardiac fibrosis by targeting TGF-beta/Smad3 signaling. *Mol. Ther.* **2014**, *22*, 974–985. [CrossRef] [PubMed]
25. van Rooij, E.; Sutherland, L.B.; Thatcher, J.E.; DiMaio, J.M.; Naseem, R.H.; Marshall, W.S.; Hill, J.A.; Olson, E.N. Dysregulation of microRNAs after myocardial infarction reveals a role of miR-29 in cardiac fibrosis. *Proc. Natl. Acad. Sci. USA* **2008**, *105*, 13027–13032. [CrossRef] [PubMed]
26. Roderburg, C.; Urban, G.W.; Bettermann, K.; Vucur, M.; Zimmermann, H.; Schmidt, S.; Janssen, J.; Koppe, C.; Knolle, P.; Castoldi, M.; et al. Micro-RNA profiling reveals a role for miR-29 in human and murine liver fibrosis. *Hepatology* **2011**, *53*, 209–218. [CrossRef] [PubMed]
27. Sekiya, Y.; Ogawa, T.; Yoshizato, K.; Ikeda, K.; Kawada, N. Suppression of hepatic stellate cell activation by microRNA-29b. *Biochem. Biophys. Res. Commun.* **2011**, *412*, 74–79. [CrossRef] [PubMed]
28. Zhang, Y.; Wu, L.; Wang, Y.; Zhang, M.; Li, L.; Zhu, D.; Li, X.; Gu, H.; Zhang, C.Y.; Zen, K. Protective role of estrogen-induced miRNA-29 expression in carbon tetrachloride-induced mouse liver injury. *J. Biol. Chem.* **2012**, *287*, 14851–14862. [CrossRef] [PubMed]
29. Maurer, B.; Stanczyk, J.; Jungel, A.; Akhmetshina, A.; Trenkmann, M.; Brock, M.; Kowal-Bielecka, O.; Gay, R.E.; Michel, B.A.; Distler, J.H.; et al. MicroRNA-29, a key regulator of collagen expression in systemic sclerosis. *Arthritis Rheum.* **2010**, *62*, 1733–1743. [CrossRef] [PubMed]
30. Cushing, L.; Kuang, P.; Lu, J. The role of miR-29 in pulmonary fibrosis. *Biochem. Cell Biol.* **2015**, *93*, 109–118. [CrossRef] [PubMed]

31. Pandit, K.V.; Milosevic, J. MicroRNA regulatory networks in idiopathic pulmonary fibrosis. *Biochem. Cell Biol.* **2015**, *93*, 129–137. [CrossRef] [PubMed]
32. Pandit, K.V.; Corcoran, D.; Yousef, H.; Yarlagadda, M.; Tzouvelekis, A.; Gibson, K.F.; Konishi, K.; Yousem, S.A.; Singh, M.; Handley, D.; et al. Inhibition and role of let-7d in idiopathic pulmonary fibrosis. *Am. J. Respir. Crit. Care Med.* **2010**, *182*, 220–229. [CrossRef] [PubMed]
33. Xiao, Y.Y.; Fan, P.J.; Lei, S.R.; Qi, M.; Yang, X.H. MiR-138/peroxisome proliferator-activated receptor beta signaling regulates human hypertrophic scar fibroblast proliferation and movement in vitro. *J. Dermatol.* **2015**, *42*, 485–495. [CrossRef] [PubMed]
34. Qu, B.; Xia, X.; Wu, H.H.; Tu, C.Q.; Pan, X.M. PDGF-regulated miRNA-138 inhibits the osteogenic differentiation of mesenchymal stem cells. *Biochem. Biophys. Res. Commun.* **2014**, *448*, 241–247. [CrossRef] [PubMed]
35. Eskildsen, T.; Taipaleenmaki, H.; Stenvang, J.; Abdallah, B.M.; Ditzel, N.; Nossent, A.Y.; Bak, M.; Kauppinen, S.; Kassem, M. MicroRNA-138 regulates osteogenic differentiation of human stromal (mesenchymal) stem cells in vivo. *Proc. Natl. Acad. Sci. USA* **2011**, *108*, 6139–6144. [CrossRef] [PubMed]
36. Liu, X.; Wang, C.; Chen, Z.; Jin, Y.; Wang, Y.; Kolokythas, A.; Dai, Y.; Zhou, X. MicroRNA-138 suppresses epithelial-mesenchymal transition in squamous cell carcinoma cell lines. *Biochem. J.* **2011**, *440*, 23–31. [CrossRef] [PubMed]
37. Geraldo, M.V.; Yamashita, A.S.; Kimura, E.T. MicroRNA miR-146b-5p regulates signal transduction of TGF-beta by repressing SMAD4 in thyroid cancer. *Oncogene* **2012**, *31*, 1910–1922. [CrossRef] [PubMed]
38. He, Y.; Huang, C.; Sun, X.; Long, X.R.; Lv, X.W.; Li, J. MicroRNA-146a modulates TGF-beta1-induced hepatic stellate cell proliferation by targeting SMAD4. *Cell Signal* **2012**, *24*, 1923–1930. [CrossRef] [PubMed]
39. Liu, Z.; Lu, C.L.; Cui, L.P.; Hu, Y.L.; Yu, Q.; Jiang, Y.; Ma, T.; Jiao, D.K.; Wang, D.; Jia, C.Y. MicroRNA-146a modulates TGF-beta1-induced phenotypic differentiation in human dermal fibroblasts by targeting SMAD4. *Arch. Dermatol. Res.* **2012**, *304*, 195–202. [CrossRef] [PubMed]
40. Morishita, Y.; Imai, T.; Yoshizawa, H.; Watanabe, M.; Ishibashi, K.; Muto, S.; Nagata, D. Delivery of microRNA-146a with polyethylenimine nanoparticles inhibits renal fibrosis in vivo. *Int. J. Nanomedicine* **2015**, *10*, 3475–3488. [CrossRef] [PubMed]
41. Milosevic, J.; Pandit, K.; Magister, M.; Rabinovich, E.; Ellwanger, D.C.; Yu, G.; Vuga, L.J.; Weksler, B.; Benos, P.V.; Gibson, K.F.; et al. Profibrotic role of miR-154 in pulmonary fibrosis. *Am. J. Respir. Cell Mol. Biol.* **2012**, *47*, 879–887. [CrossRef] [PubMed]

Somatostatin Analogue Treatment Primarily Induce miRNA Expression Changes and Up-Regulates Growth Inhibitory miR-7 and miR-148a in Neuroendocrine Cells

Kristina B. V. Døssing [1,2,3,†], Christina Kjær [4,†], Jonas Vikesâ [1], Tina Binderup [2,3], Ulrich Knigge [5,6], Michael D. Culler [7], Andreas Kjær [2,3], Birgitte Federspiel [8] and Lennart Friis-Hansen [1,9,*]

1 Center for Genomic Medicine, Rigshospitalet, Blegdamsvej 9, 2100 Copenhagen, Denmark; kbdoessing@sund.ku.dk (K.B.V.D.); jonas.vikesaa@roche.com (J.V.)
2 Department of Clinical Physiology, Nuclear Medicine and PET, Rigshospitalet, Blegdamsvej 9, 2100 Copenhagen, Denmark; tina.binderup@rh.regionh.dk (T.B.); andreas.kjaer@rh.regionh.dk (A.K.)
3 Cluster for Molecular Imaging, Faculty of Health Sciences University of Copenhagen, Blegdamsvej 3B, 2100 Copenhagen, Denmark
4 University College Copenhagen, Sigurdsgade 26, 2200 Copenhagen, Denmark; chkj@phmetropol.dk
5 Department of Surgical Gastroenterology C, Rigshospitalet, University of Copenhagen, Blegdamsvej 9, 2100 Copenhagen, Denmark; ulrich.knigge@rh.regionh.dk
6 Department of Clinical Endocrinology PE, Rigshospitalet, Blegdamsvej 9, 2100 Copenhagen, Denmark
7 Biomeasure Incorporated/IPSEN, 650 W Kendall St, Cambridge, MA 02142, USA; mdculler@comcast.net
8 Department of Pathology, Rigshospitalet, University of Copenhagen, Blegdamsvej 9, 2100 Copenhagen, Denmark; birgitte.federspiel@rh.regionh.dk
9 Department of Clinical Biochemistry, Hillerød Hospital, Dyrhavevej 29, 3400 Hillerød, Demark
* Correspondence: Lennart.jan.friis-hansen@regionh.dk
† The two authors contributed equally.

Abstract: Somatostatin (SST) analogues are used to control the proliferation and symptoms of neuroendocrine tumors (NETs). MicroRNAs (miRNA) are small non-coding RNAs that modulate posttranscriptional gene expression. We wanted to characterize the miRNAs operating under the control of SST to elucidate to what extent they mediate STT actions. NCI-H727 carcinoid cell line was treated with either a chimeric SST/dopamine analogue; a SST or dopamine analogue for proliferation assays and for identifying differentially expressed miRNAs using miRNA microarray. The miRNAs induced by SST analogue treatment are investigated in carcinoid cell lines NCI-H727 and CNDT2 using in situ hybridization, qPCR and proliferation assays. SST analogues inhibited the growth of carcinoid cells more potently compared to the dopamine analogue. Principal Component Analysis (PCA) of the samples based on miRNA expression clearly separated the samples based on treatment. Two miRNAs which were highly induced by SST analogues, miR-7 and miR-148a, were shown to inhibit the proliferation of NCI-H727 and CNDT2 cells. SST analogues also produced a general up-regulation of the let-7 family members. SST analogues control and induce distinct miRNA expression patterns among which miR-7 and miR-148a both have growth inhibitory properties.

Keywords: somatostatin analogues; neuroendocrine tumors; cancer; miRNAs; miR-148a; miR-7; let-7

---

1. Introduction

Gastro-Entero-Pancreatic neuroendocrine neoplasms (GEP-NEN) are generally slow growing tumors originating from neuroendocrine cells in the gastro-intestinal tract, the diffuse neuroendocrine system, pancreas and the bronco-pulmonary system [1]. According to the World Health Organization (WHO) 2010 classification [2] GEP-NEN can be classified according to Ki67 index and/or mitotic count into neuroendocrine tumors (NETs, NET-G1 and NET-G2) and neuroendocrine carcinomas (NEC G3) [2,3]. The bronco-pulmonary NENs are classified according to mitoses pr. 2 mm^2 and the presence or absence of necrosis according to the 2015 WHO classification [4]. This separates small-cell and large-cell, poorly differentiated based on mitosis only, and the typical carcinoids (TCs) and atypical carcinoids (ACs) by <2 mitoses per 2 mm^2/no necrosis and 2–10 mitoses per 2 mm^2/with focal necrosis. In general, NETs of the lung comprise <20% of all lung cancers, and TCs and ACs comprise 1–2% of these [4,5]. The yearly incidence of GEP-NENs and bronco-pulmonary NENs is 2–3 and 0.2–2 per 1,000,000 inhabitants respectively, but this has been increasing during the last decades [1,6–8]. TCs share homologies with G1 GEP-NENs and ACs with G2 GEP-NENS coinciding with the fact that ACs are more malignant in nature and more prone to metastasize compared to TCs [8,9]. The clinical picture depends on the site of the primary tumor and its ability to secrete neuroamines and/or peptides at supra-physiological levels (functioning tumors) as the clinical syndrome depends on the transmitter/hormone secreted [10]. However, most of the GEP-NENs and bronco-pulmonary NENs are silent or 'non- functional', as they either do not secrete any transmitters/hormones or the one they secrete does not cause clinical symptoms [10,11]. The lack of endocrine symptoms in these patients often delays the diagnosis until the presence of symptoms caused by the mass effect and/or the presence of metastases, mainly hepatic metastases [12]. In patients with localized low grade GEP-NETs, the 5- year survival rate is 60–100% whereas patients with regional disease or distant metastases have 40% and 29% 5-year survival rates respectively [13]. In bronco-pulmonary NENs the overall survival is higher in TCs versus ACs with a 5-year survival rate of 90% and 60% respectively and increases with resectable tumors [14].

A common feature for endocrine cells is that both their secretion and growth can be inhibited by somatostatin (SST) [15]. In the nervous tissue, SST acts like a neurotransmitter and a neuromodulator. In the gastrointestinal tract, SST is expressed by specialized endocrine cells and acts as a paracrine factor [16]. The somatostatin receptors (SSTRs) belong to the G-protein coupled receptor family and there are five different SSTR subtypes (SSTR1-5), all differently expressed by neuroendocrine cells [17]. Seventy to ninety percent of all GEP-NET express SSTRs and tumors expressing SSTRs often contain one or more receptor subtypes, most often SSTR1, 2, 3 and 5, whereas SSTR4 is less commonly seen in endocrine tumors [18]. This has made SST useful for both visualization and treatment [19,20]. The expression is also related to the tumor grade, since the receptors are preferably expressed in low grade tumors, whereas the expression of some receptor subtypes is reduced in the more dedifferentiated tumors [21]. The different SSTR subtypes all bind the endogenous ligands; SST14, SST 28 and cortistatin; with equally high affinity in the nanomolar (nM) range. The widespread expression of SSTRs by a large number of human tumors and the biological actions of SSTRs form the basis for in vivo tumor targeting. However, the short half-life in circulation (1–3 min) of the endogenous SST peptides makes them unsuited for therapy [22]. Therefore a number of synthetic analogues with longer half-lives, such as somatostatin SMS201-995 (Octreotide), RC-160 (Vapreotide), BIM-23014 (Lanreotide), MK678 (Seglitide) and SOM-230 (Pasireotide LAR), were developed and are currently in clinical use [23,24]. Surgical removal is the optimal treatment for GEP-NENs and bronco-pulmonary NENs, and somatostatin analogue (SSA) treatment constitutes the gold standard for symptomatic and anti-proliferative control [8,25].

MicroRNAs (miRNAs) are small non-coding RNA molecules (on average 20–23 nucleotide (nt) long) that modulate gene expression by binding to complementary sequences on target messenger RNA (mRNAs). This predominantly results in a decrease of target mRNA levels [26]. An increasing number of studies indicate that miRNAs are involved in many important biological processes, including proliferation, apoptosis, differentiation, angiogenesis and immune response. miRNA deregulation leads to aberrant gene expression in various diseases and dysregulation of miRNA expression have been shown to be involved in cancer development [27]. We therefore wanted to examine which miRNA operate under the control of SST and if they contribute to the growth- inhibitory effects of SST.

2. Materials and Methods

2.1. Cell Lines and Tissue Culture

Four human carcinoid cell lines, two intestinal (CNDT2 and HC45) and two of pulmonary origin (NCI-H720 and NCI-H727) were used for experiments.

CNDT2 is a human midgut carcinoid cell line kindly provided by Lee M. Ellis M.D. Anderson Center Texas USA [28] and grown in DMEM/F12 with 15 mM HEPES (Life Technologies, Carlsbad, CA, USA) supplemented with 10% FBS (Th. Geyer GmbH, Stuttgart, Germany), penicillin 100 U/mL and streptomycin 100 µg/mL (Life Technologies), 5 mL Sodium pyruvate 100 mM (Sigma-Aldrich, St. Louis, MO, USA), 5 mL MEM NEAA 100x (Life Technologies), 5 mL L-Glutamine 200 mM 100x (Life) and 10 ng/mL NGF (Life Technologies) and kept at 37 °C/5% CO_2. HC45 is a human ileal carcinoid cell line kindly provided by Ricardo V. Lloyd Mayo Clinic [29] and kept in RPMI 1640/Glutamax (GIBCO, Waltham, MA, USA) supplemented with 10% FBS, 1% P/S and 10 ng/mL Insulin (Invitrogen, Carlsbad, CA, USA) at 37 °C and 5% CO_2.

NCI-H720 is an atypical and NCI-H727 is a typical human pulmonary carcinoid cell line obtained from ATCC (Boras, Sweden). Both cell lines were cultured in RPMI 1640 Glutamax supplemented with 10% FBS, penicillin 100 U/mL and streptomycin 100 µg/mL, 1 mM Sodium Pyruvate (Life Technologies) at 37 °C and 5% CO_2.

For experiments involving seeding cells into new plates, cells were always allowed to adhere overnight.

2.2. Tumor Tissue

Five formalin-fixed and paraffin-embedded (FFPE) tissue samples of carcinoid tumors (see Table 1), obtained from the Department of Pathology (Rigshospitalet, Copenhagen, Denmark), were used for Laser Capture Microdissection, qPCR and in situ hybridization, see descriptions for the procedures below.

Table 1. Tissue used for immunohistochemistry, in situ hybridization and qPCR.

Patient	Age/Sex	Location	Ki67 Index	Treatment
1	65/F	Colon	1%	Surgery
2	69/F	Small intestine	3%	Surgery/SSA
3	62/M	Ileum	2%	Surgery
4	55/F	Small intestine	7–8%	Surgery
5	64/F	Ileocecal	1%	Surgery/SSA

2.3. Somatostatin Analogues

Two SST analogues (BIM-23014 (Lanreotide) and BIM-23023), a combined SST-dopamine 2 receptor analogue (BIM-23A760) and a pure dopamine receptor D2 (DRD2) analogue (BIM-53097) were used. The compounds' binding affinities for the SST and dopamine receptors are shown in (Table 2). The compounds were initially dissolved in 100 µL 99% EtOH then in 0.1 N acetic acid/0.1% BSA to a final stock concentration of 10^{-3} M and used for experiments at suitable concentrations in complete growth medium.

Table 2. Human Somatostatin Receptor Subtype Specificity (IC50-nM) [30,31].

Compound	Somatostatin Receptor Subtype					Dopamine
	1	2	3	4	5	
Somatostatin 14	1.95	0.25	1.2	1.77	1.4	
Somatostatin 28	1.86	0.31	1.3	5.4	0.4	
BIM-23014 (Lanreotide)	>1000	0.75	98	>1000	12.7	
BIM-23023	>1000	0.42	87	2.7	4.2	
BIM-53097						22.1
BIM-23A760	622	0.03	160	>1000	42.0	15

2.4. *Transfection Studies and Cell Growth Analyses*

A total of 4 × 10^6 NCI-H727 or 1.5 × 10^6 CNDT2 cells were seeded and used for each transfection. To 965 µL Opti-MEM (Invitrogen) 10 µL Negative control 1 or 2 (Thermo Scientific, Waltham, MA, USA) or mature miRNA miR-7/miR-148a (Applied Biosystems, Carlsbad, CA, USA) or inhibitor—miR-7 LNA/miR- 148a LNA (Exiqon, Vedbæk, Denmark) was added to a final concentration of 50 nM together with 25 µL Turbofect transfection reagent (Fermentas, Leon-Rot, Germany) in 5 mL complete growth medium and left to incubate for 15–20 min at room temperature (RT) before being added drop-wise to the cells.

For growth analyses, 4 × 10^4 cells (NCI-H727) or 1.5 × 10^4 cells (CNDT2) were seeded in each well into E-plates for use in the xCELLigence system (Roche/ACEA, San Diego, CA, USA) for proliferation studies. The xCELLigence analyzer, which is an electronic cell sensor array technology, allows label-free and real-time monitoring of cell proliferation. The presence of the cells on top of the electrodes will affect the local ionic environment at the electrode/solution interface, leading to an increase in the electrode impedance. The more cells that are attached to the electrodes, the larger the increases in electrode impedance. For further details, see references [32,33]. The difference in cell number seeding for growth assays is due to differences in size and proliferative rate between the two cell lines. A series of analogue concentrations were used to find the optimal concentration for the actual growth experiments. The analogues were added daily directly to the wells of the E-plates in complete growth medium used for the normal passage of cells, without changing it for the duration of the entire growth experiment. The concentrations of the analogues for growth experiments were BIM-23a760 10^{-7} M, BIM-23023 10^{-7} M, BIM-23014 10^{-9} M and finally BIM-53097 10^{-9} M.

2.5. *RNA Extraction*

Total RNA was extracted using Trizol reagent (Life Technologies) according to the manufactures specifications. The RNA concentration was measured on the NanoDrop (Thermo Fisher Scientific, Wilmington, DE, USA) and the integrity determined using the Agilent 2100 Bioanalyzer (Agilent Technologies, Santa Clara, CA, USA).

2.6. *Somatostatin Receptors mRNA Quantification*

The expressions of the SSTRs were quantitated as previously described [34] and normalized to the expression of β-actin.

2.7. *Laser Capture Microdissection*

Tumor and normal cells from FFPE tissue, see tissue specifications above, were Laser Capture Microdissected (LCM) using an Arcturus LCM system (ThermoFisher Scientific, Waltham, MA, USA). 10 µm sections of tissue were mounted on Pen Membrane slides (Applied Biosystems, Foster City, CA, USA). Sections were stained with cresyl violet, using the LCM staining kit (Ambion, Foster City, CA, USA/Life Technologies, Carlsbad, CA, USA) with cresyl violet according to the manufacturer's instructions. After tissue sections had been collected and transferred to the collection cap, the cap was immediately transferred and clicked on to an Eppendorf tube containing the 100 μL of the Lysis Solution used in the first step of the RNAqueous®-Micro Kit (ThermoFisher Scientific). The tube was inverted to ensure complete coverage of the dissected cells in the buffer. RNA isolation was performed according to the manufacturer's instructions.

2.8. *qPCR of microRNA Expression*

The expression of miR-7 and miR-148a was quantitated using TaqMan miRNA assay (Applied Biosystems). Briefly, 100 ng of the total RNA was reversely transcribed into cDNA using gene specific primers and the TaqMan MicroRNA Reverse Transcription Kit (Applied Biosystems) according to the manufacturer's protocol. Samples were run on the ABI PRISM 7900 HT Sequence Detection System (Applied Biosystems). For normalization of the miRNA expression data the geometric mean of hsa-miR-191 and RNU-44 were used [35]. Primer sequences are listed in Table 3.

Table 3. Primer sequences.

MiRNA	Mature Sequence	Product ID
miR-7	UGGAAGACUAGUGAUUUUGUUGU	000268
miR-148a	UCAGUGCACUACAGAACUUUGU	000470
MiR-191	CAACGGAAUCCCAAAAGCAGCUG	002299
RNU-44	CCTGGATGATGATAGCAAATGCTG-ACTGAACATGAAGGTCTT AATTAGCTCTAACTGACT	001094

2.9. *In Situ Hybridization*

Formalin-fixed and paraffin-embedded (FFPE) tissue samples of carcinoid tumors were obtained from the Department of Pathology (Rigshospitalet, Copenhagen, Denmark). A double-DIG-labeled LNA-modified oligos miR-7 (Exiqon, Munich, Germany), probe sequence 5′-ACAACAAAATCACTAGTCTTCCA-3′, RNA-T*m* 80 °C and double-DIG-labeled LNA-modified oligos miR-148a (Exiqon), probe sequence 5′-ACAAAGTTCTGTAGTGCACTGA-3′, RNA-Tm 80 °C were used for detection as described [36]. Probe concentration was 100 nM and slides were hybridized at 50 °C. Sections were counterstained with Nuclear Fast Red. Pictures of representative areas of the slides were taken with a Zeiss Axio Imager (Zeiss, Jena, Germany), original magnification × 20/10. Cells with intense blue nuclear stain were scored as positive. The level of expression within a positive cell was not scored. A LNA probe against snRNA U6 (Exiqon) was used as positive control and a scramble probe (Exiqon) as negative control.

2.10. *MicroRNA Microarray*

For microarray analysis 1 μg of total RNA was labeled using the Flashtag RNA labeling kit for Affymetrix (Genisphere LLC., Hatfield, PA, USA) according to the manufacturer's instructions. The labeled samples were hybridized to GeneChip miRNA Array (Affymetrix, Santa Clara, CA, USA). The Affymetrix miRNA array assay miRNAs includes small nucleolar RNAs (snoRNAs) and small Cajal Body specific RNAs (scaRNAs) in human. The 847 human miRNAs on the array are derived from Sanger miRbase miRNA database V11. Four copies of each miRNA probe are distributed on the array.

2.11. *RNA Profiling*

Arrays were washed and stained with phycoerytrin conjugated streptavidin (SAPE) using the Affymetrix Fluidics Station® 450, and the arrays were scanned in the Affymetrix GeneArray® 3000 scanner to generate fluorescent images, as described in the Affymetrix Gene Chip® protocol. To minimize batch variation, equal numbers of treatment groups were included in each batch.

2.12. Data Analysis

Raw data files were imported into Affymetrix's miRNA QC Tool (Affymetrix) and normalized using the quantiles normalization and median Polish summarization following a background correction that corrects for the GC content of the each particular probe. Log2 intensities of the 847 human miRNAs were imported into the Data Analysis software package Qlucore Omnics Explorer v2.1. Principal Component Analysis (PCA) visualization of the clustering of samples using the genes selected in the class comparison was performed using the build-in PCA tool in Qlucore Omnics Explorer v2.1 (Qlucore AB, Lund, Sweden).

Class comparison analysis was performed using Students *t*-test. A multiple comparison test was used for all three tissue-types and a two-group comparison was used for comparing AP versus NF. The miRNA was defined as being differentially expressed between the compared groups if the p value was less than 0.05 and the fold change above 1.5.

2.13. Statistical Analyses

Students' unpaired *t*-test was used and differences with a $p \leq 0.05$ were considered significant and indicated by *. Unless otherwise stated results are given as median $\pm$ standard deviation (SD).

3. Results

3.1. Somatostatin and Dopamine Analogues Inhibit the Growth of a Carcinoid Cell Line NCI-H727

We first examined the expression of the SSTRs in four different carcinoid cell lines, HC45 and CNDT2 (both intestinal) and the atypical NCI-H720 and typical NCI-H727 pulmonary cell lines, in order to choose an optimal cell line as a model system for examining the effect of SSAs on miRNA expression in NETs. All the carcinoid cell lines expressed SSTR subtypes 2 and 5 Figure 1A, and for further analyses we selected two of the cell lines with highest SSTR2 mRNA level and each representing common NET origins in lung (NCI-H727) and intestinal (CNDT2). Furthermore, the HC45 proved very difficult to grow even after having been immortalized by retroviral transfection with a constitutive active human Telomerase Reverse Transcriptase TERT expression vector and we discontinued using this cell line. We have also previously shown that the CNDT2 and NCI-H727 cell lines are good model systems when examining NETs [37] and proceed with these two cell lines as our model.

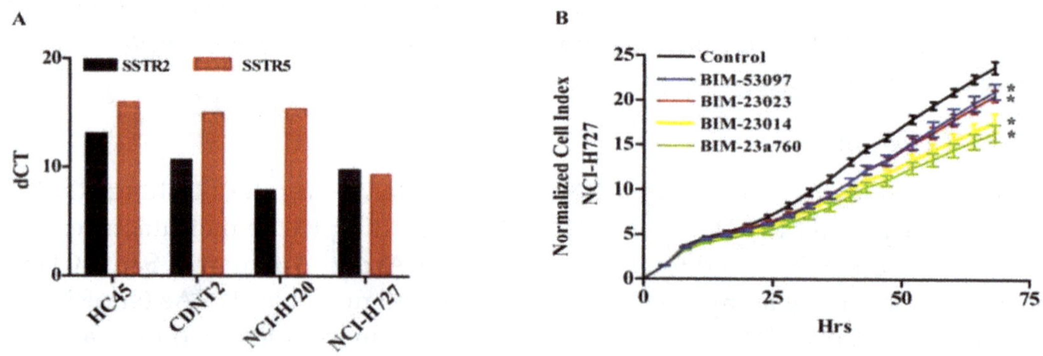

Figure 1. Carcinoid cell lines express all somatostatin receptors (SSRTs), and activation of both SSTRs and DRD2 gives the most potent growth inhibition of a carcinoid cell line (**A**) All the four examined carcinoid cell lines express SSTR subtype 2 and 5 when analyzed by qPCR (**B**) The growth of NCI-H727 is inhibited by somatostatin, dopamine and the chimeric somatostatin-dopamine agonists. The dopamine analogue BIM-53097 (blue) and the somatostatin analogues (SSAs) BIM—23023 (red) were the weaker inhibitors of cell growth. The somatostatin agonist BIM-23014 (yellow) and the chimeric somatostatin-dopamine compound BIM-23A760 (green) were the stronger inhibitors of carcinoid cell growth. The control (black) is vehicle without agonist.

Having shown that NCI-H727 cells expressed SSTR2- and five subtypes we examined the effect of the SST and dopamine analogues on their growth. We found that all compounds inhibited the growth of the carcinoid cell line (Figure 1B). The most potent inhibitor of proliferation was the combined SSTR and dopamine receptor DR agonist BIM-23A760 (green) followed by the SSTR agonist BIM-23014 (yellow) and BIM-23023 (red). The least effective of the SSAs was the DR agonist BIM-53097 (blue) Figure 1B.

3.2. Somatostatin Receptor Activation Primarily Induces microRNA Expression Changes

After having seen that SST analogues inhibit growth we treated NCI-H727 cells with the most potent inhibitor—BIM-23A760—and compared the mRNA and miRNA profiles of treated cells with that of untreated cells (Figure 2). PCA plots based on either mRNAs or miRNAs of cells treated with SST analogue showed that PCA based miRNA expression profiles clearly separated controls from treated cells. In contrast, PCA based mRNA profiles did not (Figure 2). Furthermore, when calculating the number of transcripts changed and regulated by SST 24 h after treatment it is clear that treating with SST regulates miRNAs to a higher degree than mRNAs (Figure 2).

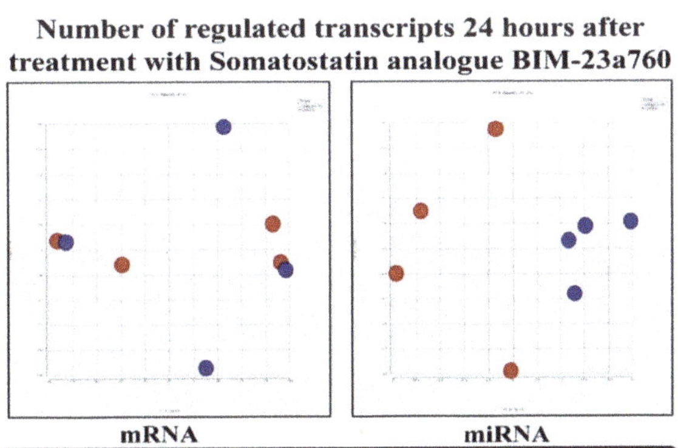

mRNA		miRNA	
15 (0.07%)	p<0.01	12 (1.4%)	p<0.01
133 (0.59%)	p<0.001	13 (1.5%)	p<0.001
Total	22371	Total	847

Figure 2. The dual somatostatin-dopamine agonist BIM-23a760 induces greater changes in microRNA (miRNA) than in messenger RNA (mRNA). Carcinoid NCI-H727 cells were treated with the dual somatostatin-dopamine agonist BIM-23a760 for 24 h and the changes in mRNA and miRNA were examined using microarray. Principal component analysis (PCA) of the changes in mRNA and miRNA demonstrated that mRNA based PCA did not clearly separate the BIM-23a760 treated cells (shown in red) from the control cells treated with vehicle without BIM-23a76 (shown in blue). In contrast a PCA based on the changes in miRNA expression clearly separated the treated cells (shown in red) from the untreated cells (shown in blue). This suggests that the miRNA changes induced by BIM-23a760 are more specific than the changes in mRNA expression. This was also supported by the fact, that a higher proportion of miRNA transcripts than of mRNA are changed by BIM-23a760 treatment.

3.3. Somatostatin Induces Distinct Receptor Based/Activated microRNA Expression Profiles and Particularly Up-Regulates miR-7 and miR-148a

Having demonstrated that SST and dopamine analogues inhibited the growth of the carcinoid cell line NCI-H727 and that activation of these receptors primarily affected miRNA expression, we hypothesized that at least some of their growth inhibitory effects depended on receptor activation and miRNA expression regulation. Also, after identifying the most potent inhibitor of growth among the different SST analogues we expanded the study and examined all the different SST analogues' effect on miRNA expression by miRNA analysis of NCI-H727 cells treated with BIM-23A760, BIM-23014, BIM-23023, BIM-53097 and with a control. A PCA of the miRNA expression profiles showed that, depending on which analogue the cells were treated with, they could be separated into groups according to the analogue used and receptor(s) activated. In the first dimension the samples treated with BIM-23a760 (SSTR/DRD2) and BIM-23014 (SSTR-only) analogues separated from the rest and in the third dimension there was a difference in miRNA expression depending on receptor type activation (Figure 3A).

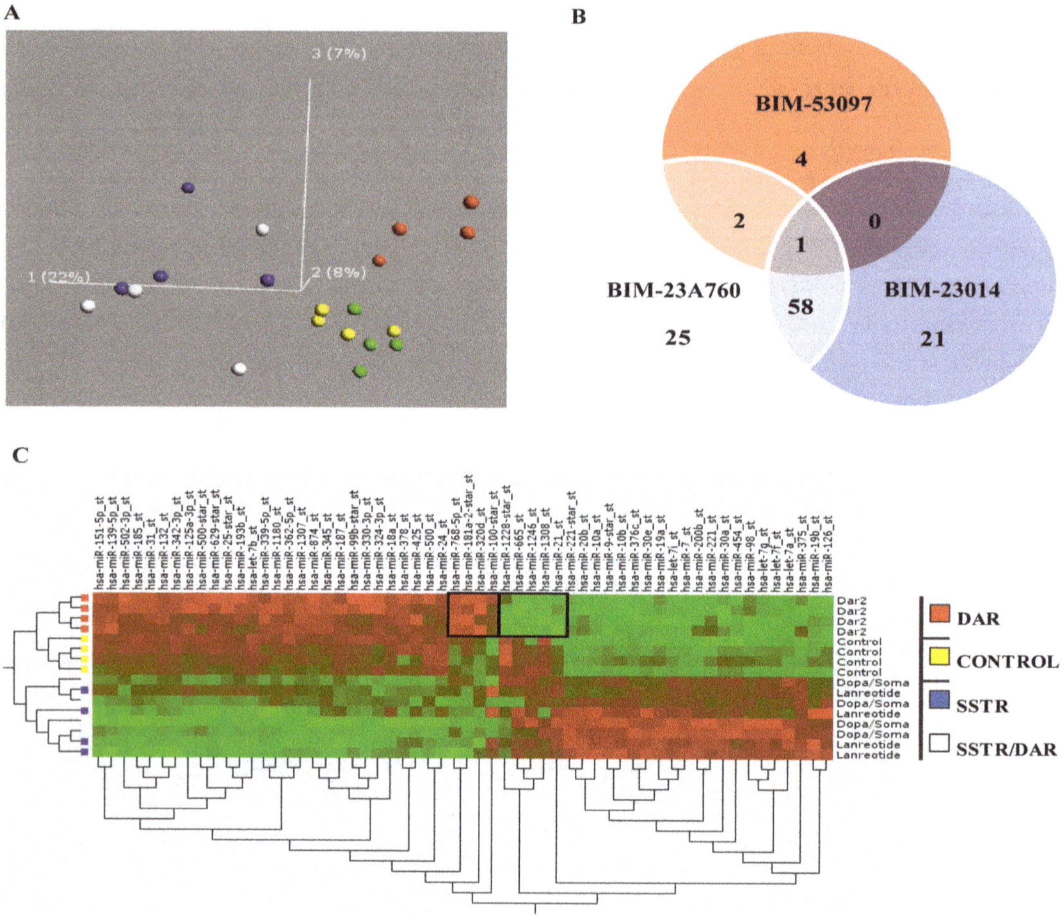

Figure 3. PCA shows specific miRNA expression depending on receptor activation. (**A**) The unbiased PCA separates the samples based on variation in the expression of miRNA in each sample. In the first dimension samples treated with BIM-23014 (blue—SSTR agonist) and BIM- 23A760 (white—SSTR and dopamine receptor D2 (DRD2) chimeric agonist) separated from the others. BIM-23023 (green—SSTR) did not separate from the control (yellow). Also in the first dimension BIM-53097 (red—DRD2) did not separate from the controls. However, in the third dimension the BIM-53097 (red—DRD2) treated samples clearly separated from the controls. The percentage indicates how much of the total variation is present in each dimension. In this experiment 22% of the total variation of miRNA expression is found in the first dimension, 8% in the second and 7% in the third. The two most potent inhibitors of growth BIM-23A760 and BIM-23014 group together, the DRD2 agonist and the least effective growth inhibitor separates by itself. (**B**) The Venn diagram shows the miRNAs shared between the different analogue treatments and that the number of miRNAs shared between BIM-23A760 (SSTR-DRD2) and BIM-23014 (SSTR) are greater than BIM-53097 (DRD2). (**C**) On the right side of the map is the clustering of treatment where DRD2 and controls separate from chimeric compound BIM-23A760 (SSTR-DRD2) and BIM-23014 (SSTR). Furthermore it seems to separate controls and DRD2. Notice in the middle of the control/DRD2 treatment, there is a small fraction of miRNA which seems to be controlled by dopamine alone and different from the control group.

A Venn diagram shows the overlap between miRNA expression shared between the analogues and analogue treatment. The results show that a higher number of miRNAs are affected and changed after treatments with BIM-23a760 and BIM-23014, which target SSTR either alone or in combination with DRD2, compared to BIM-52097 which only targets DRD2. The highest degree of miRNA expressional changes was induced by SSTR activation compared to selective DRD2 activation (Figure 3B). However, a more specific miRNA expression change was observed in the small fraction of miRNAs changed only in response to BIM-53097 and DRD2 receptor activation. Since BIM-23023 (SSTR) did not separate from the control group in the PCA plot, it was excluded (Figure 3B). Based on the miRNA array results we created a heat map to better visualize the changes in miRNA expression between the treatments with the different compounds, again data with BIM-23023 (SSTR) being excluded (Figure 3C). We also created a list of the most significantly down- or up regulated miRNAs from the two compounds which showed the biggest inhibitory effect on cell proliferation BIM-23A760 (SSTR/DRD2), BIM-23014 (SSTR) and the least effective growth inhibitor BIM-53097 (DRD2) (Table 4). From this list we chose to focus on miR-7 and miR-148a.

Table 4. Up- and down-regulated miRNAs based on Somatostatin analogue (SSA) treatment.

miRNA	Change	BIM-23014		BIM-23A760		BIM-53097	
		FC	*p*-Value	FC	*p*-Value	FC	*p*-Value
miR-769-3p		−2.8	0.001	−2.6	0.002	−1.1	NS
miR-663b		−2.0	0.009	−2.5	0.001	−2.9	0.00
miR-663	↓	−2.0	0.007	−2.3	0.002	−3.3	0.00
miR-30b-star	↓	−2.0	0.001	−2.2	0.000	−1.3	NS
miR-297	↓	−1.7	0.008	−2.0	0.001	−1.7	0.01
miR-483-5p	↓	−1.9	0.003	−2.0	0.002	−1.0	NS
miR-376c		2.1	0.000	2.0	0.000	−1.2	NS
Let-7f		1.8	0.001	2.1	0.000	−1.2	NS
miR-10a-star	↑	1.9	0.001	2.1	0.000	1.1	NS
miR-495	↑	2.0	0.001	2.2	0.000	1.6	0.01
miR-7	↑	2.3	0.000	2.2	0.000	−1.1	NS
miR-9-star	↑	2.3	0.000	2.3	0.000	−1.4	NS
miR-454		1.9	0.001	2.3	0.000	−1.0	NS
miR-26b		2.2	0.010	2.6	0.003	1.0	NS
miR-429		2.9	0.004	2.8	0.004	−1.4	NS
miR-9		1.5	NS	2.9	0.002	−1.5	NS
miR-148a		2.5	0.006	2.9	0.002	2.0	0.03
miR-30e-star		2.8	0.000	3.1	0.000	1.5	NS

The up and down regulated miRNAs in NIH-H727 cells after 24 h treatment with somatostatin analogue BIM-23014, dopamine analogue BIM-53097 and the chimeric somatostatin-dopamine analogue BIM-23a760. Cut-off $p \leq 0.001$, N = 4. NS—*p*-value not statistically significant; Fold Change—FC.

3.4. In Situ Hybridization and qPCR on Neuroendocrine Tumors and Neuroendocrine Tumor Laser Capture Microdissected Tissue Confirms the Presence of miR-7 and miR-148a

To characterize the localization and to visualize the expression of miR-7 and miR-148a in NETs we performed in situ hybridization on five NETs and found miR-7 expressed exclusively by the endocrine cells, both tumor and normal cells; miR-148a was also expressed primarily by the endocrine cells also both tumor and normal cells but to a lesser extent (Figure 4A). Expression analysis of LCM NET tissue confirmed the up-regulation of particularly miR-7, but also shows the presence of miR-148a (Figure 4B).

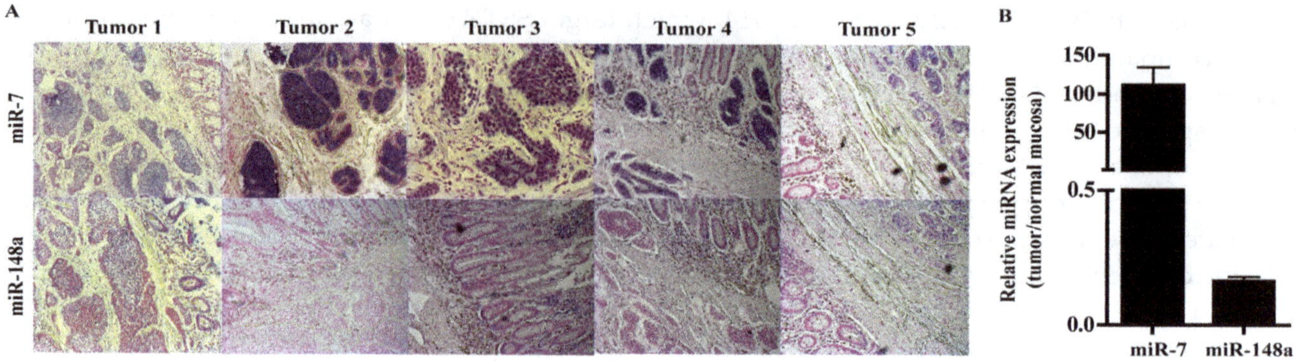

Figure 4. miR-7 and miR-148a are primarily expressed in neuroendocrine cells and inhibit the growth of a carcinoid cell line. (**A**) In situ hybridization of miR-7 show robust expression of miR-7 specifically located to the neuroendocrine cells. The expression of miR-148a is also predominantly seen in the neuroendocrine cells, although to a lesser extent than miR-7. (**B**) qPCR of laser capture micro dissected cells confirmed that miR-7 is robustly expressed in NETs compared to normal mucosa. In contrast, the expression of miR-148a is lower in NETs than in the normal mucosa that contains endocrine cells.

3.5. miR-7 and miR-148a Modulate the Growth of NCI-H727 and CNDT2 Carcinoid Cell Lines

Having demonstrated that SST analogues induced the expression of miR-7 and miR-148a and that both these miRNAs are expressed in NETs, we examined how these miRNAs would affect the growth of both the NCI-H727 and CNDT2 carcinoid cell lines by either over expressing or inhibiting them. Both miR-7 and miR-148a significantly inhibit the growth of carcinoid cells in vitro indicating that these miRNA could mediate some of the growth inhibitory effect induced by SST analogues (Figure 5A,B). We subsequently blocked the effect of miR-7 and miR-148a by transfecting LNA-inhibitors that increased the growth of both carcinoid cell lines (Figure 5C,D).

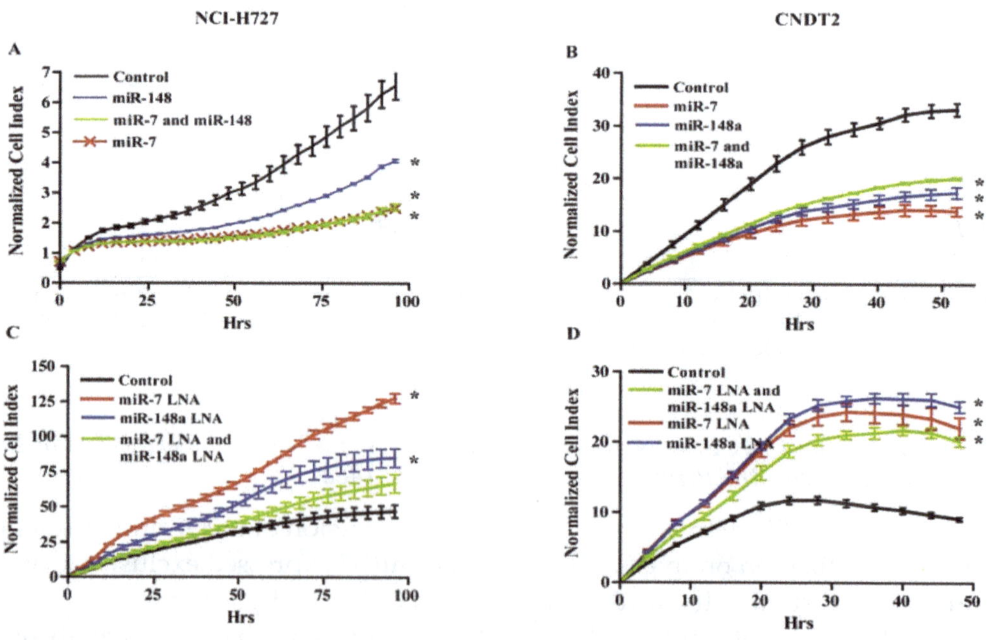

Figure 5. miR-7 and miR-148a regulates growth of the carcinoid cell lines NCI-H727 and CNDT2. (**A**) and (**B**) Both miR-7 and miR-148a reduces cellular growth of the carcinoid cell lines NCI-H727 and CNDT2 where the highest inhibitory effect is caused by transfecting with miR-7. (**C**) and (**D**) Inhibiting miR-7 and miR-148a by transfecting the carcinoid cell lines NCI-H727 and CNDT2 with miRNA inhibitors alleviate growth repression and cause the cells to grow more than the controls. * growth rates statistically significant compared to control.

3.6. Somatostatin Modulates the Expression of the Let-7 Family

We have previously shown that the expression of several Let-7 family members is reduced during NET carcinogenesis [37] and we therefore specifically examined how the SSTR and DRD2 analogues affected the expression of the let-7 miRNA family. We found that the expression of 4 of the Let-7 members increased after treatment with SSTR analogues, the expression of one was slightly reduced and the expression of 4 was unaffected (Table 5).

Table 5. Differentially regulated let-7 family members based on SSA treatment.

miRNA	Change	BIM-23014		BIM-23A760		BIM-53097	
		FC	*p*-Value	FC	*p*-Value	FC	*p*-Value
let-7a	→	1.2	0.000	1.2	0.000	−1.0	NS
let-7b	↓	−1.4	0.000	−1.4	0.001	−1.0	NS
let-7c	→	−1.0	NS	1.0	NS	−1.1	NS
let-7d	→	−1.0	NS	1.0	NS	1.1	NS
let-7e	→	−1.2	NS	−1.1	NS	1.1	NS
let-7f	↑	1.8	0.002	2.1	0.000	−1.2	NS
let-7g	↑	1.7	0.001	1.9	0.000	−1.1	NS
let-7i	↑	1.3	0.000	1.4	0.000	−1.1	NS
miR-98	↑	1.8	0.004	2.2	0.000	−1.1	NS

The alteration in the expression of the let-7 family of miRNAs in NIH-H727 cells after 24 h treatment with somatostatin analogue BIM-23014, dopamine analogue BIM-53097 and the chimeric somatostatin-dopamine analogue BIM-23a760. N = 4, NS—*p*-value not significant.

Thus, SST analogues had the ability to reintroduce the expression of the let-7 family and possibly revert some of the pathways otherwise involved in NET carcinogenesis and malignancy.

4. Discussion

We have shown that in our NET model system SST analogue treatment primarily induces changes in miRNA expression profiles in carcinoid cell lines. SST analogues are widely used to treat patients with NETs, as the SST analogues both alleviate the carcinoid syndrome and inhibit growth of the tumors [19,38–40]. Here we show that SST analogues inhibit the growth of two carcinoid cell lines and that the inhibitory potential depends on the analogue used. This is in good concordance with other studies that also show that SST analogues inhibit the growth of the NCI-H727 carcinoid cell line [30,41] and the CNDT2 carcinoid cell line [42] and that a chimeric compound targeting SSTRs and DRD2s has the most potent growth inhibitory effect [30,41,42]. The growth inhibitory effects of SST analogues are mediated directly by binding of SST to the SSTRs, which leads to cell cycle arrest or apoptosis and indirectly by the inhibition of growth factors and the suppression of oncogenic signal transducing pathways. However, the molecular mechanisms linking neuroendocrine proliferation and tumor progression are not yet fully understood. The PI3K/MAPK/mTOR pathway is known to play an important role in NETs [43,44]. Already existing therapies used to treat patients with NETs consist of agents targeting this specific signaling pathway [45–49]. SSTR2 has been shown to bind directly to the p85 subunit of PI3K, which belongs to the class I_A PI3Ks and is the class specifically involved in promoting cell survival, growth and proliferation and the most important subclass involved in human cancers [50,51]. SST analogue treatment of a pancreatic cell line can inhibit the binding between SSTR2 and p85 which is critical for the down-regulation of PI3K activity and resulting decreased cell survival [50].

We found miR-7 and miR-148a among the most up-regulated miRNAs after treatment with SST analogues and increased expression of the let-7 family members, all of which have growth inhibitory effects. miR-7 has been shown to be down-regulated in several cancers including colorectal cancer [52] and gastric cancer [53] and to be endocrine specific [54], with evidence going both ways to whether or not miR-7 acts as a tumor suppressor [55]. Here we demonstrated miR-7 to be highly present in NETs.

Both miR-7 and SST targets the PI3K/MAPK/mTOR pathway underlining the significance for the up-regulation of this specific miRNA by SST. In a study in lung cancers, including the carcinoid cell line NCI-H727 showed miR-7 to directly target PI3KR3, the regulatory subunit of PI3K, and to reduce the metastatic potential by reducing the effect of TLR9 signaling [56]. In hepatocellular carcinoma miR-7 has been shown to directly target and repress PI3KCD an integral component of the PI3K signaling pathway, thereby both inhibiting cellular growth, invasion and migration in vitro and more importantly tumorigenesis and metastasis in vivo [57].

Another important aspect of the induction of miR-7 expression in cancer lies in the fact that miR-7, through its inhibitory actions on central cancerous signaling pathways, can increase the sensitivity and improve otherwise chemo-or radiotherapy resistant tumor cells [55]. SST mediates up- regulation of miR-7 thereby increasing the cells sensitivity to the growth inhibitory actions of SST itself resulting in a positive feed-forward loop.

miR-148a has been shown to be down-regulated in early gastric- [58], pancreatic- [59] and colon cancer [60], as well as in breast cancer [61], and, when down-regulated, suggested as a biomarker for the diagnosis and prognosis of gastrointestinal cancers [58,62,63]. Shivapurkar et al. showed that a panel of six miRNAs including miR-148a could predict the risk of recurrence of colon cancer [60]; thus the up-regulation of miR-148a by SST analogue treatment might also be used as a biomarker in NETs. We found miR-148a localized to endocrine cells and tumor tissue, all of which points towards it being a specific and important miRNA in endocrine tumor formation and initiation. Our finding that miR-148a has a growth inhibitory effect on carcinoid cell lines makes it an important miRNA to up-regulate for SST analogues as it will give the SST analogues the ability to "pack that extra punch" in stopping cell growth, and which is possibly why we see that SST analogue treatment primarily gives a change in miRNA expression rather than mRNA expression.

SST itself attenuates Insulin Growth Factor 1 (IGF-1) signaling [64] and IGF-1 Receptor (IGF-1R) is important in GEP-NET tumor growth factor biology [51]. Since miR-148a targets IGF-1, SST analogue induced miR-148a expression could again lead to the interference with the PI3K signaling pathway by the inhibition of growth factors like IGF-1 and its receptor IGF-1R, which frequently are overexpressed in NETs [43,51]. Inhibition of IGF-1R signaling has been shown to decrease PI3K signaling and the induction of cell cycle arrest and apoptosis [51]. miR-148a directly targets and down-regulates IGF-1R in breast cancer and over-expression of miR-148a decreased phosphorylated Akt, a component of the PI3K signaling pathway [61]. miR-7 also targets IGF-1R in a gastric cancer model with significant influence on inhibiting the metastatic potential in this model [53]. However, while SSTR agonists have been shown to have anti-tumor growth activity [30,41,42], IGF-1R inhibition proved to be ineffective in treatment of NETs [65]. One explanation could be that IGF-1R inhibition only targets the receptor activity but not directly the downstream signaling events. In contrast, SSTR activation and induction of the miRNAs could potentially inhibit IGF-1R signaling at multiple levels and hence be more effective. We have previously shown the let-7 family to be among the most down-regulated miRNA in NETs. We also identified and that it targets HMGA2, BACH1 and MMP1 and reduce the expression of these oncogenes all present in NETs [37]. Here we find that treatment with SST analogues modulates the expression of the let-7 family and leads to the up-regulation of several family members and the down-regulation of only one.

Recently, several reports have established the let-7 miRNAs as key players in metabolic pathways, especially in the glucose metabolic pathway through the inhibition of IGF-1R, which is a key target in the PI3K/mTOR signaling pathway [66]. Others have shown the let-7 family to directly target and inhibit IGF-1 and IGF-1R, which harbors three let-7 binding sites in its 3'UTR. This leads to the elimination of PI3K activation, and hence abrogates the signaling pathway leading towards cell division, differentiation and survival. placing let-7 up-stream of IGF-1/IGF-1R with downstream effects on the PI3K signaling cascade [67,68]. The positive modulation of the let-7 family by SST analogues targets and inhibits the IGF-1/PI3K signaling pathway in addition to maybe inhibiting glucose nutrient in aiding the rapid growth of the cancer cells. By this mechanism, they are targeted on both survival signaling as well as their supply of nutrients. In conclusion miR-7, miR-148a and the let-7 family most up-regulated by SST have been shown to play a role in inhibiting the PI3K signaling pathway at different levels reducing the cancer cell's ability to escape and circumvent inhibition of a single step. The fact that the miRNAs up-regulated by the SST analogues, the analogues themselves and the already existing therapies all seem to target the same signaling pathway underlines the importance of targeting this pathway when treating NETs.

Author Contributions: K.B.V.D. and L.F.-H. conceived and designed the experiments; K.B.V.D.; C.K. and T.B. performed the experiments; K.B.V.D.; L.F.-H.; C.K., J.V. and T.B. interpreted and analyzed the data; all authors contributed reagents and materials; K.B.V.D. and L.F.-H. wrote the paper; obtained funding (K.B.V.D., B.F., A.K., L.F.-H.); study supervision (L.F.-H.). All authors critically revised the manuscript for important intellectual content and approved the paper.

Acknowledgments: Mette Moldaschl is thanked for excellent technical assistance. We also thank Lee M. Ellis, Department of Surgical Oncology, The University of Texas M. D. Anderson Cancer Center, Houston, TX 77230 for the CDNT2 cells.

References

1. Modlin, I.M.; Oberg, K.; Chung, D.C.; Jensen, R.T.; de Herder, W.W.; Thakker, R.V.; Caplin, M.; Delle Fave, G.; Kaltsas, G.A.; Krenning, E.P.; et al. Gastroenteropancreatic neuroendocrine tumours. *Lancet Oncol.* **2008**, *9*, 61–72. [CrossRef]
2. Kloppel, G. Classification and pathology of gastroenteropancreatic neuroendocrine neoplasms. *Endocr. Relat. Cancer* **2011**, *18* (Suppl. 1), S1–S16. [CrossRef] [PubMed]
3. Janson, E.T.; Sorbye, H.; Welin, S.; Federspiel, B.; Gronbaek, H.; Hellman, P.; Ladekarl, M.; Langer, S.W.; Mortensen, J.; Schalin-Jantti, C.; et al. Nordic guidelines 2014 for diagnosis and treatment of gastroenteropancreatic neuroendocrine neoplasms. *Acta Oncol.* **2014**, *53*, 1284–1297. [CrossRef] [PubMed]
4. Travis, W.D. The 2015 World Health Organization classification of lung tumors. *J. Thorac. Oncol.* **2015**, *10*, 1243–1260. [CrossRef] [PubMed]
5. Ramirez, R.A.; Beyer, D.T.; Diebold, A.E.; Voros, B.A.; Chester, M.M.; Wang, Y.Z.; Boudreaux, J.P.; Woltering, E.A.; Uhlhorn, A.P.; Ryan, P.; et al. Prognostic factors in typical and atypical pulmonary carcinoids. *Ochsner. J.* **2017**, *17*, 335–340. [PubMed]
6. Lepage, C.; Rachet, B.; Coleman, M.P. Survival from malignant digestive endocrine tumors in England and Wales: A population-based study. *Gastroenterology* **2007**, *132*, 899–904. [CrossRef] [PubMed]
7. Niederle, M.B.; Hackl, M.; Kaserer, K.; Niederle, B. Gastroenteropancreatic neuroendocrine tumours: The current incidence and staging based on the Who and European neuroendocrine tumour society classification: An analysis based on prospectively collected parameters. *Endocr. Relat. Cancer* **2010**, *17*, 909–918. [CrossRef] [PubMed]
8. Caplin, M.E.; Baudin, E.; Ferolla, P.; Filosso, P.; Garcia-Yuste, M.; Lim, E.; Oberg, K.; Pelosi, G.; Perren, A.; Rossi, R.E.; et al. Pulmonary neuroendocrine (carcinoid) tumors: European neuroendocrine tumor society expert consensus and recommendations for best practice for typical and atypical pulmonary carcinoids. *Ann. Oncol.* **2015**, *26*, 1604–1620. [CrossRef] [PubMed]
9. Hashmi, H.; Vanberkel, V.; Bade, B.C.; Kloecker, G. Clinical presentation, diagnosis, and management of typical and atypical bronchopulmonary carcinoid. *J. Community Support* **2017**, *15*, E303–E308. [CrossRef]
10. Eriksson, B.; Arnberg, H.; Lindgren, P.G.; Lorelius, L.E.; Magnusson, A.; Lundqvist, G.; Skogseid, B.; Wide, L.; Wilander, E.; Oberg, K. Neuroendocrine pancreatic tumours: Clinical presentation, biochemical and histopathological findings in 84 patients. *J. Intern. Med.* **1990**, *228*, 103–113. [CrossRef] [PubMed]

11. Hendifar, A.E.; Marchevsky, A.M.; Tuli, R. Neuroendocrine tumors of the lung: Current challenges and advances in the diagnosis and management of well-differentiated disease. *J. Thorac. Oncol.* **2017**, *12*, 425–436. [CrossRef] [PubMed]
12. Ramage, J.K.; Davies, A.H.; Ardill, J.; Bax, N.; Caplin, M.; Grossman, A.; Hawkins, R.; McNicol, A.M.; Reed, N.; Sutton, R.; et al. Guidelines for the management of gastroenteropancreatic neuroendocrine (including carcinoid) tumours. *Gut* **2005**, *54* (Suppl. 4), iv1–iv16. [CrossRef] [PubMed]
13. Plockinger, U.; Wiedenmann, B. Treatment of gastroenteropancreatic neuroendocrine tumors. *Virchows Arch.* **2007**, *451* (Suppl. 1), S71–S80. [CrossRef] [PubMed]
14. Cameselle-Teijeiro, J.M.; Mato Mato, J.A.; Fernandez Calvo, O.; Garcia Mata, J. Neuroendocrine pulmonary tumors of low, intermediate and high grade: Anatomopathological diagnosis-prognostic and predictive factors. *Mol. Diagn. Ther.* **2018**, *22*, 169–177. [CrossRef] [PubMed]
15. Reichlin, S. Secretion of somatostatin and its physiologic function. *J. Lab. Clin. Med.* **1987**, *109*, 320–326. [PubMed]
16. Larsson, L.I.; Goltermann, N.; de Magistris, L.; Rehfeld, J.F.; Schwartz, T.W. Somatostatin cell processes as pathways for paracrine secretion. *Science* **1979**, *205*, 1393–1395. [CrossRef] [PubMed]
17. Yamada, Y.; Post, S.R.; Wang, K.; Tager, H.S.; Bell, G.I.; Seino, S. Cloning and functional characterization of a family of human and mouse somatostatin receptors expressed in brain, gastrointestinal tract, and kidney. *Proc. Natl. Acad. Sci. USA* **1992**, *89*, 251–255. [CrossRef] [PubMed]
18. Papotti, M.; Bongiovanni, M.; Volante, M.; Allia, E.; Landolfi, S.; Helboe, L.; Schindler, M.; Cole, S.L.; Bussolati, G. Expression of somatostatin receptor types 1–5 in 81 cases of gastrointestinal and pancreatic endocrine tumors. A correlative immunohistochemical and reverse-transcriptase polymerase chain reaction analysis. *Virchows Arch.* **2002**, *440*, 461–475. [CrossRef] [PubMed]
19. Modlin, I.M.; Pavel, M.; Kidd, M.; Gustafsson, B.I. Review article: Somatostatin analogues in the treatment of gastroenteropancreatic neuroendocrine (carcinoid) tumours. *Aliment. Pharmacol. Ther.* **2010**, *31*, 169–188. [PubMed]
20. Binderup, T.; Knigge, U.; Loft, A.; Mortensen, J.; Pfeifer, A.; Federspiel, B.; Hansen, C.P.; Hojgaard, L.; Kjaer, A. Functional imaging of neuroendocrine tumors: A head-to-head comparison of somatostatin receptor scintigraphy, 123i-MIBG scintigraphy, and 18F-FDG PET. *J. Nucl. Med.* **2010**, *51*, 704–712. [CrossRef] [PubMed]
21. Buscail, L.; Saint-Laurent, N.; Chastre, E.; Vaillant, J.C.; Gespach, C.; Capella, G.; Kalthoff, H.; Lluis, F.; Vaysse, N.; Susini, C. Loss of sst2 somatostatin receptor gene expression in human pancreatic and colorectal cancer. *Cancer Res.* **1996**, *56*, 1823–1827. [PubMed]
22. Sheppard, M.; Shapiro, B.; Pimstone, B.; Kronheim, S.; Berelowitz, M.; Gregory, M. Metabolic clearance and plasma half-disappearance time of exogenous somatostatin in man. *J. Clin. Endocrinol. Metab.* **1979**, *48*, 50–53. [CrossRef] [PubMed]
23. Appetecchia, M.; Baldelli, R. Somatostatin analogues in the treatment of gastroenteropancreatic neuroendocrine tumours, current aspects and new perspectives. *J. Exp. Clin. Cancer Res.* **2010**, *29*, 19. [CrossRef] [PubMed]
24. Cives, M.; Kunz, P.L.; Morse, B.; Coppola, D.; Schell, M.J.; Campos, T.; Nguyen, P.T.; Nandoskar, P.; Khandelwal, V.; Strosberg, J.R. Phase II clinical trial of pasireotide long-acting repeatable in patients with metastatic neuroendocrine tumors. *Endocr. Relat. Cancer* **2015**, *22*, 1–9. [CrossRef] [PubMed]
25. Alexandraki, K.I.; Karapanagioti, A.; Karoumpalis, I.; Boutzios, G.; Kaltsas, G.A. Advances and current concepts in the medical management of gastroenteropancreatic neuroendocrine neoplasms. *Biomed Res. Int.* **2017**, *2017*, 9856140. [CrossRef] [PubMed]
26. Guo, H.; Ingolia, N.T.; Weissman, J.S.; Bartel, D.P. Mammalian microRNAs predominantly act to decrease target mRNA levels. *Nature* **2010**, *466*, 835–840. [CrossRef] [PubMed]
27. Ha, T.Y. MicroRNAs in human diseases: From cancer to cardiovascular disease. *Immune Netw.* **2011**, *11*, 135–154. [CrossRef] [PubMed]
28. Van Buren, G., 2nd; Rashid, A.; Yang, A.D.; Abdalla, E.K.; Gray, M.J.; Liu, W.; Somcio, R.; Fan, F.; Camp, E.R.; Yao, J.C.; et al. The development and characterization of a human midgut carcinoid cell line. *Clin. Cancer Res.* **2007**, *13*, 4704–4712. [CrossRef] [PubMed]

29. Stilling, G.A.; Zhang, H.; Ruebel, K.H.; Leontovich, A.A.; Jin, L.; Tanizaki, Y.; Zhang, S.; Erickson, L.A.; Hobday, T.; Lloyd, R.V. Characterization of the functional and growth properties of cell lines established from ileal and rectal carcinoid tumors. *Endocr. Pathol.* **2007**, *18*, 223–232. [CrossRef] [PubMed]
30. Kidd, M.; Drozdov, I.; Joseph, R.; Pfragner, R.; Culler, M.; Modlin, I. Differential cytotoxicity of novel somatostatin and dopamine chimeric compounds on bronchopulmonary and small intestinal neuroendocrine tumor cell lines. *Cancer* **2008**, *113*, 690–700. [CrossRef] [PubMed]
31. Arvigo, M.; Gatto, F.; Ruscica, M.; Ameri, P.; Dozio, E.; Albertelli, M.; Culler, M.D.; Motta, M.; Minuto, F.; Magni, P.; et al. Somatostatin and dopamine receptor interaction in prostate and lung cancer cell lines. *J. Endocrinol.* **2010**, *207*, 309–317. [CrossRef] [PubMed]
32. Kustermann, S.; Boess, F.; Buness, A.; Schmitz, M.; Watzele, M.; Weiser, T.; Singer, T.; Suter, L.; Roth, A. A label-free, impedance-based real time assay to identify drug-induced toxicities and differentiate cytostatic from cytotoxic effects. *Toxicol. In Vitro* **2013**, *27*, 1589–1595. [CrossRef] [PubMed]
33. Ke, N.; Wang, X.; Xu, X.; Abassi, Y.A. The xCELLigence system for real-time and label-free monitoring of cell viability. *Methods Mol. Biol.* **2011**, *740*, 33–43. [PubMed]
34. Binderup, T.; Knigge, U.; Mellon Mogensen, A.; Palnaes Hansen, C.; Kjaer, A. Quantitative gene expression of somatostatin receptors and noradrenaline transporter underlying scintigraphic results in patients with neuroendocrine tumors. *Neuroendocrinology* **2008**, *87*, 223–232. [CrossRef] [PubMed]
35. Peltier, H.J.; Latham, G.J. Normalization of microRNA expression levels in quantitative RT-PCR assays: Identification of suitable reference RNA targets in normal and cancerous human solid tissues. *RNA* **2008**, *14*, 844–852. [CrossRef] [PubMed]
36. Jorgensen, S.; Baker, A.; Moller, S.; Nielsen, B.S. Robust one-day in situ hybridization protocol for detection of microRNAs in paraffin samples using LNA probes. *Methods* **2010**, *52*, 375–381. [CrossRef] [PubMed]
37. Dossing, K.B.; Binderup, T.; Kaczkowski, B.; Jacobsen, A.; Rossing, M.; Winther, O.; Federspiel, B.; Knigge, U.; Kjaer, A.; Friis-Hansen, L. Down-regulation of miR-129-5p and the let-7 family in neuroendocrine tumors and metastases leads to up-regulation of their targets Egr1, G3bp1, Hmga2 and Bach1. *Genes* **2014**, *6*, 1–21. [CrossRef] [PubMed]
38. Strosberg, J.R.; Fisher, G.A.; Benson, A.B.; Malin, J.L.; Panel, G.T.C.; Cherepanov, D.; Broder, M.S.; Anthony, L.B.; Arslan, B.; Fisher, G.A.; et al. Systemic treatment in unresectable metastatic well-differentiated carcinoid tumors: Consensus results from a modified delphi process. *Pancreas* **2013**, *42*, 397–404. [CrossRef] [PubMed]
39. Rinke, A.; Muller, H.H.; Schade-Brittinger, C.; Klose, K.J.; Barth, P.; Wied, M.; Mayer, C.; Aminossadati, B.; Pape, U.F.; Blaker, M.; et al. Placebo-controlled, double-blind, prospective, randomized study on the effect of octreotide lar in the control of tumor growth in patients with metastatic neuroendocrine midgut tumors: A report from the PROMID study group. *J. Clin. Oncol.* **2009**, *27*, 4656–4663. [CrossRef] [PubMed]
40. Caplin, M.E.; Pavel, M.; Ruszniewski, P. Lanreotide in metastatic enteropancreatic neuroendocrine tumors. *N. Engl. J. Med.* **2014**, *371*, 1556–1557. [CrossRef] [PubMed]
41. Kidd, M.; Schally, A.V.; Pfragner, R.; Malfertheiner, M.V.; Modlin, I.M. Inhibition of proliferation of small intestinal and bronchopulmonary neuroendocrine cell lines by using peptide analogs targeting receptors. *Cancer* **2008**, *112*, 1404–1414. [CrossRef] [PubMed]
42. Li, S.C.; Martijn, C.; Cui, T.; Essaghir, A.; Luque, R.M.; Demoulin, J.B.; Castano, J.P.; Oberg, K.; Giandomenico, V. The somatostatin analogue octreotide inhibits growth of small intestine neuroendocrine tumour cells. *PLoS ONE* **2012**, *7*, e48411. [CrossRef] [PubMed]
43. Kharmate, G.; Rajput, P.S.; Lin, Y.C.; Kumar, U. Inhibition of tumor promoting signals by activation of SSTR2 and opioid receptors in human breast cancer cells. *Cancer Cell Int.* **2013**, *13*, 93. [CrossRef] [PubMed]
44. Fernandes, I.; Pacheco, T.R.; Costa, A.; Santos, A.C.; Fernandes, A.R.; Santos, M.; Oliveira, A.G.; Casimiro, S.; Quintela, A.; Fernandes, A.; et al. Prognostic significance of AKT/mTOR signaling in advanced neuroendocrine tumors treated with somatostatin analogs. *Onco Targets Ther.* **2012**, *5*, 409–416. [CrossRef] [PubMed]
45. Porta, C.; Paglino, C.; Mosca, A. Targeting PI3K/AKT/mTOR signaling in cancer. *Front. Oncol.* **2014**, *4*, 64. [CrossRef] [PubMed]
46. Johnbeck, C.B.; Munk Jensen, M.; Haagen Nielsen, C.; Fisker Hag, A.M.; Knigge, U.; Kjaer, A. ^{18}F-FDG and ^{18}F-FLT-pet imaging for monitoring everolimus effect on tumor-growth in neuroendocrine tumors: Studies in human tumor xenografts in mice. *PLoS ONE* **2014**, *9*, e91387.

47. Raymond, E.; Dahan, L.; Raoul, J.L.; Bang, Y.J.; Borbath, I.; Lombard-Bohas, C.; Valle, J.; Metrakos, P.; Smith, D.; Vinik, A.; et al. Sunitinib malate for the treatment of pancreatic neuroendocrine tumors. *N. Engl. J. Med.* **2011**, *364*, 501–513. [CrossRef] [PubMed]
48. Yao, J.C.; Shah, M.H.; Ito, T.; Bohas, C.L.; Wolin, E.M.; Van Cutsem, E.; Hobday, T.J.; Okusaka, T.; Capdevila, J.; de Vries, E.G.; et al. Everolimus for advanced pancreatic neuroendocrine tumors. *N. Engl. J. Med.* **2011**, *364*, 514–523. [CrossRef] [PubMed]
49. Pavel, M.E.; Hainsworth, J.D.; Baudin, E.; Peeters, M.; Horsch, D.; Winkler, R.E.; Klimovsky, J.; Lebwohl, D.; Jehl, V.; Wolin, E.M.; et al. Everolimus plus octreotide long-acting repeatable for the treatment of advanced neuroendocrine tumours associated with carcinoid syndrome (RADIANT-2): A randomised, placebo-controlled, phase 3 study. *Lancet* **2011**, *378*, 2005–2012. [CrossRef]
50. Bousquet, C.; Guillermet-Guibert, J.; Saint-Laurent, N.; Archer-Lahlou, E.; Lopez, F.; Fanjul, M.; Ferrand, A.; Fourmy, D.; Pichereaux, C.; Monsarrat, B.; et al. Direct binding of p85 to sst2 somatostatin receptor reveals a novel mechanism for inhibiting PI3K pathway. *EMBO J.* **2006**, *25*, 3943–3954. [CrossRef] [PubMed]
51. Briest, F.; Grabowski, P. PI3K-AKT-mTOR-signaling and beyond: The complex network in gastroenteropancreatic neuroendocrine neoplasms. *Theranostics* **2014**, *4*, 336–365. [CrossRef] [PubMed]
52. Zhang, N.; Li, X.; Wu, C.W.; Dong, Y.; Cai, M.; Mok, M.T.; Wang, H.; Chen, J.; Ng, S.S.; Chen, M.; et al. MicroRNA-7 is a novel inhibitor of YY1 contributing to colorectal tumorigenesis. *Oncogene* **2013**, *32*, 5078–5088. [CrossRef] [PubMed]
53. Zhao, X.; Dou, W.; He, L.; Liang, S.; Tie, J.; Liu, C.; Li, T.; Lu, Y.; Mo, P.; Shi, Y.; et al. MicroRNA-7 functions as an anti-metastatic microRNA in gastric cancer by targeting insulin-like growth factor-1 receptor. *Oncogene* **2013**, *32*, 1363–1372. [CrossRef] [PubMed]
54. Kredo-Russo, S.; Mandelbaum, A.D.; Ness, A.; Alon, I.; Lennox, K.A.; Behlke, M.A.; Hornstein, E. Pancreas-enriched miRNA refines endocrine cell differentiation. *Development* **2012**, *139*, 3021–3031. [CrossRef] [PubMed]
55. Gu, D.N.; Huang, Q.; Tian, L. The molecular mechanisms and therapeutic potential of microRNA-7 in cancer. *Expert Opin. Ther. Targets* **2015**, *19*, 415–426. [CrossRef] [PubMed]
56. Xu, L.; Wen, Z.; Zhou, Y.; Liu, Z.; Li, Q.; Fei, G.; Luo, J.; Ren, T. MicroRNA-7-regulated TLR9 signaling-enhanced growth and metastatic potential of human lung cancer cells by altering the phosphoinositide-3-kinase, regulatory subunit 3/Akt pathway. *Mol. Biol. Cell* **2013**, *24*, 42–55. [CrossRef] [PubMed]
57. Fang, Y.; Xue, J.L.; Shen, Q.; Chen, J.; Tian, L. MicroRNA-7 inhibits tumor growth and metastasis by targeting the phosphoinositide 3-kinase/Akt pathway in hepatocellular carcinoma. *Hepatology* **2012**, *55*, 1852–1862. [CrossRef] [PubMed]
58. Zheng, G.; Xiong, Y.; Xu, W.; Wang, Y.; Chen, F.; Wang, Z.; Yan, Z. A two-microRNA signature as a potential biomarker for early gastric cancer. *Oncol. Lett.* **2014**, *7*, 679–684. [CrossRef] [PubMed]
59. Hanoun, N.; Delpu, Y.; Suriawinata, A.A.; Bournet, B.; Bureau, C.; Selves, J.; Tsongalis, G.J.; Dufresne, M.; Buscail, L.; Cordelier, P.; et al. The silencing of microRNA 148a production by DNA hypermethylation is an early event in pancreatic carcinogenesis. *Clin. Chem.* **2010**, *56*, 1107–1118. [CrossRef] [PubMed]
60. Shivapurkar, N.; Weiner, L.M.; Marshall, J.L.; Madhavan, S.; Deslattes Mays, A.; Juhl, H.; Wellstein, A. Recurrence of early stage colon cancer predicted by expression pattern of circulating microRNAs. *PLoS ONE* **2014**, *9*, e84686. [CrossRef]
61. Xu, Q.; Jiang, Y.; Yin, Y.; Li, Q.; He, J.; Jing, Y.; Qi, Y.T.; Xu, Q.; Li, W.; Lu, B.; et al. A regulatory circuit of miR-148a/152 and DNMT1 in modulating cell transformation and tumor angiogenesis through IGF-IR and IRS1. *J. Mol. Cell Biol.* **2013**, *5*, 3–13. [CrossRef] [PubMed]
62. Sakamoto, N.; Naito, Y.; Oue, N.; Sentani, K.; Uraoka, N.; Zarni Oo, H.; Yanagihara, K.; Aoyagi, K.; Sasaki, H.; Yasui, W. MicroRNA-148a is downregulated in gastric cancer, targets MMP7, and indicates tumor invasiveness and poor prognosis. *Cancer Sci.* **2014**, *105*, 236–243. [CrossRef] [PubMed]
63. Sun, J.; Song, Y.; Wang, Z.; Wang, G.; Gao, P.; Chen, X.; Gao, Z.; Xu, H. Clinical significance of promoter region hypermethylation of microRNA-148a in gastrointestinal cancers. *Onco Targets Ther.* **2014**, *7*, 853–863. [PubMed]
64. Murray, R.D.; Kim, K.; Ren, S.G.; Chelly, M.; Umehara, Y.; Melmed, S. Central and peripheral actions of somatostatin on the growth hormone-IGF-I axis. *J. Clin. Investig.* **2004**, *114*, 349–356. [CrossRef] [PubMed]

65. Libutti, S.K. Therapy: Blockade of IGF-1R-not effective in neuroendocrine tumours. *Nat. Rev. Endocrinol.* **2013**, *9*, 389–390. [CrossRef] [PubMed]
66. Thornton, J.E.; Gregory, R.I. How does Lin28 let-7 control development and disease? *Trends Cell Biol.* **2012**, *22*, 474–482. [CrossRef] [PubMed]
67. Shen, G.; Wu, R.; Liu, B.; Dong, W.; Tu, Z.; Yang, J.; Xu, Z.; Pan, T. Upstream and downstream mechanisms for the promoting effects of IGF-1 on differentiation of spermatogonia to primary spermatocytes. *Life Sci.* **2014**, *101*, 49–55. [CrossRef] [PubMed]
68. Alajez, N.M.; Shi, W.; Wong, D.; Lenarduzzi, M.; Waldron, J.; Weinreb, I.; Liu, F.F. Lin28b promotes head and neck cancer progression via modulation of the insulin-like growth factor survival pathway. *Oncotarget* **2012**, *3*, 1641–1652. [CrossRef] [PubMed]

14

Serum and Lipoprotein Particle miRNA Profile in Uremia Patients

Markus Axmann [1], Sabine M. Meier [1], Andreas Karner [2], Witta Strobl [1], Herbert Stangl [1,*] and Birgit Plochberger [2,*]

[1] Center for Pathobiochemistry and Genetics, Institute of Medical Chemistry and Pathobiochemistry, Medical University Vienna, 1090 Vienna, Austria; markus.axmann@meduniwien.ac.at (M.A.); sabine.meier@meduniwien.ac.at (S.M.M.); witta.strobl@meduniwien.ac.at (W.S.)

[2] University of Applied Sciences Upper Austria, School of Medical Engineering and Applied Social Sciences, 4020 Linz, Austria; andreas.karner@fh-linz.at

* Correspondence: herbert.stangl@meduniwien.ac.at (H.S.); birgit.plochberger@fh-linz.at (B.P.)

Abstract: microRNAs (miRNAs) are post-transcriptional regulators of messenger RNA (mRNA), and transported through the whole organism by—but not limited to—lipoprotein particles. Here, we address the miRNA profile in serum and lipoprotein particles of healthy individuals in comparison with patients with uremia. Moreover, we quantitatively determined the cellular lipoprotein-particle-uptake dependence on the density of lipoprotein particle receptors and present a method for enhancement of the transfer efficiency. We observed a significant increase of the cellular miRNA level using reconstituted high-density lipoprotein (HDL) particles artificially loaded with miRNA, whereas incubation with native HDL particles yielded no measurable effect. Thus, we conclude that no relevant effect of lipoprotein-particle-mediated miRNA-transfer exists under in vivo conditions though the miRNA profile of lipoprotein particles can be used as a diagnostic marker.

Keywords: microRNA; lipoprotein; transfer; uremia

---

1. Introduction

In general, intercellular communication occurs via cell-to-cell adhesion or by secretory messengers via e.g., hormones. A majority of circulating microRNAs (miRNAs) are derived from shedding of plasma membrane vesicles [1–4]. However, the protein Argonaute2 was also reported to carry a population of circulating miRNA without the need for vesicles [5]. Recently, high-density lipoprotein (HDL) particles were shown to transfer proteins and miRNAs from donor to recipient cells [6,7]. miRNAs are a class of small, non-coding RNAs which have emerged as important post-transcriptional regulators [8]. They intracellularly target specific messenger RNAs (mRNAs) to degrade or repress the translation process [8–11]. In extracellular space, miRNAs were thought to be relative unstable molecules; however, surprisingly, they were found to circulate in a highly stable form in various body fluids like blood, milk, saliva and urine [12]. Turchinovich et al. [13] proposed that extracellular miRNA carried by HDL is associated with the Argonaute family of proteins. However, proteome analysis did not reveal these proteins as a component of HDL [14–19]. The profile of circulating miRNAs display characteristic changes when pathogenic alterations occur. This dysregulation is linked to diseases such as cancer, cardiovascular [20–24] and kidney diseases [25–28]. Altered miRNA profiles in the HDL fraction of patients suffering from familial hypercholesterolemia were seen [6]. Thus, they may serve as markers for various diseases [29] and their progression [12] or the development of new therapeutic agents [8,30]. Thereby, mode and specificity of HDL uptake routes may influence the functional role of secreted miRNAs. This is one of the least understood issues in the field of miRNAs.

Here, we present data that miRNA is present—besides serum—in all lipoprotein particles and find that the miRNA profile is different in patients with uremia. Finally, we aimed to follow the uptake of miRNA to cells using quantitative real-time PCR (qPCR). For this purpose, we derived a protocol to increase the uptake for miRNA from HDL particles significantly. Taken together, our data indicate that—despite the presence of miRNA on lipoprotein particles—its abundance is rather low, which makes any influence on the cell metabolism via lipoprotein-mediated transfer in vivo generally highly unlikely.

2. Materials and Methods

The manuscript and the used nomenclature are written according to the minimum information for publication of quantitative real-time PCR (qPCR) experiments (MIQE) guidelines [31].

2.1. Patients

For the miRNA TaqMan™ (Applied Biosystems, Carlsbad, CA, USA) array analyses, 17 adult chronic renal failure (CRF) patients suffering from CRF Kidney Disease (Outcomes Quality Initiative stage 3–5 without hemodialysis) and 14 adult CRF patients on maintenance hemodialysis and matched controls were recruited (for lipid analyses see Table 1 of Meier et al. [32]). This study was approved by the Ethics Committee, Medical University of Vienna (EK-Nr. 511/2007). Written informed consent was obtained from all participants. Patients with an age over 55 years, with diabetes mellitus, nephrotic syndrome, severe hyperlipidemia, inflammatory diseases, malignancy and infections within the last three months, and patients on corticosteroids, lipid lowering or immunosuppressive drugs were excluded. Venous blood was collected after an overnight fast. Sera were frozen in aliquots at −20 °C. Lipoprotein particles were isolated immediately and stored frozen in aliquots at −20 °C until analysis. For detailed lipid and apolipoprotein analyses see Meier et al. [32].

2.2. microRNA in Serum and Lipoprotein Particles

As there is no known endogenous control for miRNAs in serum or lipoprotein particles, we added cel-miR-39 from the *C. elegans* species as internal control and for reference [33]. Total miRNA was reverse transcribed using the TaqMan™ microRNA Reverse Transcription Kit and the appropriate reverse transcription (RT) primers (Applied Biosystems), according to the manufacturer's protocol. Resulting samples were subjected to RT-qPCR using TaqMan™ Gene Expression Universal Master Mix (Applied Biosystems) and TaqMan™ miRNA Assays (Applied Biosystems) following the manufacturer's instructions. The amplification was conducted in a StepOne Real-Time PCR-System (48-well, Applied Biosystems); data was collected using the StepOne Software v2.1 (Applied Biosystems).

2.3. TaqMan™ Arrays

Equal volumes of patient samples were pooled and miRNA was isolated from 100 µL serum or 500 µg lipoprotein particles using the miRNeasy Kit (QIAGEN GmbH, Hilden, Germany), according to the supplier's instructions, in two independent experiments. RNA quantity and purity was measured using a NanoDrop ND-1000 Spectrophotometer (peqlab Biotechnologie GmbH, Erlangen, Germany) and RNA was stored at −80 °C. Total miRNA was reverse transcribed using the TaqMan™ microRNA RT Kit with MegaPlex RT Primers (Applied Biosystems). To ensure the sample content for the TaqMan™ arrays, miRNAs from lipoprotein particles and serum were further processed by applying preamplification using the appropriate primers and the PreAmp Master Mix (Applied Biosystems). Afterwards, samples were loaded into TaqMan™ Array Cards A+B (in total 754 miRNAs were detected) and analyzed using a 7900HT Fast Real-Time PCR System (Applied Biosystems). Data was collected and analyzed using appropriate software from Applied Biosystems

(SDS 2.4 and Data assist v3.0) yielding $\Delta\Delta c_q$. The value RQ (relative quantification) is defined as $RQ = 2^{-\Delta\Delta\, c_q}$.

2.4. Synthetic miRNA

Human mature miRNAs hsa-miR145 (5′-GUC CAG UUU UCC CAG GAA UCC CU-3′), hsa-miR155 (5′-UUA AUG CUA AUC GUC AUA GGG GU-3′) and hsa-miR223 (5′-UGU CAG UUU GUC AAA UAC CCC A-3′) were synthesized by Microsynth (Microsynth, Vienna, Austria). The manufacturer did purification with HPLC & dialysis. miRNAs were solubilized in 10 mM tris(hydroxymethyl) aminomethane (TRIS) buffer, pH 7.5 (Thermo Fisher Scientific, Vienna, Austria) and stored at −20 °C in aliquots of 100 µL (final storage concentration 10 µM).

2.5. Lipoprotein Particle Isolation

For the reconstitution/labeling experiments, human plasma was collected from two normolipidemic healthy volunteers twice (time between donations was roughly one month), in accordance with the medical and ethical guidelines of the Medical University of Vienna. This part of the study was approved by the Ethics Committee, Medical University of Vienna (EK-Nr. 1414/2016). Written informed consent was obtained from all participants. Individual lipoprotein particle (HDL and low-density lipoprotein (LDL)) fractions were isolated by serial ultra-centrifugation at a density of 1.21 g/mL or 1.06 g/mL, respectively [34]. Final protein concentration was determined photometrically (Bradford assay) and samples were stored under an inert atmosphere at +4 °C. For the preparation of lipoprotein particle deficient serum (LPDS), human sera from both donors were spun using ultra-centrifugation at a density of 1.21 g/mL, dialyzed and stored at −20 °C.

2.6. Reconstitution of HDL Particles

HDL particles were reconstituted by a modified protocol, previously published in [35]. In short, lipids from HDL particles were extracted two times with ethanol : diethyl ether (3:2) at −20 °C for 2 h. Precipitate was dried under nitrogen gas flow and resuspended in buffer A (150 mM NaCl, 0.1‰ ethylenediaminetetraacetic acid (EDTA), 10 mM TRIS/HCl, pH 8.0, all Sigma Aldrich, Vienna, Austria). Protein concentration was determined photometrically (Bradford assay). A lipid mixture, consisting of L-α-phosphatidylcholine, cholesterol oleate and cholesterol (all Sigma Aldrich) at a molar ratio of 100:22:4.8 dissolved in chloroform : methanol (2:1), was dried under nitrogen gas and resuspended in buffer A. Aliquots of synthetic miRNAs (100 µL, 10 µM) were mixed with freshly prepared spermine solution (final concentration 15 mM, Sigma Aldrich) for 30 min at 30 °C. Lipid suspension and miRNA/spermine solution were mixed and sodium deoxycholate (Sigma Aldrich) was added for lipid solubilization at a final concentration of 15 mM. In negative control experiments, HDL particles were reconstituted without addition of miRNA and/or spermine. The mixture was stirred at +4 °C for 2 h. Delipidated HDL was added at a final molar ratio of L-α-phosphatidylcholine, cholesterol oleate, cholesterol and protein (HDL) of 100:22:4.8:1, and the mixture was stirred at +4 °C overnight. Extensive dialysis using Slide-A-Lyzer™ dialysis cassettes (cut-off 20 kDa, Thermo Fisher Scientific) against phosphate-buffered saline (PBS, Roth, Graz, Austria) + Amberlite XAD-2 polymeric adsorbent (15 g/L, Sigma Aldrich) was performed to separate generated reconstituted HDL (rHDL) particles form free miRNA, spermine and detergent. Final protein concentration was determined photometrically (Bradford assay) and samples were stored under inert atmosphere at +4 °C.

2.7. Labeling of LDL Particles

LDL particles were labeled with miRNA by a modified protocol, previously published in [36]. In short, LDL particle solution was incubated with 0.3 mM EDTA/0.1 mM ethylene glycol-bis(β-aminoethyl ether)-*N*,*N*,*N*′,*N*′-tetraacetic acid (EGTA, Roth) for 10 min at +4 °C. Aliquots of miRNA were mixed with freshly prepared spermine solution (final concentration 15 mM, Sigma Aldrich) for 30 min at 30 °C. After addition of an equal volume of dimethyl sulfoxide

DMSO (Thermo Fisher Scientific), the mixture was diluted five-fold with buffer containing 150 mM NaCl/0.3 mM EDTA/0.1 mM EGTA. LDL particle solution and miRNA mixture were combined and incubated for 2 h at 40 °C. In negative control experiments, LDL particles were labeled without addition of miRNA and/or spermine. Extensive dialysis using Slide-A-Lyzer™ dialysis cassettes (cut-off 20 kDa, Thermo Fisher Scientific) against PBS (Roth) + Amberlite XAD-2 polymeric adsorbent (15 g/L, Sigma Aldrich) was performed to separate labeled LDL particles from free miRNA, spermine and organic solvent. Final protein concentration was determined photometrically (Bradford assay) and samples were stored under inert atmosphere at +4 °C.

2.8. *Quality Control of Reconstituted/Labeled Lipoprotein Particles*

High-speed atomic force microscopy (HS-AFM) (see Figure A1) was performed in tapping mode with free amplitudes of 1.5 nm–2.5 nm. The amplitude setpoint was larger than 90% of the free oscillation amplitude. USC-F1.2-k0.15 cantilevers (Nanoworld AG, Neuchâtel, Switzerland) were used. HDL and LDL particle aliquots were diluted $1{:}10^3$ in PBS (in order to observe individual particles) and incubated for 5 min on freshly cleaved mica. After rinsing with PBS, the samples were imaged in PBS at room temperature. Image processing and particle analysis were done in Gwyddion 2.49 (CMI, Brno, Czech Republic) and SPIP (Image Metrology, Hørsholm, Denmark). The probability density functions (pdfs) representing the height distribution of particles were calculated using the 'ks-density' algorithm in MATLAB (Mathworks, Natick, MA, USA).

2.9. *Cells*

Chinese hamster ovary (CHO) cell lines CHOK1, ldlA7 (reduced expression of LDL-receptor), and ldlA7-SR-B1 (overexpression of scavenger receptor class B type 1 (SR-B1)) were cultured as previously described in [37,38]. In short, cells were maintained in Dulbecco's modified Eagle Medium/Nutrient Mixture F-12 (Sigma Aldrich) supplemented with 10% fetal calf serum (FCS, Sigma Aldrich), Penicillin/Streptomycin (100 U/mL and 0.1 mg/mL final concentration, Sigma Aldrich) and 2 mM L-Glutamine (Roth). Geneticin sulphate (Thermo Fisher Scientific) was added to ldlA7-SR-B1 cells at a final concentration of 0.5 mg/mL as selection antibiotic.

Cells were grown until just reaching confluency in individual LabTek™ chambers (eight chambers per slide, each with a 4 mm × 4 mm surface area and ~700 μL total volume, Thermo Fisher Scientific). Cells were washed three times with Hanks' balanced salt solution (HBSS, Roth) and incubated in FCS-free culture medium. Next, HDL particle solution (final concentration 50 μg/mL or 5 μg/mL) or LDL particle solution (final concentration 5 μg/mL) from the different preparations (native/reconstituted/labeled) was added. Incubation was performed at 37 °C and 5% CO_2. After 16 h, cells were washed three times with HBSS and covered with 100 μL of medium without FCS. For negative control experiments, chambers without cells and with cells without any addition of particles were used. For blocking experiments (testing of unspecific binding), native lipoprotein particle solution or bovine serum albumin (BSA, Sigma Aldrich) was used. Cell number was determined for each experiment in two independent chambers, which were similarly treated as above (medium containing 50 μg/mL HDL or 5 μg/mL LDL particle solution) and the average cell number of both chambers was used for normalization.

2.10. *miRNA-Extraction*

miRNA-extraction was performed according to the manufacturer's protocol of the miRNeasy Mini Kit (Qiagen, Vienna, Austria). In short, a sample volume (ranging from 1 μL to 100 μL (lowest protein concentration)) of native/reconstituted HDL particles or native/labeled LDL particles normalized to the lowest protein concentration was lysed and homogenized. The cell sample volume used for miRNA-extraction contained the pooled cells from two independent chambers. For the negative control sample, 100 μL of RNase-free water (Roth) was used. A phenol/chloroform extraction was performed subsequently to separate proteins/DNA from RNA.

2.11. Reverse Transcription

Reverse transcription was performed according to the manufacturer's protocol of the TaqMan™ microRNA Reverse Transcription Kit (Thermo Fisher Scientific). Briefly, 7 μL of the kit's mastermix solution were added to 5 μL of the extracted miRNA-sample and 3 μL of miRNA-specific primer (Assay ID 002623, 000526, and 002278, Thermo Fisher Scientific). In order to use the same cell number for each cell line, the sample volume of the cell line with a higher overall cell number was reduced accordingly; the residual volume to reach the total sample volume of 5 μL was RNase-free water (thus, sample volume from the individual cell lines varied between 5 μL and 2 μL). As a sample for standard curve preparation, an aliquot of miRNA was diluted sequentially in RNase-free water. Reverse transcription was performed using a Thermocycler (LabCycler Sensoquest, Göttingen, Germany) with the following program: (1) 30 min at 16 °C, (2) 30 min at 42 °C, (3) 5 min at 85 °C, (4) ∞ at 4 °C. Usually, the qPCR step was done immediately after the reverse transcription—otherwise the complementary DNA synthesized from the miRNA samples was stored at −20 °C.

2.12. qPCR

qPCR of the reverse transcribed samples was performed according to the manufacturer's protocol of the TaqMan™ assay (Thermo Fisher Scientific). In short, per sample, 7.5 μL of mastermix solution (iTaq™, Biorad, Vienna, Austria) was diluted with 4.75 μL RNase-free water and 0.756 μL hydrolysis probe (TaqMan™ Assay ID 002623, 000526, and 002278, Thermo Fisher Scientific). Each sample contained further 2 μL of cDNA sample or RNase-free water and was measured twice. qPCR was performed using a Corbett RG-6000 PCR machine (Corbett Research, Cambridge, UK) with the following program: (1) 2 min at 50 °C, (2) 10 min at 95 °C, (3) 15 s at 95 °C, (4) 60 s at 60 °C. Steps (3) and (4) were repeated up to 50 times. For the analysis, RotorGene 6000 Software Version 1.7 (Qiagen, Vienna, Austria) was used with activated 'DynamicTube Normalization' (for compensation of different background levels using the second derivative of each sample trace) and 'Noise Slope Correction' (normalization to the noise level). The threshold for calculation of the quantification cycle c_q for each miRNA was determined from the standard curves of each miRNA individually, via the 'Auto-Find Threshold' function of the software package—hereby, the software maximizes the R-value of the fit of the standard curve—and kept equal for each specific miRNA. The c_q value of the negative control was always at least six cycles higher than the highest sample c_q value (in the range between 30–40 cycles). In each run, at least one of the standard dilutions of miRNA was added as a sample to calibrate each individual run for the same reaction efficiency.

2.13. Calculation of miRNA Strands Number

For standard curves (Figure A2), the number of miRNA strands of each sequential dilution step was calculated from the initial concentration and the subsequent dilution steps during reverse transcription and PCR. Thereby, we assumed that the reverse transcription performed linearly over the low concentration range used here and in a 1:1 stoichiometry. Accordingly, the number of strands corresponding to the c_q value of each sample run was calculated from the standard curve regression line and normalized to the number of cells or the lipoprotein particle number, respectively. Lipoprotein particle number was estimated from the initial protein concentration and its average molecular weight (molecular weight MW_{HDL} ~250 kDa, MW_{LDL} ~3 MDa). We assumed no lipid contribution to the molecular weight—thus, we slightly overestimated the number of miRNA strands per lipoprotein particle. We further assumed a 100% recovery of miRNA during the miRNA extraction step.

3. Results

3.1. miRNA in Lipoprotein Particle Fractions

First, we tested the presence of miRNA in serum and all lipoprotein particle fractions of healthy volunteers. To do so, we separated serum and lipoprotein particle fractions and purified the contained miRNAs. For analysis, we selected two miRNAs known to be involved in renal and cardiovascular disease, namely miR-21 [39] and miR-223 [40]. Both miRNAs were detectable in all fractions (Table 1) with the highest relative amount in HDL particles; however, the abundance was low (absolute c_q values varied between 26 and 36 cycles). This data indicates a lipoprotein-particle-selective association of miRNAs. To demonstrate that HDL particles did not contain remaining serum constituents, we diluted the HDL particle fraction with excess PBS and centrifuged it a second time. Still, this fraction contained the highest miRNA amount among the lipoprotein particle classes ($<\Delta\Delta c_q> = -2.5$ (SD = 0.2) and -1.5 (SD = 0.1) for miR-223 and miR-21, respectively), indicating that miRNAs are relatively enriched at least in the HDL particle fraction.

Table 1. miRNAs circulating in human serum and associated to lipoprotein particles.

Sample	miRNA	$<\Delta\Delta c_q>$	SD $<\Delta\Delta c_q>$
LPDS	223	0.0	0.1
	21	0.0	0.2
VLDL	223	5.7	0.2
	21	4.6	0.2
LDL	223	2.8	0.2
	21	1.8	0.2
HDL	223	−3.5	0.1
	21	−2.5	0.1

Lipoprotein particle deficient serum (LPDS) and lipoprotein particle fractions (very-low density lipoprotein (VLDL), low density lipoprotein (LDL) and high density lipoprotein (HDL) particles) were analyzed for two distinct microRNAs (miRNAs). Values are in relation to the levels of cel-miR-39 and subsequently LPDS. ($< x >$ = average value of x, SD = standard deviation, n = 3 independent experiments).

Next, we investigated miRNA profiles in serum and HDL particles of uremic patients recruited for a study on the influence of chronic renal failure (CRF) on cholesterol efflux from macrophages [32]. Figure 1 depicts the expression profile of miRNAs found in the serum/HDL particles of patients suffering from CRF (top) and from the more advanced disease course (hemodialysis, bottom) relative to their matched healthy controls. In both settings, the miRNA profile difference between serum and HDL particles was diverse: some miRNAs were represented in HDL particles and serum in a similar portion (see Figure 1 overlap of the blue and orange line); however, some miRNAs were increased at different rates in serum and HDL particles. The most regulated miRNA in CRF patients—miR-192—was about 3.8 times increased in HDL particles while it was decreased in serum to one-third. Nevertheless, there was a substantial overlap in the miRNA signature between serum and HDL particles in both CRF and hemodialysis patients.

Furthermore, we assessed the differences in the serum and HDL signatures between the two patient cohorts; most of the changes in the miRNA levels were in a similar direction (Figure 2). The miR-122 level increased sharply in the HDL fraction with disease progression (ΔRQ ~+6.6), while it hardly decreased in the serum (ΔRQ ~−0.7). On the other hand, some miRNA levels like miR-24, miR574-3p, miR-222, miR-27b and miR-29a were increased in the serum with disease progression, whereas in HDL particles, the levels remained nearly constant. While these miRNA levels were significantly different between HDL particles and serum, a substantial overlap in the miRNA profile exists.

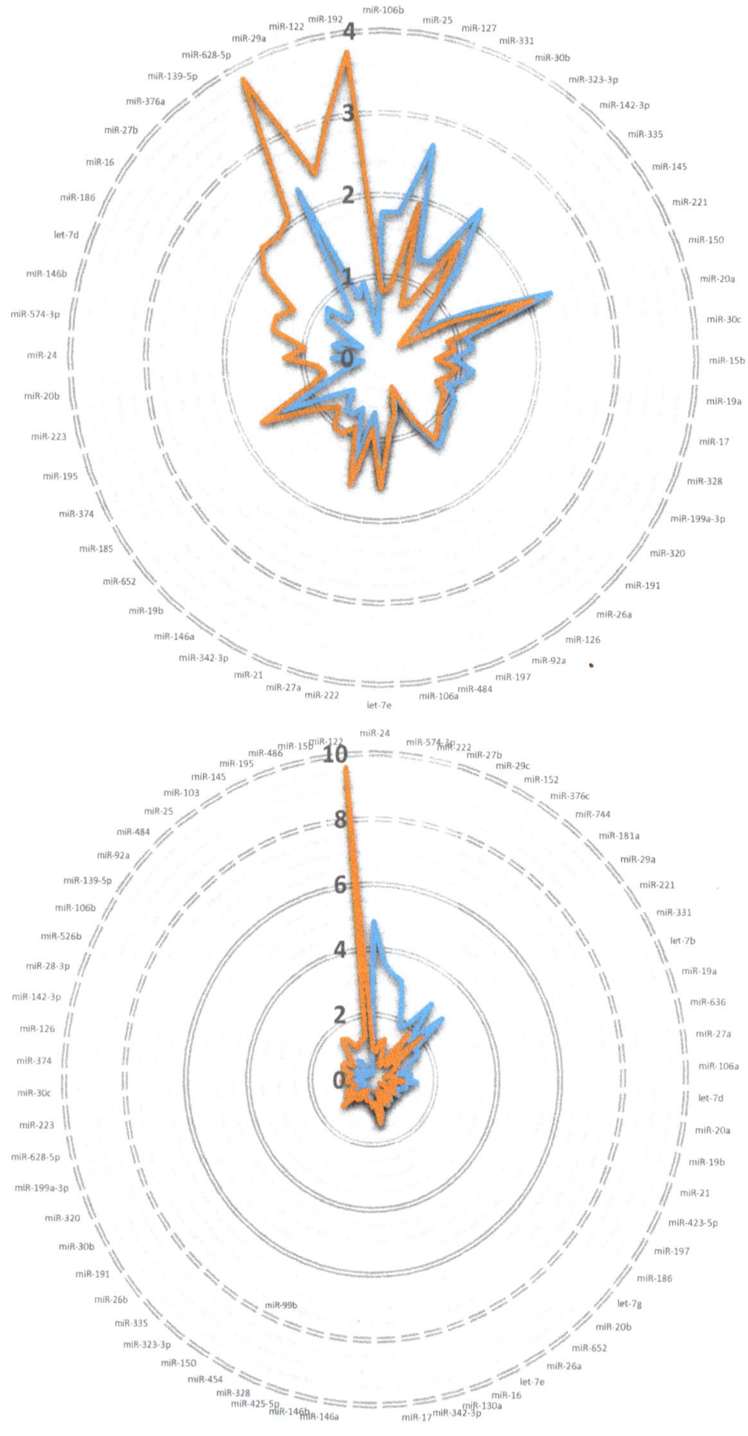

Figure 1. miRNA profile in serum and high-density lipoprotein (HDL) particles of chronic renal failure (CRF) and hemodialysis patients. Pooled serum and HDL particles of adult CRF (top) and hemodialysis patients (bottom) and their matched controls was analyzed by the TaqMan™ miRNA array A (duplicates). From the 381 different miRNAs, 57 (serum) and 90 (HDL particles) for CRF patients and 75 (serum) and 89 (HDL particles) for hemodialysis patients were detectable (plotted were only miRNAs detectable in serum and HDL particles) and the relative quantitation (RQ) was calculated subsequently (the corresponding control miRNAs were set to RQ = 1). RQ-values were arranged in the following manner: starting at the twelve o'clock position with the maximum negative difference between the RQ values of HDL particles (orange) and serum (blue), the values increase clockwise (reaching nearly equality at the six o'clock position) and reach the maximum positive difference before closing the circle again (see Table A2 (in Appendix A) for the complete data set).

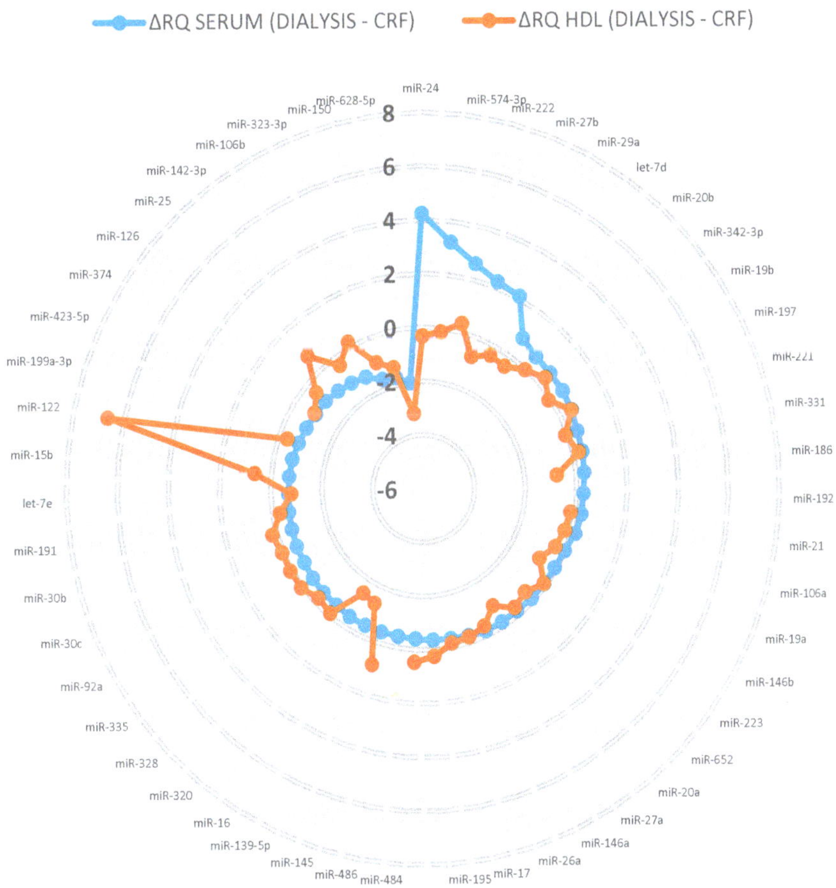

Figure 2. Differences in the miRNA profile in serum and HDL particles during disease progression from CRF to hemodialysis. Data is arranged in the following manner: starting at the twelve o'clock position with the maximum difference between the RQ values of serum (blue) during disease progression from CRF to hemodialysis (=ΔRQ), the values decrease clockwise (reaching nearly no change at the four o'clock position) and reach the maximum negative difference before closing the circle again (plotted are only miRNAs detectable in both disease progression steps). For comparison, ΔRQ values were superimposed for the same miRNAs (if available) in the HDL particle fraction (orange).

3.2. miRNA Associated to Lipoprotein Particles

After having shown that miRNA also circulates with lipoprotein particles and that their expression profile was altered in the disease progression, we assessed if we can reconstitute this system in order to observe and follow cellular miRNA uptake. Therefore, we tested the uptake of native HDL/LDL and reconstituted HDL/labeled LDL particles with two selected miRNAs: miR-223, which shows a high abundance and miR-155, which is rare in lipoprotein particles [41]. To assess the validity of the reconstituted and labeled lipoprotein particles in comparison to their native counterpart, we imaged single lipoprotein particles by HS-AFM and analyzed their size (Figure A1). We observed globular particles comparable to the native forms. However, the reconstituted HDL particle variants were slightly smaller than the native ones; reasonably due to the partial reconstitution process. For LDL particles, the size distributions of the native and labeled particles largely overlap; obviously, spermine addition accounts for the observed size increase. According to our data, the ratio of tested miRNA strands to the number of native lipoprotein particles was in the range of one strand of a specific miRNA to around 1 billion lipoprotein particles (Tables 2 and 3).

Table 2. * miRNA content of HDL particle variants.

Sample	Donor	miRNA	< # miRNA/particle >	SD < # miRNA/particle >
Native HDL	1	223	6.10^{-9}	7.10^{-10}
		155	3.10^{-9}	9.10^{-10}
	2	223	7.10^{-9}	4.10^{-9}
		155	4.10^{-8}	2.10^{-8}
rHDL	1	223	4.10^{-9}	8.10^{-10}
		155	5.10^{-9}	3.10^{-9}
	2	223	7.10^{-9}	4.10^{-9}
		155	1.10^{-8}	1.10^{-8}
rHDL + miR-155	1	155	1.10^{-4}	1.10^{-4}
	2	155	8.10^{-5}	5.10^{-5}
rHDL + miR-155 & spermine	1	155	1.10^{-4}	5.10^{-5}
	2	155	7.10^{-5}	2.10^{-5}

* Each condition was tested twice with two independently obtained samples. rHDL: reconstituted HDL.

Table 3. * miRNA content of LDL particle variants.

Sample	Donor	miRNA	< # miRNA/particle >	SD < # miRNA/particle >
Native LDL	1	223	3.10^{-8}	2.10^{-8}
		155	9.10^{-8}	5.10^{-8}
	2	223	5.10^{-9}	3.10^{-9}
		155	9.10^{-8}	2.10^{-8}
Labeled LDL	1	223	5.10^{-8}	3.10^{-8}
		155	9.10^{-8}	5.10^{-8}
	2	223	4.10^{-9}	3.10^{-9}
		155	9.10^{-8}	2.10^{-8}
Labeled LDL + miR-155	1	155	1.10^{-4}	6.10^{-5}
	2	155	2.10^{-3}	3.10^{-4}
Labeled LDL + miR-155 & spermine	1	155	5.10^{-4}	3.10^{-4}
	2	155	5.10^{-3}	6.10^{-4}

* Each condition was tested twice with two independently obtained samples.

Nevertheless, we were able to increase the miRNA/particle ratio via reconstitution or labeling of the respective lipoprotein particle. The reconstitution (which involves thorough delipidation of the particle) or the labeling process of the lipoprotein particle itself did not influence the miRNA/particle ratio (compare miRNA/particle ratio of native and reconstituted/labeled particles). Thus, we found neither sequence-specificity (no significant difference between miRNA/particle ratio of miR-223 and miR-155), nor any significant effect of electrical-charge-compensation of the poly-anionic nature of the miRNA with addition of the natural poly-cationic nucleotide-sequence-stabilizer-agent spermine. However, our approach increased the miRNA/particle ratio by around five orders of magnitude (from 10^{-9} to 10^{-4} for HDL particles and 10^{-8} to 10^{-3} for LDL particles as shown in Tables 2 and 3).

Next, we tested whether the uptake of lipoprotein particles in cells depends on the cell membrane density of their respective lipoprotein receptor. In order to reach a steady state between uptake and potential degradation of miRNA, we incubated cells for 16 h with lipoprotein particles. We observed no increase of the cellular miRNA-content after incubation with native lipoprotein particles at a concentration in the physiological range (5 µg/mL and 50 µg/mL for HDL or LDL particles) independent of the receptor density (Tables 4 and 5). We did neither observe a significant increase in the number of miRNA strands per cell for HDL particles (overexpression of the HDL-receptor SR-B1 in ldlA7-SR-B1 cells in comparison with native HDL-receptor density in ldlA7 cells) nor for LDL particles (a low LDL-receptor density in ldlA7 cells in comparison with native LDL-receptor density in CHOK1).

However, incubation using lipoprotein particles with artificially increased miRNA content yielded—at least for HDL particles—a significant and receptor-density-dependent increase of the number of miRNA strands per cell (approximately by a factor 4 or 10 at native or overexpressed receptor densities). In contrast, incubation with LDL particles did not influence the miRNA level.

Table 4. * Influence of lipoprotein receptor density on cellular miRNA-content before/after HDL particle uptake.

Cell Line	HDL Conditions and Concentration	< # miRNA/cell >	SD < # miRNA/cell >
ldlA7	no HDL addition	240	35
	native, 50 μg/mL	250	25
	miR-155 and spermine, 50 μg/mL	980	55
ldlA7-SR-B1	no HDL addition	260	20
	native, 50 μg/mL	330	55
	miR-155 and spermine, 50 μg/mL	2500	190

* Each condition was tested twice with two independently obtained samples. The HDL particles with the highest < # miRNA/particle > = 1×10^{-4} were used for these experiments. The used sample volume in the qPCR step equated for both cell lines to 3100 cells.

Table 5. * Influence of lipoprotein particle receptor density on cellular miRNA-content before/after LDL particle uptake.

Cell Line	LDL Conditions and Concentration	< # miR/cell >	SD < # miR/cell >
ldlA7	no LDL addition	190	10
	native, 5 μg/mL	240	20
	miR-155 and spermine, 5 μg/mL	210	40
CHOK1	no LDL addition	120	10
	native, 5 μg/mL	130	20
	miR-155 and spermine, 5 μg/mL	160	70

* Each condition was tested twice with two independently obtained samples. The LDL particles with the highest < # miRNA/particle > = 5×10^{-3} were used for these experiments. The used sample volume in the qPCR step equated for both cell lines to 5800 cells.

In order to exclude the possibility of a non-cell-mediated unspecific interaction of HDL particles with the cell culture vessel itself, we performed control experiments with chambers without cells (see Table A1). Here, the determined numbers for the miRNA amount corresponding to an individual cell (comparable to numbers from Tables 4 and 5) were around 1% and were, therefore, neglected. An incubation-concentration-dependence of the unspecific binding could be seen, and it was further observed that pre-blockage with BSA or native HDL particles had no influence.

4. Discussion

We have shown that, for some miRNAs, there is a significant difference in the profile between the HDL particle fraction and serum derived from CRF as well as hemodialysis patients. For other miRNAs, their values are nearly identical. Moreover, during disease progression, a similar bimodal behavior was observable. Especially miR-24, which has been reported to be upregulated in patients with kidney transplants [42], showed an increase of more than four RQ values in the serum, while its value even decreased in the HDL particle fraction by more than three values. Contrarily, RQ values of miR-122 increased by more than six values during disease progression in the HDL particle fraction while the serum RQ value hardly decreased. Interestingly, miR-21 and miR-223—two miRNAs known to be involved in renal and cardiovascular disease—did not change, neither in serum nor in the HDL particle fraction. Thus, we postulate that the miRNA profile may be suitable to identify certain diseases and follow their progression [43]; however, one has to be careful regarding its origin, as some miRNA levels differ between serum and HDL particle fraction.

Vickers et al. showed that HDL particles deliver miR-223 to SR-B1–transfected baby hamster kidney (BHK) cells, leading to repression of the Renilla-SR-B1-3′UTR luciferase reporters [6]. Therefore, HDL particles can be expected to transport miRNAs to target cells, leading to altered gene expression. Tabet et al. demonstrated that HDL-transported miR-223 down-regulated ICAM-1 expression in endothelial cells [7]. In the present study, we verified the principal functionality of HDL particles as miRNA-transfer-vehicle. Nevertheless, the number of miRNAs transported via native HDL particles may be insignificant due to the extremely low miRNA/particle ratio. This number may slightly vary depending on the miRNA and individual; however, any relevant influence on the cellular mRNA profile due to an uptake of miRNA-containing lipoprotein particles seems highly unlikely—at least in normolipidemic healthy individuals. We like to note that our numbers (strands/particle) are comparable to data published by Dimmler et al. (10^4 strands/µg lipoproteins are comparable to 10^{-9} strands/particle) [41].

Moreover, we have estimated that for the observed increase of miRNA strands per cell at standard receptor density, each cell has to interact, on average, with at least 7.5 million HDL particles (assuming a miRNA/particle ratio of 1:10^4, see Table 1) during the incubation period of 16 h. Under the assumption of no intracellular degradation of miRNA, this value represents only a lower limit. This yields a minimum of roughly 60 uptake events of HDL particles per second—a calculation previously unavailable, which enables a rough estimation of the absolute cargo transfer. According to our measured ratio of 10^{-9} miRNA strands per native HDL particle, on average, it would take more than half a year until a single miRNA strand has been taken up by a single cell. In contrast to HDL-mediated miRNA-uptake, we did not observe a significant increase of the miRNA level using LDL particles—an observation in agreement with the different lysosomal uptake pathway of LDL, which leads to degradation of the cargo itself.

Furthermore, our observation that reconstitution and the associated delipidation step did not influence the miRNA/particle ratio leads us to speculate that miRNA itself is not dominantly lipid-associated, and thus, not associated to the particle surface. Experiments using fluorescently labeled miRNA added to planar supported lipid bilayers (data not shown) yielded no observable signal—independent of the tested conditions (lipid composition, miRNA-concentration, temperature, pH-value, label). Moreover, extra-particular electrostatic association seems unlikely, due to the negative surface potential of lipoprotein particles and the poly-anionic nature of RNA strands. In addition, miRNAs are usually incorporated into a protein of the Argonaute family (AGO1-4) and have been proposed to circulate with HDL [13]. More than 150 proteins were found within the HDL particle fraction using proteome analysis; 95 of these proteins were identified to be specifically bound to the HDL particle. However, neither of the AGO proteins were detected [14–19]. Additionally, using Western Blot analysis and fluorescence imaging, we were unable to detect any AGO-2 protein (unpublished data). Therefore, it seems to be unlikely that this protein family is carrying the miRNA of HDL particles. The protein moiety enabling some of the HDL particles to bind miRNA still remains elusive.

In summary, we conclude that miRNA is transported and transferred to cells via lipoprotein particles, however, most likely plays no relevant role in vivo regarding the alteration of the cellular miRNA, and thus, the mRNA profile. However, lipoprotein particles may serve as diagnostic tools for certain disease-specific miRNAs—as exemplarily demonstrated here by us—or moreover, as an artificial drug delivery system.

Author Contributions: Conceptualization, M.A., S.M.M., W.S., H.S. and B.P.; Methodology, M.A., S.M.M., W.S., H.S. and B.P.; Validation, M.A., H.S. and B.P.; Formal Analysis, M.A., A.K.; Investigation, M.A., S.M.M. and A.K.; Resources, H.S. and B.P.; Writing-Original Draft Preparation, M.A., H.S. and B.P.; Visualization, M.A., A.K. and B.P.; Supervision, H.S. and B.P.; Project Administration, H.S. and B.P.; Funding Acquisition, S.M.M., H.S. and B.P.

Acknowledgments: We thank C. Röhrl and M. Ogris for helpful discussions. We thank J. Strasser for providing the miRNA-image used in the graphical abstract.

Appendix A

Table A1. Unspecific binding of HDL particles to the chamber.

Chamber Conditions	< # miRNA >	SD < # miRNA >
Blocking with 5% BSA for 1 h + medium for 16 h	2600	5
Blocking with 500 μg/mL native HDL for 1 h + medium for 16 h	2200	90
Blocking with 5% BSA for 1 h + 50 μg/mL rHDL + miR-155 & spermine for 16 h	14500	2800
Blocking with 500 μg/mL native HDL for 1 h + 50 μg/mL rHDL + miR-155 & spermine for 16 h	11000	2800
50 μg/mL rHDL + miR-155 & spermine for 16 h	11000	2100
5 μg/mL rHDL + miR-155 & spermine for 16 h	2800	140

The average number of miRNA < # miRNA > corresponds to the same volume as is contained in the chamber with ldlA7-SR-B1 cells (Table 4). For the authors, the only reasonable way to compare these values to the numbers from Table 4 is to divide them by the number of cells used (there ~3100 cells).

Table A2. RQ Values from Figure 1 of CRF (top) and hemodialysis patients (bottom).

Assay	RQ_{SERUM}	RQ_{HDL}	Assay	RQ_{SERUM}	RQ_{HDL}	Assay	RQ_{SERUM}	RQ_{HDL}
miR-106b	1.78	0.82	**miR-19b**	0.52	0.80	**miR-425-5p**	N.D.	0.83
miR-25	1.80	0.87	**miR-652**	0.64	0.93	**miR-148b**	N.D.	0.85
miR-127	2.68	1.94	**miR-185**	0.86	1.16	**miR-485-3p**	N.D.	0.87
miR-331	1.29	0.70	**miR-374**	1.39	1.71	**miR-99b**	N.D.	0.88
miR-30b	1.34	0.78	**miR-195**	0.62	0.97	**miR-28-3p**	N.D.	0.96
miR-323-3p	2.21	1.73	**miR-223**	0.38	0.74	**miR-376c**	N.D.	0.97
miR-142-3p	1.46	1.04	**miR-20b**	0.23	0.82	**let-7b**	N.D.	1.12
miR-335	0.79	0.39	**miR-24**	0.62	1.23	**let-7g**	N.D.	1.34
miR-145	0.63	0.27	**miR-574-3p**	0.35	0.98	**miR-29c**	N.D.	1.39
miR-221	1.22	0.95	**miR-146b**	0.63	1.41	**miR-26b**	N.D.	1.45
miR-150	2.30	2.05	**let-7d**	0.55	1.42	**miR-155**	N.D.	1.66
miR-20a	1.11	0.87	**miR-186**	0.28	1.26	**miR-494**	N.D.	1.85
miR-30c	1.17	0.94	**miR-16**	0.83	1.82	**miR-152**	N.D.	1.97
miR-15b	0.93	0.70	**miR-27b**	0.91	2.03	**miR-454**	N.D.	2.05
miR-19a	1.17	0.97	**miR-376a**	0.63	1.97	**miR-133a**	N.D.	2.12
miR-17	1.07	0.88	**miR-139-5p**	0.65	2.09	**miR-539**	N.D.	2.37
miR-328	0.93	0.75	**miR-628-5p**	2.32	3.83	**miR-148a**	N.D.	2.85
miR-199a-3p	1.06	0.91	**miR-29a**	0.80	2.40	**miR-518f**	N.D.	2.99
miR-320	1.07	0.93	**miR-122**	0.96	2.97	**miR-627**	N.D.	3.02
miR-191	1.12	0.98	**miR-192**	0.32	3.76	**miR-636**	N.D.	3.73
miR-26a	1.20	1.09	**miR-142-5p**	N.D.	0.35	**miR-301b**	N.D.	3.83
miR-126	1.30	1.20	**miR-526b**	N.D.	0.38	**miR-107**	N.D.	5.30
miR-92a	0.90	0.81	**miR-103**	N.D.	0.45	**miR-410**	N.D.	5.77
miR-197	0.46	0.37	**miR-744**	N.D.	0.15	**miR-424**	N.D.	6.91
miR-484	0.75	0.71	**miR-340**	N.D.	0.48	**miR-487b**	N.D.	6.92
miR-106a	0.92	0.90	**let-7a**	N.D.	0.50	**let-7c**	N.D.	13.26
let-7e	1.53	1.58	**miR-130b**	N.D.	0.56	**miR-486**	0.61	N.D.
miR-222	0.67	0.87	**miR-130a**	N.D.	0.58	**miR-532**	1.11	N.D.
miR-27a	1.38	1.59	**miR-301**	N.D.	0.65	**miR-375**	1.20	N.D.
miR-21	0.66	0.90	**miR-128a**	N.D.	0.72	**miR-423-5p**	1.24	N.D.
miR-342-3p	0.78	1.04	**miR-28**	N.D.	0.75	**miR-212**	1.63	N.D.
miR-146a	0.77	1.03	**miR-345**	N.D.	0.82			

Table A2. *Cont.*

Assay	RQ_{SERUM}	RQ_{HDL}	Assay	RQ_{SERUM}	RQ_{HDL}	Assay	RQ_{SERUM}	RQ_{HDL}
miR-24	4.86	0.92	**miR-99b**	0.82	0.84	**miR-483-5p**	0.43	N.D.
miR-574-3p	3.59	0.86	**miR-146a**	0.58	0.65	**miR-192**	0.59	N.D.
miR-222	3.29	1.20	**miR-146b**	0.55	0.65	**miR-99a**	2.80	N.D.
miR-27b	3.16	1.30	**miR-425-5p**	0.56	0.66	**miR-660**	3.94	N.D.
miR-29c	2.36	0.52	**miR-328**	0.44	0.54	**miR-193a-5p**	6.52	N.D.
miR-152	1.84	0.50	**miR-454**	0.23	0.34	**miR-125b**	16.31	N.D.
miR-376c	2.05	0.73	**miR-150**	0.57	0.71	**miR-193b**	17.65	N.D.
miR-744	2.98	1.81	**miR-323-3p**	0.59	0.78	**miR-200c**	90.15	N.D.
miR-181a	1.67	0.61	**miR-335**	0.26	0.45	**miR-203**	1626.14	N.D.
miR-29a	2.90	2.03	**miR-26b**	1.03	1.24	**miR-532**	N.D.	0.39
miR-221	1.61	0.86	**miR-191**	0.46	0.69	**miR-370**	N.D.	0.47
miR-331	1.67	0.92	**miR-30b**	0.72	0.96	**miR-339-3p**	N.D.	0.52
let-7b	1.30	0.68	**miR-320**	0.58	0.83	**miR-301**	N.D.	0.54
miR-19a	1.17	0.56	**miR-199a-3p**	0.24	0.57	**miR-125a-5p**	N.D.	0.54
miR-636	0.87	0.37	**miR-628-5p**	0.29	0.65	**miR-148b**	N.D.	0.55
miR-27a	1.20	0.72	**miR-223**	0.27	0.63	**miR-590-5p**	N.D.	0.58
miR-106a	1.09	0.64	**miR-30c**	0.59	0.98	**miR-142-5p**	N.D.	0.62
let-7d	1.38	0.99	**miR-374**	0.45	0.83	**miR-597**	N.D.	0.64
miR-20a	0.94	0.55	**miR-126**	0.34	0.73	**miR-133a**	N.D.	0.65
miR-19b	1.09	0.71	**miR-142-3p**	0.30	0.69	**miR-539**	N.D.	0.70
miR-21	0.91	0.71	**miR-28-3p**	0.52	0.93	**let-7a**	N.D.	0.73
miR-423-5p	0.36	0.23	**miR-526b**	0.57	1.05	**miR-127**	N.D.	0.79
miR-197	0.95	0.85	**miR-106b**	0.52	1.03	**miR-18a**	N.D.	0.87
miR-186	0.60	0.51	**miR-139-5p**	0.22	0.76	**miR-411**	N.D.	0.92
let-7g	0.68	0.60	**miR-92a**	0.32	0.88	**miR-324-3p**	N.D.	1.10
miR-20b	0.85	0.78	**miR-484**	0.34	1.15	**miR-185**	N.D.	1.28
miR-652	0.53	0.46	**miR-25**	0.75	1.58	**miR-148a**	N.D.	1.31
miR-26a	0.93	0.87	**miR-103**	0.63	1.46	**miR-130b**	N.D.	1.35
let-7e	0.83	0.78	**miR-145**	0.20	1.11	**miR-155**	N.D.	1.65
miR-16	0.39	0.35	**miR-195**	0.25	1.20	**miR-132**	N.D.	2.02
miR-130a	1.07	1.04	**miR-486**	0.18	1.28	**miR-487b**	N.D.	2.11
miR-342-3p	1.37	1.35	**miR-15b**	0.23	1.35	**miR-28**	N.D.	2.22
miR-17	0.73	0.71	**miR-122**	0.23	9.61			

Arrangement of data is equal to the clockwise order shown in Figure 1; for miRNA data with only one available value (not shown in Figure 1), the miRNAs are arranged in an ascending order.

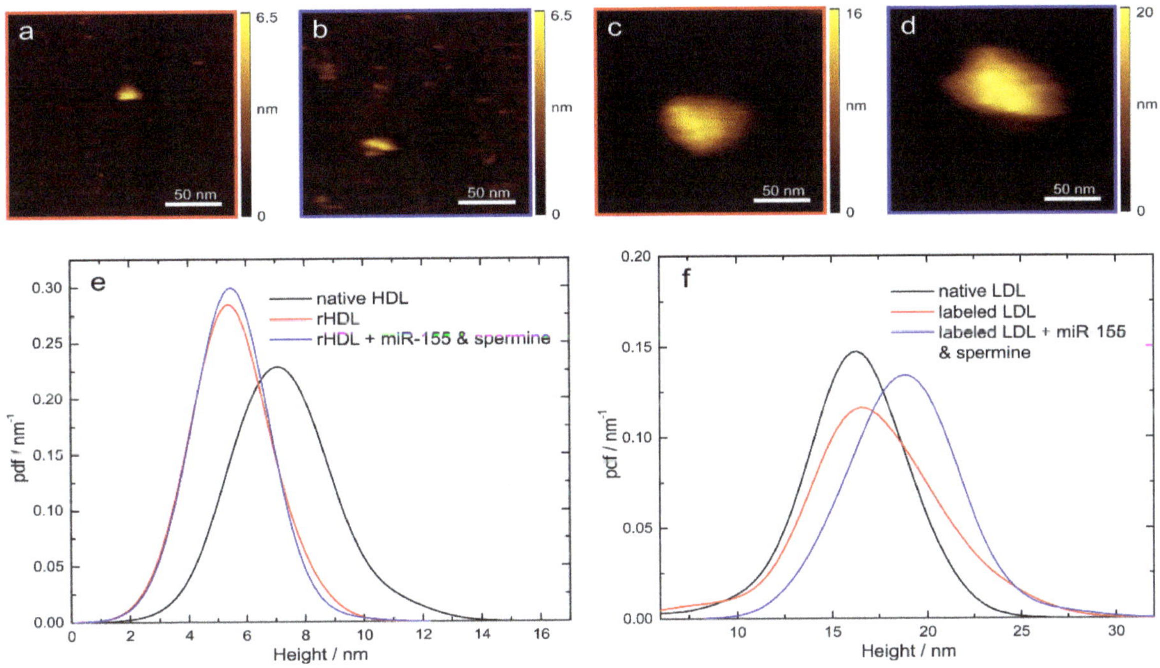

Figure A1. Spatial characterization of reconstituted/labeled lipoprotein particles. High-speed atomic force microscopy (HS-AFM) images of reconstituted HDL particles (**a**), rHDL particles with miR-155 & spermine (**b**), labeled LDL particles (**c**) and labeled LDL particles with miR-155 & spermine (**d**). The size analysis yielded distributions of particle heights for native HDL particles (N = 125 particles, <height> = 7.3 nm ± 1.5 nm), rHDL particles (N =110, <height> = 5.5 nm ± 1.0 nm) and rHDL particles + miR-155 & spermine (N = 122, <height> = 5.5 nm ± 0.9 nm) (**e**), as well as for native LDL particles (N = 79, <height> = 16.1 nm ± 2.4 nm), labeled LDL particles (N = 89, <height> = 17.2 nm ± 3.4 nm) and LDL particles + miR-155 & spermine (N = 57, <height> = 18.8 nm ± 2.4 nm) (**f**).

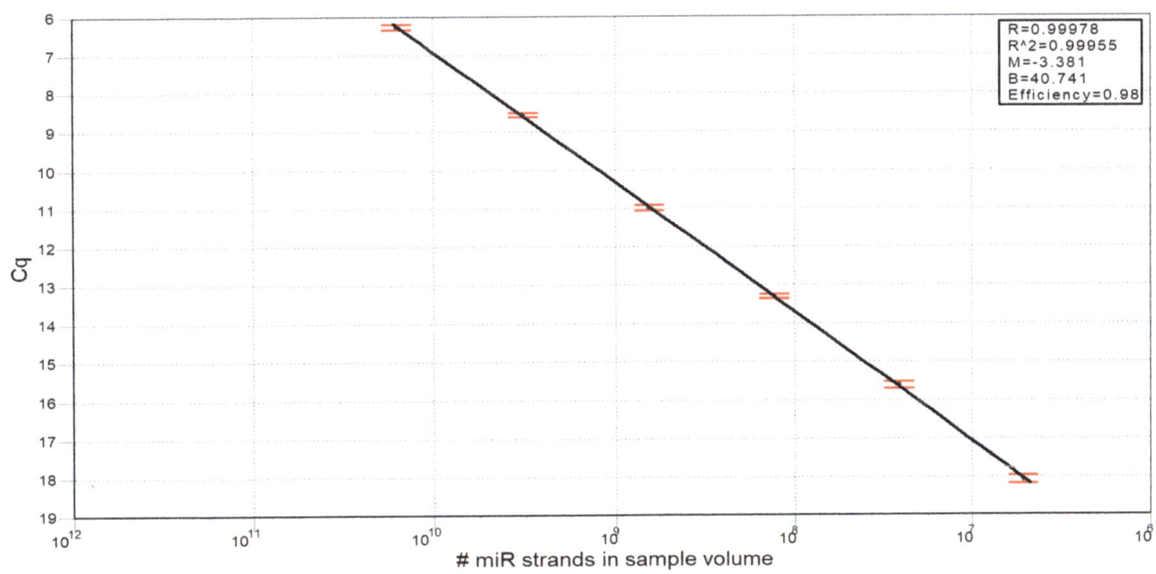

Figure A2. *Cont.*

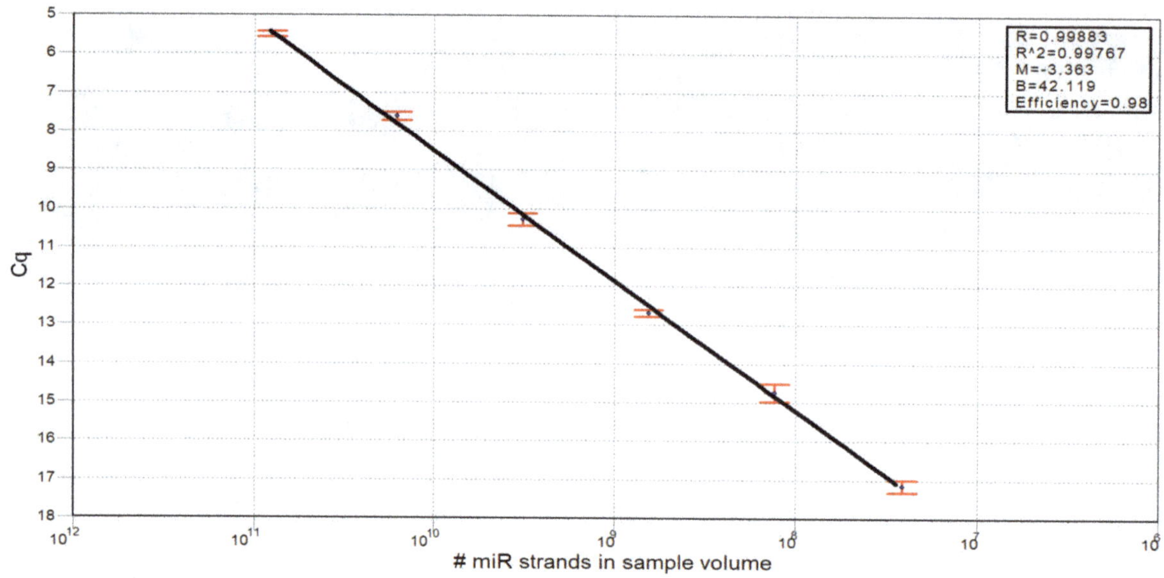

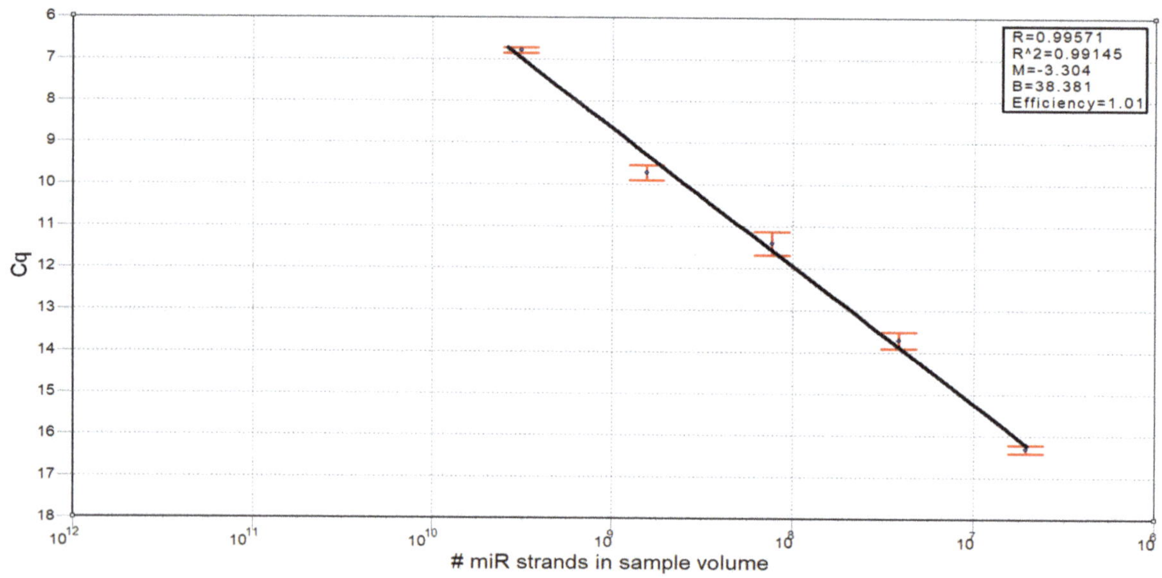

Figure A2. Standard curves for miR-145 (top), miR-155 (center) and miR-223 (bottom).

References

1. Simons, M.; Raposo, G. Exosomes—Vesicular carriers for intercellular communication. *Curr. Opin. Cell Biol.* **2009**, *21*, 575–581. [CrossRef] [PubMed]
2. Valadi, H.; Ekstrom, K.; Bossios, A.; Sjostrand, M.; Lee, J.J.; Lotvall, J.O. Exosome-mediated transfer of mrnas and microRNAs is a novel mechanism of genetic exchange between cells. *Nat. Cell Biol.* **2007**, *9*, 654–659. [CrossRef] [PubMed]
3. Hunter, M.P.; Ismail, N.; Zhang, X.; Aguda, B.D.; Lee, E.J.; Yu, L.; Xiao, T.; Schafer, J.; Lee, M.L.; Schmittgen, T.D.; et al. Detection of microRNA expression in human peripheral blood microvesicles. *PLoS ONE* **2008**, *3*, e3694. [CrossRef] [PubMed]
4. Ratajczak, J.; Wysoczynski, M.; Hayek, F.; Janowska-Wieczorek, A.; Ratajczak, M.Z. Membrane-derived microvesicles: Important and underappreciated mediators of cell-to-cell communication. *Leukemia* **2006**, *20*, 1487–1495. [CrossRef] [PubMed]
5. Arroyo, J.D.; Chevillet, J.R.; Kroh, E.M.; Ruf, I.K.; Pritchard, C.C.; Gibson, D.F.; Mitchell, P.S.; Bennett, C.F.; Pogosova-Agadjanyan, E.L.; Stirewalt, D.L.; et al. Argonaute2 complexes carry a population of circulating microRNAs independent of vesicles in human plasma. *Proc. Natl. Acad. Sci. USA* **2011**, *108*, 5003–5008. [CrossRef] [PubMed]

6. Vickers, K.C.; Palmisano, B.T.; Shoucri, B.M.; Shamburek, R.D.; Remaley, A.T. MicroRNAs are transported in plasma and delivered to recipient cells by high-density lipoproteins. *Nat. Cell Biol.* **2011**, *13*, 423–433. [CrossRef] [PubMed]
7. Tabet, F.; Vickers, K.C.; Cuesta Torres, L.F.; Wiese, C.B.; Shoucri, B.M.; Lambert, G.; Catherinet, C.; Prado-Lourenco, L.; Levin, M.G.; Thacker, S.; et al. Hdl-transferred microRNA-223 regulates icam-1 expression in endothelial cells. *Nat. Commun.* **2014**, *5*, 3292. [CrossRef] [PubMed]
8. Wahid, F.; Shehzad, A.; Khan, T.; Kim, Y.Y. MicroRNAs: Synthesis, mechanism, function, and recent clinical trials. *Biochim. Biophys. Acta* **2010**, *1803*, 1231–1243. [CrossRef] [PubMed]
9. Carthew, R.W.; Sontheimer, E.J. Origins and mechanisms of miRNAs and siRNAs. *Cell* **2009**, *136*, 642–655. [CrossRef] [PubMed]
10. Fabian, M.R.; Sonenberg, N.; Filipowicz, W. Regulation of mRNA translation and stability by micrornas. *Ann. Rev. Biochem.* **2010**, *79*, 351–379. [CrossRef] [PubMed]
11. Filipowicz, W.; Bhattacharyya, S.N.; Sonenberg, N. Mechanisms of post-transcriptional regulation by microRNAs: Are the answers in sight? *Nat. Rev. Genet.* **2008**, *9*, 102–114. [CrossRef] [PubMed]
12. Chen, X.; Liang, H.; Zhang, J.; Zen, K.; Zhang, C.Y. Secreted microRNAs: A new form of intercellular communication. *Trends Cell Biol.* **2012**, *22*, 125–132. [CrossRef] [PubMed]
13. Turchinovich, A.; Samatov, T.R.; Tonevitsky, A.G.; Burwinkel, B. Circulating miRNAs: Cell-cell communication function? *Front. Genet.* **2013**, *4*, 119. [CrossRef] [PubMed]
14. Vaisar, T.; Pennathur, S.; Green, P.S.; Gharib, S.A.; Hoofnagle, A.N.; Cheung, M.C.; Byun, J.; Vuletic, S.; Kassim, S.; Singh, P.; et al. Shotgun proteomics implicates protease inhibition and complement activation in the antiinflammatory properties of HDL. *J. Clin. Investig.* **2007**, *117*, 746–756. [CrossRef] [PubMed]
15. Weichhart, T.; Kopecky, C.; Kubicek, M.; Haidinger, M.; Doller, D.; Katholnig, K.; Suarna, C.; Eller, P.; Tolle, M.; Gerner, C.; et al. Serum amyloid a in uremic HDL promotes inflammation. *J. Am. Soc. Nephrol.* **2012**, *23*, 934–947. [CrossRef] [PubMed]
16. Holzer, M.; Wolf, P.; Curcic, S.; Birner-Gruenberger, R.; Weger, W.; Inzinger, M.; El-Gamal, D.; Wadsack, C.; Heinemann, A.; Marsche, G. Psoriasis alters HDL composition and cholesterol efflux capacity. *J. Lipid Res.* **2012**, *53*, 1618–1624. [CrossRef] [PubMed]
17. Holzer, M.; Birner-Gruenberger, R.; Stojakovic, T.; El-Gamal, D.; Binder, V.; Wadsack, C.; Heinemann, A.; Marsche, G. Uremia alters HDL composition and function. *J. Am. Soc. Nephrol.* **2011**, *22*, 1631–1641. [CrossRef] [PubMed]
18. Singh, S.A.; Andraski, A.B.; Pieper, B.; Goh, W.; Mendivil, C.O.; Sacks, F.M.; Aikawa, M. Multiple apolipoprotein kinetics measured in human HDL by high-resolution/accurate mass parallel reaction monitoring. *J. Lipid Res.* **2016**, *57*, 714–728. [CrossRef] [PubMed]
19. Singh, S.A.; Aikawa, M. Unbiased and targeted mass spectrometry for the HDL proteome. *Curr. Opin. Lipidol.* **2017**, *28*, 68–77. [CrossRef] [PubMed]
20. Ha, T.Y. MicroRNAs in human diseases: From cancer to cardiovascular disease. *Immune Netw.* **2011**, *11*, 135–154. [CrossRef] [PubMed]
21. Jamaluddin, M.S.; Weakley, S.M.; Zhang, L.; Kougias, P.; Lin, P.H.; Yao, Q.; Chen, C. miRNAs: Roles and clinical applications in vascular disease. *Expert Rev. Mol. Diagn.* **2011**, *11*, 79–89. [CrossRef] [PubMed]
22. Kosaka, N.; Iguchi, H.; Yoshioka, Y.; Takeshita, F.; Matsuki, Y.; Ochiya, T. Secretory mechanisms and intercellular transfer of microRNAs in living cells. *J. Biol. Chem.* **2010**, *285*, 17442–17452. [CrossRef] [PubMed]
23. Qin, S.; Zhang, C. MicroRNAs in vascular disease. *J. Cardiovasc. Pharm.* **2011**, *57*, 8–12. [CrossRef] [PubMed]
24. Vickers, K.C.; Remaley, A.T. MicroRNAs in atherosclerosis and lipoprotein metabolism. *Curr. Opin. Endocrinol.* **2010**, *17*, 150–155. [CrossRef] [PubMed]
25. Akkina, S.; Becker, B.N. MicroRNAs in kidney function and disease. *Transl. Res.* **2011**, *157*, 236–240. [CrossRef] [PubMed]
26. Bhatt, K.; Mi, Q.S.; Dong, Z. MicroRNAs in kidneys: Biogenesis, regulation, and pathophysiological roles. *Am. J. Physiol. Renal Physiol.* **2011**, *300*, F602–F610. [CrossRef] [PubMed]
27. Li, J.Y.; Yong, T.Y.; Michael, M.Z.; Gleadle, J.M. Review: The role of microRNAs in kidney disease. *Nephrology* **2010**, *15*, 599–608. [CrossRef] [PubMed]

28. Neal, C.S.; Michael, M.Z.; Pimlott, L.K.; Yong, T.Y.; Li, J.Y.; Gleadle, J.M. Circulating microRNA expression is reduced in chronic kidney disease. *Nephrol. Dial. Transplant.* **2011**, *26*, 3794–3802. [CrossRef] [PubMed]
29. Mitchell, P.S.; Parkin, R.K.; Kroh, E.M.; Fritz, B.R.; Wyman, S.K.; Pogosova-Agadjanyan, E.L.; Peterson, A.; Noteboom, J.; O'Briant, K.C.; Allen, A.; et al. Circulating microRNAs as stable blood-based markers for cancer detection. *Proc. Natl. Acad. Sci. USA* **2008**, *105*, 10513–10518. [CrossRef] [PubMed]
30. Olson, E.N. MicroRNAs as therapeutic targets and biomarkers of cardiovascular disease. *Sci. Transl. Med.* **2014**, *6*, 239ps3. [CrossRef] [PubMed]
31. Bustin, S.A.; Benes, V.; Garson, J.A.; Hellemans, J.; Huggett, J.; Kubista, M.; Mueller, R.; Nolan, T.; Pfaffl, M.W.; Shipley, G.L.; et al. The MIQE guidelines: Minimum information for publication of quantitative real-time PCR experiments. *Clin. Chem.* **2009**, *55*, 611–622. [CrossRef] [PubMed]
32. Meier, S.M.; Wultsch, A.; Hollaus, M.; Ammann, M.; Pemberger, E.; Liebscher, F.; Lambers, B.; Fruhwurth, S.; Stojakovic, T.; Scharnagl, H.; et al. Effect of chronic kidney disease on macrophage cholesterol efflux. *Life Sci.* **2015**, *136*, 1–6. [CrossRef] [PubMed]
33. Fichtlscherer, S.; De Rosa, S.; Fox, H.; Schwietz, T.; Fischer, A.; Liebetrau, C.; Weber, M.; Hamm, C.W.; Roxe, T.; Muller-Ardogan, M.; et al. Circulating microRNAs in patients with coronary artery disease. *Circ. Res.* **2010**, *107*, 677–684. [CrossRef] [PubMed]
34. Schumaker, V.N.; Puppione, D.L. Sequential flotation ultracentrifugation. *Method Enzymol.* **1986**, *128*, 155–170.
35. Jonas, A. Reconstitution of high-density lipoproteins. *Method Enzymol.* **1986**, *128*, 553–582.
36. Krieger, M. Reconstitution of the hydrophobic core of low-density lipoprotein. *Method Enzymol.* **1986**, *128*, 608–613.
37. Stangl, H.; Cao, G.; Wyne, K.L.; Hobbs, H.H. Scavenger receptor, class b, type I-dependent stimulation of cholesterol esterification by high density lipoproteins, low density lipoproteins, and nonlipoprotein cholesterol. *J. Biol. Chem.* **1998**, *273*, 31002–31008. [CrossRef] [PubMed]
38. Stangl, H.; Hyatt, M.; Hobbs, H.H. Transport of lipids from high- and low-density lipoproteins via scavenger receptor-bi. *J. Biol. Chem.* **1999**, *274*, 32692–32698. [CrossRef] [PubMed]
39. Zhou, T.B.; Jiang, Z.P. Role of miR-21 and its signaling pathways in renal diseases. *J. Recept. Signal Transduct.* **2014**, *34*, 335–337. [CrossRef] [PubMed]
40. Yuan, X.; Wang, X.; Chen, C.; Zhou, J.; Han, M. Bone mesenchymal stem cells ameliorate ischemia/reperfusion-induced damage in renal epithelial cells via microRNA-223. *Stem Cell Res. Ther.* **2017**, *8*, 146. [CrossRef] [PubMed]
41. Wagner, J.; Riwanto, M.; Besler, C.; Knau, A.; Fichtlscherer, S.; Roxe, T.; Zeiher, A.M.; Landmesser, U.; Dimmeler, S. Characterization of levels and cellular transfer of circulating lipoprotein-bound microRNAs. *Arterioscler. Thromb. Vasc.* **2013**, *33*, 1392–1400. [CrossRef] [PubMed]
42. Lorenzen, J.M.; Kaucsar, T.; Schauerte, C.; Schmitt, R.; Rong, S.; Hubner, A.; Scherf, K.; Fiedler, J.; Martino, F.; Kumarswamy, R.; et al. MicroRNA-24 antagonism prevents renal ischemia reperfusion injury. *J. Am. Soc. Nephrol.* **2014**, *25*, 2717–2729. [CrossRef] [PubMed]
43. Desgagne, V.; Guerin, R.; Guay, S.P.; Corbin, F.; Couture, P.; Lamarche, B.; Bouchard, L. Changes in high-density lipoprotein-carried miRNA contribution to the plasmatic pool after consumption of dietary trans fat in healthy men. *Epigenomics* **2017**, *9*, 669–688. [CrossRef] [PubMed]

Permissions

All chapters in this book were first published by MDPI; hereby published with permission under the Creative Commons Attribution License or equivalent. Every chapter published in this book has been scrutinized by our experts. Their significance has been extensively debated. The topics covered herein carry significant findings which will fuel the growth of the discipline. They may even be implemented as practical applications or may be referred to as a beginning point for another development.

The contributors of this book come from diverse backgrounds, making this book a truly international effort. This book will bring forth new frontiers with its revolutionizing research information and detailed analysis of the nascent developments around the world.

We would like to thank all the contributing authors for lending their expertise to make the book truly unique. They have played a crucial role in the development of this book. Without their invaluable contributions this book wouldn't have been possible. They have made vital efforts to compile up to date information on the varied aspects of this subject to make this book a valuable addition to the collection of many professionals and students.

This book was conceptualized with the vision of imparting up-to-date information and advanced data in this field. To ensure the same, a matchless editorial board was set up. Every individual on the board went through rigorous rounds of assessment to prove their worth. After which they invested a large part of their time researching and compiling the most relevant data for our readers.

The editorial board has been involved in producing this book since its inception. They have spent rigorous hours researching and exploring the diverse topics which have resulted in the successful publishing of this book. They have passed on their knowledge of decades through this book. To expedite this challenging task, the publisher supported the team at every step. A small team of assistant editors was also appointed to further simplify the editing procedure and attain best results for the readers.

Apart from the editorial board, the designing team has also invested a significant amount of their time in understanding the subject and creating the most relevant covers. They scrutinized every image to scout for the most suitable representation of the subject and create an appropriate cover for the book.

The publishing team has been an ardent support to the editorial, designing and production team. Their endless efforts to recruit the best for this project, has resulted in the accomplishment of this book. They are a veteran in the field of academics and their pool of knowledge is as vast as their experience in printing. Their expertise and guidance has proved useful at every step. Their uncompromising quality standards have made this book an exceptional effort. Their encouragement from time to time has been an inspiration for everyone.

The publisher and the editorial board hope that this book will prove to be a valuable piece of knowledge for researchers, students, practitioners and scholars across the globe.

List of Contributors

Rishipal R. Bansode, Janak R. Khatiwada and Leonard L. Williams
Center for Excellence in Post-Harvest Technologies, North Carolina Research Campus, North Carolina Agricultural and Technical State University, Kannapolis, NC 28081, USA

Jack N. Losso
School of Nutrition & Food Sciences, Louisiana State University, Baton Rouge, LA 70803, USA

Zainab Jahangir
John P. Stevens High School, North Edison, NJ 08820, USA

Jahangir Iqbal
King Abdullah International Medical Research Center, Ministry of National Guard Health Affairs, Al Ahsa 31982, Saudi Arabia

Ahmed Bakillah
Department of Medicine, Downstate Medical Center, State University of New York, 450 Clarkson Ave, Brooklyn, NY 11203, USA
King Abdullah International Medical Research Center, Ministry of National Guard Health Affairs, Al Ahsa 31982, Saudi Arabia

Barbara Zellinger
radART—Institute for Research and Development on Advanced Radiation Technologies, Paracelsus Medical University, Müllner Hauptstrasse 48, 5020 Salzburg, Austria
Department of Pathology, Paracelsus Medical University, SALK, Müllner Hauptstrasse 48, 5020 Salzburg, Austria

Cornelia Kronberger and Karl Sotlar
Department of Pathology, Paracelsus Medical University, SALK, Müllner Hauptstrasse 48, 5020 Salzburg, Austria

Ulrich Bodenhofer
School of Informatics, Communications and Media, University of Applied Sciences Upper Austria, Softwarepark 11, 4232 Hagenberg, Austria
Institute for Machine Learning, Campus Science Park 3, Johannes Kepler University, Altenbergerstrasse 69, 4040 Linz, Austria

Immanuela A. Engländer, Felix Sedlmayer and Franz Zehentmayr
radART—Institute for Research and Development on Advanced Radiation Technologies, Paracelsus Medical University, Müllner Hauptstrasse 48, 5020 Salzburg, Austria
Department of Radiation Oncology, Paracelsus Medical University, SALK, Müllner Hauptstrasse 48, 5020 Salzburg, Austria

Brane Grambozov, Gerd Fastner and Markus Stana
Department of Radiation Oncology, Paracelsus Medical University, SALK, Müllner Hauptstrasse 48, 5020 Salzburg, Austria

Peter Strasser
Department of Laboratory Medicine, Paracelsus Medical University, SALK, Müllner Hauptstrasse 48, 5020 Salzburg, Austria

Roland Reitsamer
Department of Gynecology and Obstetrics, Paracelsus Medical University, SALK, Müllner Hauptstrasse 48, 5020 Salzburg, Austria

Mohamed Zaiou
School of Pharmacy, University of Lorraine, 5 rue Albert Lebrun, 54000 Nancy, France

Viorel Simion, Basak Icli, Daniel Pérez-Cremades, Henry S. Cheng and Mark W. Feinberg
Department of Medicine, Cardiovascular Division, Brigham and Women's Hospital, Harvard Medical School, Boston, MA 02115, USA

Jacob B. Pierce
Department of Medicine, Cardiovascular Division, Brigham and Women's Hospital, Harvard Medical School, Boston, MA 02115, USA
Feinberg School of Medicine, Northwestern University, Chicago, IL 60611, USA

Chang Liu, Yue Hu and Wenyuan Shen
Laboratory of Medical Molecular Virology, School of Medicine, Nankai University, Tianjin 300071, China

Ze Chen
State Key Laboratory of Veterinary Etiological Biology and Key Laboratory of Veterinary Parasitology of Gansu Province, Lanzhou Veterinary Research Institute, Chinese Academy of Agricultural Science, Lanzhou 730046, China
Co-Innovation Center for Prevention and Control of Important Animal Infectious Diseases and Zoonoses, Yangzhou 225009, China

Haishuo Ji and Shan Gao
College of Life Sciences, Nankai University, Tianjin 300071, China
Institute of Statistics, Nankai University, Tianjin 300071, China

Deshui Yu, Siyu Li and Wenjun Bu
College of Life Sciences, Nankai University, Tianjin 300071, China

Jishou Ruan
School of Mathematical Sciences, Nankai University, Tianjin 300071, China

Miguel Hueso, Josep M. Cruzado and Joan Torras
Department of Nephrology, Hospital Universitari Bellvitge and Bellvitge Research Institute (IDIBELL), L'Hospitalet de Llobregat, 08907 Barcelona, Spain

Estanis Navarro
Independent Researcher, Esplugues de Llobregat, 08950 Barcelona, Spain

Ce Yuan and Subbaya Subramanian
Bioinformatics and Computational Biology Program, University of Minnesota, Minneapolis, MN 55455, USA
Department of Surgery, University of Minnesota Medical School, Minneapolis, MN 55455, USA

Clifford J. Steer
Department of Medicine, University of Minnesota Medical School, Minneapolis, MN 55455, USA
Department of Genetics, Cell Biology, and Development, University of Minnesota Medical School, Minneapolis, MN 55455, USA

Kadri Rekker, Marina Suhorutshenko and Maire Peters
Institute of Clinical Medicine, Department of Obstetrics and Gynecology, University of Tartu, 50406 Tartu, Estonia
Competence Centre on Health Technologies, 50410 Tartu, Estonia

Signe Altmäe
Competence Centre on Health Technologies, 50410 Tartu, Estonia
Department of Biochemistry and Molecular Biology, Faculty of Sciences, University of Granada, 18071 Granada, Spain

Juan F. Martinez-Blanch and Francisco M. Codoñer
Lifesequencing SL, 46980 Valencia, Spain

Felipe Vilella
Igenomix Foundation/INCLIVA, 46010 Valencia, Spain
Igenomix SL, 46980 Valencia, Spain

Carlos Simón
Igenomix Foundation/INCLIVA, 46010 Valencia, Spain
Igenomix SL, 46980 Valencia, Spain
Department of Obstetrics and Gynecology, Valencia University, 46010 Valencia, Spain

Andres Salumets
Institute of Clinical Medicine, Department of Obstetrics and Gynecology, University of Tartu, 50406 Tartu, Estonia
Competence Centre on Health Technologies, 50410 Tartu, Estonia
Institute of Biomedicine and Translational Medicine, Department of Biomedicine, University of Tartu, 50412 Tartu, Estonia
Department of Obstetrics and Gynecology, University of Helsinki and Helsinki University Hospital, FI-00014 Helsinki, Finland

Agne Velthut-Meikas
Competence Centre on Health Technologies, 50410 Tartu, Estonia
Department of Chemistry and Biotechnology, Tallinn University of Technology, 12618 Tallinn, Estonia

Gabrielle Bradshaw, Larisa M. Haupt, Eunise M. Aquino, Rodney A. Lea, Heidi G. Sutherland and Lyn R. Griffiths
Genomics Research Centre, School of Biomedical Sciences, Institute of Health and Biomedical Innovation, Queensland University of Technology, Brisbane, QLD 4001, Australia

Steven Mullenbrock, Fei Liu, Suzanne Szak, Xiaoping Hronowski, Benbo Gao, Peter Juhasz, Chao Sun, Mei Liu, Helen McLaughlin, Qiurong Xiao and Timothy S. Zheng
Biogen, Cambridge, MA 02142, USA

Carol Feghali-Bostwick
Division of Rheumatology and Immunology, Department of Medicine, The Medical University of South Carolina, Charleston, SC 29425, USA

Kristina B. V. Døssing
Center for Genomic Medicine, Rigshospitalet, Blegdamsvej 9, 2100 Copenhagen, Denmark
Department of Clinical Physiology, Nuclear Medicine and PET, Rigshospitalet, Blegdamsvej 9, 2100 Copenhagen, Denmark
Cluster for Molecular Imaging, Faculty of Health Sciences University of Copenhagen, Blegdamsvej 3B, 2100 Copenhagen, Denmark

Christina Kjær
University College Copenhagen, Sigurdsgade 26, 2200 Copenhagen, Denmark

Jonas Vikesa
Center for Genomic Medicine, Rigshospitalet, Blegdamsvej 9, 2100 Copenhagen, Denmark

Tina Binderup
Department of Clinical Physiology, Nuclear Medicine and PET, Rigshospitalet, Blegdamsvej 9, 2100 Copenhagen, Denmark
Cluster for Molecular Imaging, Faculty of Health Sciences University of Copenhagen, Blegdamsvej 3B, 2100 Copenhagen, Denmark

Ulrich Knigge
Department of Surgical Gastroenterology C, Rigshospitalet, University of Copenhagen, Blegdamsvej 9, 2100 Copenhagen, Denmark
Department of Clinical Endocrinology PE, Rigshospitalet, Blegdamsvej 9, 2100 Copenhagen, Denmark

Michael D. Culler
Biomeasure Incorporated/IPSEN, 650 W Kendall St, Cambridge, MA 02142, USA

Andreas Kjær
Department of Clinical Physiology, Nuclear Medicine and PET, Rigshospitalet, Blegdamsvej 9, 2100 Copenhagen, Denmark
Cluster for Molecular Imaging, Faculty of Health Sciences University of Copenhagen, Blegdamsvej 3B, 2100 Copenhagen, Denmark

Birgitte Federspiel
Department of Pathology, Rigshospitalet, University of Copenhagen, Blegdamsvej 9, 2100 Copenhagen, Denmark

Lennart Friis-Hansen
Center for Genomic Medicine, Rigshospitalet, Blegdamsvej 9, 2100 Copenhagen, Denmark
Department of Clinical Biochemistry, Hillerød Hospital, Dyrhavevej 29, 3400 Hillerød, Demark

Markus Axmann, Sabine M. Meier, Witta Strobl and Herbert Stangl
Center for Pathobiochemistry and Genetics, Institute of Medical Chemistry and Pathobiochemistry, Medical University Vienna, 1090 Vienna, Austria

Andreas Karner and Birgit Plochberger
University of Applied Sciences Upper Austria, School of Medical Engineering and Applied Social Sciences, 4020 Linz, Austria

Index

Printed in the USA
CPSIA information can be obtained
at www.ICGtesting.com
LVHW080053211023
761596LV00010BA/2044

9 781639 877904